Lecture Notes in Computer Science   11202

Commenced Publication in 1973
Founding and Former Series Editors:
Gerhard Goos, Juris Hartmanis, and Jan van Leeuwen

Corina Cîrstea (Ed.)

Coalgebraic Methods in Computer Science

14th IFIP WG 1.3 International Workshop, CMCS 2018
Colocated with ETAPS 2018
Thessaloniki, Greece, April 14–15, 2018
Revised Selected Papers

 Springer

Editor
Corina Cîrstea 🆔
University of Southampton
Southampton
UK

ISSN 0302-9743 ISSN 1611-3349 (electronic)
Lecture Notes in Computer Science
ISBN 978-3-030-00388-3 ISBN 978-3-030-00389-0 (eBook)
https://doi.org/10.1007/978-3-030-00389-0

Library of Congress Control Number: 2018954549

LNCS Sublibrary: SL1 – Theoretical Computer Science and General Issues

This Springer imprint is published by the registered company Springer Nature Switzerland AG
The registered company address is: Gewerbestrasse 11, 6330 Cham, Switzerland

Preface

The 14th International Workshop on Coalgebraic Methods in Computer Science, CMCS 2018, was held during April 14–15, 2018, in Thessaloniki, Greece, as a satellite event of the Joint Conference on Theory and Practice of Software, ETAPS 2018. Coalgebras allow for a uniform treatment of a large variety of state-based dynamical systems, such as transition systems, automata (including weighted and probabilistic variants), Markov chains, and game-based systems. Over the last two decades, coalgebra has developed into a field of its own interest, presenting a deep mathematical foundation, a growing field of applications, and interactions with various other fields such as reactive and interactive system theory, object-oriented and concurrent programming, formal system specification, modal and description logics, artificial intelligence, dynamical systems, control systems, category theory, algebra, analysis, etc. The aim of the workshop is to bring together researchers with a common interest in the theory of coalgebras, their logics, and their applications.

Previous workshops of the CMCS series were held in Lisbon (1998), Amsterdam (1999), Berlin (2000), Genova (2001), Grenoble (2002), Warsaw (2003), Barcelona (2004), Vienna (2006), Budapest (2008), Paphos (2010), Tallinn (2012), Grenoble (2014), and Eindhoven (2016). Since 2004, CMCS has been a biennial workshop, alternating with the International Conference on Algebra and Coalgebra in Computer Science (CALCO).

The CMCS 2018 program featured a keynote talk by Samson Abramsky (University of Oxford, UK), an invited talk by Clemens Kupke (University of Strathclyde, UK), and an invited talk by Daniela Petrişan (Université Paris Diderot, France). In addition, a special session on categorical quantum computation was held, featuring invited tutorials by Bob Coecke (University of Oxford, UK) and Aleks Kissinger (Radboud University Nijmegen, The Netherlands).

This volume contains revised regular contributions (10 accepted out of 17 submissions) and the abstracts of two keynote/invited talks. All regular contributions were refereed by three reviewers. Special thanks go to all the authors for the high quality of their contributions, to the reviewers and Program Committee members for their thorough reviewing and help in improving the papers presented at CMCS 2018, and to all the participants for active discussions.

May 2018   Corina Cîrstea

Organization

CMCS 2018 was organized as a satellite event of the Joint Conference on Theory and Practice of Software (ETAPS 2018).

Steering Committee

Filippo Bonchi	University of Pisa, Italy
Marcello Bonsangue	Leiden University, The Netherlands
Corina Cîrstea	University of Southampton, UK
Ichiro Hasuo	National Institute of Informatics, Japan
Bart Jacobs	Radboud University Nijmegen, The Netherlands
Bartek Klin	University of Warsaw, Poland
Alexander Kurz	University of Leicester, UK
Marina Lenisa	University of Udine, Italy
Stefan Milius (Chair)	Friedrich-Alexander-Universität Erlangen-Nürnberg, Germany
Lawrence Moss	Indiana University, USA
Dirk Pattinson	Australian National University, Australia
Jan Rutten	Radboud University Nijmegen, The Netherlands
Lutz Schröder	Friedrich-Alexander-Universität Erlangen-Nürnberg, Germany
Alexandra Silva	University College London, UK

Programme Committee

Filippo Bonchi	University of Pisa, Italy
Marcello Bonsangue	Leiden University, The Netherlands
Corina Cîrstea (Chair)	University of Southampton, UK
Fredrik Dahlqvist	University College London, UK
Ugo Dal Lago	University of Bologna, Italy
Sergey Goncharov	Friedrich-Alexander-Universität Erlangen-Nürnberg, Germany
Helle Hvid Hansen	Delft University of Technology, The Netherlands
Ichiro Hasuo	National Institute of Informatics, Japan
Bart Jacobs	Radboud University Nijmegen, The Netherlands
Bartek Klin	University of Warsaw, Poland
Paul Levy	University of Birmingham, UK
Stefan Milius	Friedrich-Alexander-Universität Erlangen-Nürnberg, Germany
Lawrence Moss	Indiana University, USA
Dirk Pattinson	Australian National University, Australia
Dusko Pavlovic	University of Hawaii, USA
Daniela Petrisan	Université Diderot Paris 7, France
Damien Pous	CNRS and ENS Lyon, France

Juriaan Rot	Radboud University Nijmegen, The Netherlands
Jan Rutten	Radboud University Nijmegen, The Netherlands
Lutz Schröder	Friedrich-Alexander-Universität Erlangen-Nürnberg, Germany
Alexandra Silva	University College London, UK
Ana Sokolova	University of Salzburg, Austria
Henning Urbat	Technische Universität Braunschweig, Germany
Jamie Vicary	University of Oxford, UK

Publicity Chair

Henning Basold	CNRS and ENS Lyon, France

Additional Reviewers

Henning Basold	Jens Seeber
Antonin Delpeuch	David Sprunger
Stefano Gogioso	Natsuki Urabe
Reuben Rowe	Thorsten Wißmann
Alexis Saurin	

Sponsoring Institutions

IFIP WG 1.3
Logical Methods in Computer Science e.V.

Contents

Relating Structure and Power: Comonadic Semantics for Computational Resources
Extended Abstract

Samson Abramsky$^{(\boxtimes)}$ and Nihil Shah

Department of Computer Science, University of Oxford, Oxford, UK
samson.abramsky@cs.ox.ac.uk, nihil@berkeley.edu

Abstract. Combinatorial games are widely used in finite model theory, constraint satisfaction, modal logic and concurrency theory to characterize logical equivalences between structures. In particular, Ehrenfeucht-Fraïssé games, pebble games, and bisimulation games play a central role. We show how each of these types of games can be described in terms of an indexed family of comonads on the category of relational structures and homomorphisms. The index k is a resource parameter which bounds the degree of access to the underlying structure. The coKleisli categories for these comonads can be used to give syntax-free characterizations of a wide range of important logical equivalences. Moreover, the coalgebras for these indexed comonads can be used to characterize key combinatorial parameters: tree-depth for the Ehrenfeucht-Fraïssé comonad, tree-width for the pebbling comonad, and synchronization-tree depth for the modal unfolding comonad. These results pave the way for systematic connections between two major branches of the field of logic in computer science which hitherto have been almost disjoint: categorical semantics, and finite and algorithmic model theory.

1 Introduction

There is a remarkable divide in the field of logic in Computer Science, between two distinct strands: one focussing on semantics and compositionality ("Structure"), the other on expressiveness and efficiency ("Power"). It is remarkable because these two fundamental aspects of our field are studied using almost disjoint technical languages and methods, by almost disjoint research communiities. We believe that bridging this divide is a major issue in Computer Science, and may hold the key to fundamental advances in the field.

In this paper, we develop a novel approach to relating categorical semantics, which exemplifies the first strand, to finite model theory, which exemplifies the second. It builds on the ideas introduced in [1], but goes much further, showing clearly that there is a strong and robust connection, which can serve as a basis for many further developments.

© IFIP International Federation for Information Processing 2018
Published by Springer Nature Switzerland AG 2018. All Rights Reserved
C. Cîrstea (Ed.): CMCS 2018, LNCS 11202, pp. 1–5, 2018.
https://doi.org/10.1007/978-3-030-00389-0_1

1.1   The Setting

Relational structures and the homomorphisms between them play a fundamental rôle in finite model theory, constraint satisfaction and database theory. The existence of a homomorphism $A \to B$ is an equivalent formulation of constraint satisfaction, and also equivalent to the preservation of existential positive sentences [4]. This setting also generalizes what has become a central perspective in graph theory [5].

1.2   Model Theory and Deception

In a sense, the purpose of model theory is "deception". It allows us to see structures not "as they really are", *i.e.* up to isomorphism, but only up to *definable properties*, where definability is relative to a logical language $\mathcal{L}$. The key notion is *logical equivalence* $\equiv^{\mathcal{L}}$. Given structures $\mathcal{A}$, $\mathcal{B}$ over the same vocabulary:

$$\mathcal{A} \equiv^{\mathcal{L}} \mathcal{B} \iff^{\Delta} \forall \varphi \in \mathcal{L}. \ \mathcal{A} \models \varphi \iff \mathcal{B} \models \varphi.$$

If a class of structures $\mathcal{K}$ is definable in $\mathcal{L}$, then it must be saturated under $\equiv^{\mathcal{L}}$. Moreover, for a wide class of cases of interest in finite model theory, the converse holds [6].

The idea of syntax-independent characterizations of logical equivalence is quite a classical one in model theory, exemplified by the Keisler-Shelah theorem [10]. It acquires additional significance in finite model theory, where model comparison games such as Ehrenfeucht-Fraïssé games, pebble games and bisimulation games play a central role [7].

We offer a new perspective on these ideas. We shall study these games, not as external artefacts, but as semantic constructions in their own right. Each model-theoretic comparison game encodes "deception" in terms of limited access to the structure. These limitations are indexed by a parameter which quantifies the resources which control this access. For Ehrenfeucht-Fraïssé games, this is the number of rounds; for pebble games, the number of pebbles; and for bisimulation games, the modal depth.

2   Main Results

We now give a conceptual overview of our main results. Technical details are provided in [2].

We shall consider three forms of model comparison game: Ehrenfeucht-Fraïssé games, pebble games and bisimulation games [7]. For each of these notions of game G, and value of the resource parameter k, we shall define a corresponding *comonad* $\mathbb{C}_k$ on the category of relational structures and homomorphisms over some relational vocabulary. For each structure $\mathcal{A}$, $\mathbb{C}_k \mathcal{A}$ is another structure over the same vocabulary, which encodes the limited access to $\mathcal{A}$ afforded by playing the game on $\mathcal{A}$ with k resources. There is always an associated homomorphism $\varepsilon_{\mathcal{A}} : \mathbb{C}_k \mathcal{A} \to \mathcal{A}$ (the *counit* of the comonad), so that $\mathbb{C}_k \mathcal{A}$ "covers" $\mathcal{A}$. Moreover, given a homomorphism $h : \mathbb{C}_k \mathcal{A} \to \mathcal{B}$, there is a *Kleisli coextension* homomorphism $h^* : \mathbb{C}_k \mathcal{A} \to \mathbb{C}_k \mathcal{B}$. This allows us to form the *coKleisli category* $\mathsf{Kl}(\mathbb{C}_k)$ for the comonad. The objects are relational structures, while the morphisms from $\mathcal{A}$ to $\mathcal{B}$ in $\mathsf{Kl}(\mathbb{C}_k)$ are exactly the homomorphisms of the form $\mathbb{C}_k \mathcal{A} \to \mathcal{B}$. Composition of these morphisms uses the Kleisli coextension. The connection between this construction and the corresponding form of game G is expressed by the following result:

Theorem 1. *The following are equivalent:*

1. *There is a coKleisli morphism $\mathbb{C}_k \mathcal{A} \to \mathcal{B}$*
2. *Duplicator has a winning strategy for the existential G-game with k resources, played from $\mathcal{A}$ to $\mathcal{B}$.*

The existential form of the game has only a "forth" aspect, without the "back". This means that Spoiler can only play in $\mathcal{A}$, while Duplicator only plays in $\mathcal{B}$. This corresponds to the asymmetric form of the coKleisli morphisms $\mathbb{C}_k \mathcal{A} \to \mathcal{B}$. Intuitively, Spoiler plays in $\mathbb{C}_k \mathcal{A}$, which gives them limited access to $\mathcal{A}$, while Duplicator plays in $\mathcal{B}$. The Kleisli coextension guarantees that Duplicator's strategies can always be lifted to $\mathbb{C}_k \mathcal{B}$; while we can always compose a strategy $\mathbb{C}_k \mathcal{A} \to \mathbb{C}_k \mathcal{B}$ with the counit on $\mathcal{B}$ to obtain a coKleisli morphism.

This asymmetric form may seem to limit the scope of this approach, but in fact this is not the case. For each of these comonads $\mathbb{C}_k$, we have the following three equivalences:

- $\mathcal{A} \rightleftarrows_k \mathcal{B}$ iff there are coKleisli morphisms $\mathbb{C}_k \mathcal{A} \to \mathcal{B}$ and $\mathbb{C}_k \mathcal{B} \to \mathcal{A}$. Note that there need be no relationship between these morphisms.
- $\mathcal{A} \cong_{\mathsf{Kl}(\mathbb{C}_k)} \mathcal{B}$ iff $\mathcal{A}$ and $\mathcal{B}$ are isomorphic in the coKleisli category $\mathsf{Kl}(\mathbb{C}_k)$. This means that there are morphisms $\mathbb{C}_k \mathcal{A} \to \mathcal{B}$ and $\mathbb{C}_k \mathcal{B} \to \mathcal{A}$ which are inverses of each other in $\mathsf{Kl}(\mathbb{C}_k)$.

Clearly, $\cong_{\mathsf{Kl}(\mathbb{C}_k)}$ strictly implies $\rightleftarrows_k$. We can also define an intermediate "back-and-forth" equivalence $\leftrightarrow_k$, parameterized by a winning condition $\mathsf{W}_{\mathcal{A},\mathcal{B}} \subseteq \mathbb{C}_k \mathcal{A} \times \mathbb{C}_k \mathcal{B}$.

For each of our three types of game, there are corresponding fragments $\mathcal{L}_k$ of first-order logic:

- For Ehrenfeucht-Fraïssé games, $\mathcal{L}_k$ is the fragment of quantifier-rank $\leq k$.
- For pebble games, $\mathcal{L}_k$ is the k-variable fragment.
- For bismulation games over relational vocabularies with symbols of arity at most 2, $\mathcal{L}_k$ is the modal fragment [3] with modal depth $\leq k$.

In each case, we write $\exists \mathcal{L}_k$ for the existential positive fragment of $\mathcal{L}_k$, and $\mathcal{L}_k^\#$ for the extension of $\mathcal{L}_k$ with counting quantifiers [7].

We can now state our first main result, in a suitably generic form.

Theorem 2. *For finite structures $\mathcal{A}$ and $\mathcal{B}$:*

$$(1)\ \mathcal{A} \equiv^{\exists\mathcal{L}_k} \mathcal{B} \iff \mathcal{A} \rightleftarrows_k \mathcal{B}.$$

$$(2)\ \mathcal{A} \equiv^{\mathcal{L}_k} \mathcal{B} \iff \mathcal{A} \leftrightarrow_k \mathcal{B}.$$

$$(3)\ \mathcal{A} \equiv^{\mathcal{L}_k^{\#}} \mathcal{B} \iff \mathcal{A} \cong_{\mathsf{KI}(\mathbb{C}_k)} \mathcal{B}.$$

Note that this is really a family of three theorems. Thus in each case, we capture the salient logical equivalences in syntax-free, categorical form.

We now turn to the significance of indexing by the resource parameter k. When $k \leq l$, we have a natural inclusion morphism $\mathbb{C}_k\mathcal{A} \rightarrow \mathbb{C}_l\mathcal{A}$, since playing with k resources is a special case of playing with $l \geq k$ resources. This tells us that the smaller k is, the easier it is to find a morphism $\mathbb{C}_k\mathcal{A} \rightarrow \mathcal{B}$. Intuitively, the more we restrict Spoiler's abilities to access the structure of $\mathcal{A}$, the easier it is for Duplicator to win the game.

The contrary analysis applies to morphisms $A \rightarrow \mathbb{C}_k\mathcal{B}$. The smaller k is, the *harder* it is find such a morphism. Note, however, that if $\mathcal{A}$ is a finite structure of cardinality k, then $\mathcal{A} \rightleftarrows_k \mathbb{C}_k\mathcal{A}$. In this case, with k resources we can access the whole of $\mathcal{A}$. What can we say when k is strictly smaller than the cardinality of $\mathcal{A}$?

It turns out that there is a beautiful connection between these indexed comonads and combinatorial invariants of structures. This is mediated by the notion of *coalgebra*, another fundamental (and completely general) aspect of comonads. A coalgebra for a comonad $\mathbb{C}_k$ on a structure $\mathcal{A}$ is a morphism $\mathcal{A} \rightarrow \mathbb{C}_k\mathcal{A}$ satisfying certain properties. We define the *coalgebra number* of a structure $\mathcal{A}$, with respect to the indexed family of comonads $\mathbb{C}_k$, to be the least k such that there is a $\mathbb{C}_k$-coalgebra on $\mathcal{A}$.

We now come to our second main result.

Theorem 3.

- *For the pebbling comonad, the coalgebra number of $\mathcal{A}$ corresponds precisely to the* treewidth *of $\mathcal{A}$.*
- *For the Ehrenfeucht-Fraïssé comonad, the coalgebra number of $\mathcal{A}$ corresponds precisely to the* tree-depth *of $\mathcal{A}$ [8].*
- *For the modal comonad, the coalgebra number of $\mathcal{A}$ corresponds precisely to the* forest depth *of $\mathcal{A}$.*

The main idea behind these results is that coalgebras on $\mathcal{A}$ are in bijective correspondence with decompositions of $\mathcal{A}$ of the appropriate form. We thus obtain categorical characterizations of these key combinatorial invariants.

References

1. Abramsky, S., Dawar, A., Wang, P.: The pebbling comonad in finite model theory. In: 2017 32nd Annual ACM/IEEE Symposium on Logic in Computer Science (LICS), pp. 1–12. IEEE (2017)
2. Abramsky, S., Shah, N.: Relating structure to power: comonadic semantics for computational resources 2018. To appear in Proceedings of Computer Science Logic (2018)
3. Andréka, H., Németi, I., van Benthem, J.: Modal languages and bounded fragments of predicate logic. J. Philos. Log. **27**(3), 217–274 (1998)
4. Chandra, A.K., Merlin, P.M.: Optimal implementation of conjunctive queries in relational data bases. In: Proceedings of the Ninth Annual ACM Symposium on Theory of Computing, pp. 77–90. ACM (1977)
5. Hell, P., Nesetril, J.: Graphs and Homomorphisms. Oxford University Press, Oxford (2004)
6. Kolaitis, P.G., Vardi, M.Y.: Infinitary logics and 0–1 laws. Inf. Comput. **98**(2), 258–294 (1992)
7. Libkin, L.: Elements of Finite Model Theory. Texts in Theoretical Computer Science. An EATCS Series. Springer, Heidelberg (2004). https://doi.org/10.1007/978-3-662-07003-1
8. Něsetřil, J., De Mendez, P.O.: A unified approach to structural limits, and limits of graphs with bounded tree-depth. arXiv preprint, page arXiv:1303.6471 (2013)
9. Shah, N.: Game comonads in finite model theory. Master's thesis, University of Oxford (2017)
10. Shelah, S.: Every two elementarily equivalent models have isomorphic ultrapowers. Isr. J. Math. **10**(2), 224–233 (1971)

Coalgebraic Logics & Duality

Clemens Kupke[(✉)]

Department of Computer and Information Sciences, University of Strathclyde,
Glasgow, Scotland
`clemens.kupke@strath.ac.uk`

Abstract. I will provide a brief introduction to coalgebraic modal logics
and highlight a few central concepts concerning these logics. After that
I will outline my current research in the area.

This note is not a survey of coalgebraic logics such as [1,2]. Instead, I am
going to highlight ideas that continue to be important for my research within
the area. A leitmotif is the fundamental role played by duality theory.

1 Logics for Coalgebras

The concepts *behaviour* and *observation* are central for the coalgebraic modelling
of systems. Whereas behaviour is formalised within the theory of Universal Coalgebra [3] via bisimilarity and finality, it is less clear how to devise a matching
notion of observations that allows to formally specify, verify and reason about
this behaviour. Providing such a theory of observations is an important goal that
has been driving the development of coalgebraic logics.

Why Modal Logic? A simple answer to this question is that basic modal logic
is the logic of Kripke frames and Kripke frames are coalgebras for the covariant
power set functor $\mathcal{P}$. More importantly, modal logics usually express properties
that are *invariant under bisimulations* which matches our intuition that formulas
of a coalgebraic logic should allow to observe coalgebraic behaviour. In addition
to that, a more categorical answer was provided in [4–6] where it was shown
that the abstract relationship between coalgebra and modal logic dualises the
fundamental link between algebra and equational logic. A basic problem that had
to be overcome was to devise suitable (and usable!) modal languages that would
allow to specify properties about coalgebras. Probably the two most successful
proposals where on the one hand Moss' ∇-modality [7,8] (which originally was
denoted by Δ) that made the radical step to use the coalgebraic type functor
as a syntax constructor of the logic and, on the other hand, Pattinson's logic
given by predicate liftings [9]. Another important line of research was to use the
syntactic structure of polynomial functors to inductively define corresponding
modal operators [10–12]. This research helped to develop one of the key features

C. Cîrstea (Ed.): CMCS 2018, LNCS 11202, pp. 6–12, 2018.
https://doi.org/10.1007/978-3-030-00389-0_2

of coalgebraic modal logics: languages and deduction systems can be composed in an elegant, seamless fashion [13,14] that mirrors the composition of functors.

Expressive Languages. One criterion for what a suitable language for specifying coalgebra is, is the so-called Hennessy-Milner property stating that two coalgebra states are bisimilar iff they satisfy the same formulas of the language. A language that satisfies this property is often called *expressive*. It is clear that expressive languages do not exist for functors for which there is no final coalgebra [15–17]. An important positive result in coalgebraic modal logic states that for finitary set functors there is a always an expressive language of predicate liftings [18]. Its proof uses an alternative characterisation of predicate liftings via the Yoneda Lemma. Similarly, for finitary, weak pullback preserving set functors, the ∇-language is always expressive, a statement that is easily proven using terminal sequence induction. Other positive results include functors on the category of Stone spaces [19] and measure spaces [20]. In these cases the proof of expressivity goes via a logical construction of final coalgebras that proves expressivity of the language at the same time as completeness of the logic.

Logics via a Dual Adjunction. All modal languages for coalgebras can be abstractly described via a dual adjunction

$$T \, \righthalfarrow \, \mathcal{C} \underset{G}{\overset{F}{\underset{\perp}{\rightleftarrows}}} \mathcal{D}^{\mathrm{op}} \, \lefthalfarrow \, L \tag{1}$$

together with a natural transformation $\delta : L \circ F \to F \circ T$, sometimes referred to as the "one-step semantics" of the language. One of the first papers advocating this view was probably [21] but only for the restricted case of the well-known dual equivalence between the category of Stone Spaces and the category of Boolean algebras whereas the more general picture above was fully developed in [22]. The abstract approach allows on the one hand to formulate properties of the logic as properties of δ, eg., completeness of the logic is linked to δ being componentwise mono whereas expressivity to its mate $\delta^\sharp : T \circ G \to G \circ L$ having that property (cf. [22,23]). Since then several researchers have pushed this approach significantly further covering - for example - positive modal logics [24], process algebra [25] and logics for trace equivalence [26], to name just a few.

2   The Power Law for ∇

A Distributive Law For ∇. While the duality-based approach to coalgebraic logics originated for logics based on predicate liftings, it is not too difficult to see that Moss' original ∇-logic also fits into the framework [2]. Key for showing this is the following distributive law

$$\rho_\nabla : T \circ \mathcal{P} \to \mathcal{P} \circ T$$
$$\pi \mapsto \{t \in TX \mid (t, \pi) \in \overline{T}(\in)\}$$

that exists for all weak pullback preserving set functors T (the law has been called the *power law* in [27]). Here $\overline{T}$ denotes the unique extension of the set functor T to a *relator* [28]. The significance of this law, however, goes far beyond being the one-step semantics of the ∇-logic. It forms the basis of the definition of so-called redistributions [29,30]. Roughly speaking, an element $\Xi \in T\mathcal{P}X$ is called *redistribution* of some $\Pi \in \mathcal{P}TX$ if $\Pi \subseteq \rho_\nabla(\Xi)$. Redistributions allow to formulate an important *logical distributive law* for the ∇-logic

$$\bigwedge_{\pi \in \Pi} \nabla\pi \leftrightarrow \bigvee_{\Xi \in \mathrm{SRD}(\Pi)} \nabla(T\wedge)(\Xi). \tag{2}$$

where $\mathrm{SRD}(\Pi)$ the collection of all (slim) redistributions of Π. The law is the key for the work [30,31] developing a complete deduction system for the ∇-logic that is entirely parametric in the set functor T.

Coalgebraic Fixpoint Logics. The coalgebraic logics that I have discussed so far can only formulate the finite-step behaviour of a coalgebra. For properties such as liveness ("something will happen infinitely often in the future") and safety ("at no point in the future the systems will crash") we need to be able to specify the ongoing, possibly infinite behaviour. A coalgebraic treatment of fixpoint-logics is difficult as duality-based techniques cannot be applied easily [32]. This makes completeness proofs for such logics notoriously hard. Nevertheless these logics have been studied successfully on a coalgebraic level. In the first instance, the focus was on automata for coalgebraic fixpoint logics that employed the ∇-operator [33]. The above logical distributive law (2) provided the key to prove important closure properties [29] of these "coalgebra automata" and thus a general finite model property and decidability result. After these initial proof-of-concept results attention shifted to fixpoint logics using predicate liftings. Both automata [34] and tableau-systems [35] were developed - the role of the power law and of redistributions is played by the assumption that the given predicate liftings come with a sound and complete axiomatisation via so-called one-step rules [36]. Apart from these results on checking satisfiability of coalgebraic fixpoint logic research on complete axiomatisations also made gradual progress, first for so-called flat coalgebraic fixpoint logics [37,38], later for coalgebraic dynamic logics [39] and finally with a recent breakthrough result on completeness for the full coalgebraic μ-calculus [40].

3   Current Research

I am now going to list research directions within coalgebraic logic that I am currently focusing on and that I am planning to discuss in my talk.

Coalgebra Automata and Duality. Coalgebra automata play an important role for studying the coalgebraic μ-calculus: not only do they provide a tool for deciding satisfiability but they are also instrumental in completeness proofs, cf. e.g. [40]. Building on our recent work for game logic [41] we are working on devising automata for coalgebraic dynamic logics. Furthermore we plan to develop automata that operate on coalgebras over Stone Spaces. A first step in this direction was made in [42] where we obtained a characterisation of the clopen semantics of the (standard) μ-calculus in terms of parity games. The long term goal is a completeness proof for coalgebraic fixpoint logics via a duality theoretic argument.

Learning and Duality. In recent work [43] we devised a generalisation of Angluin's well-known L^*-algorithm for learning regular languages [44]. The generalisation can be summarised in the following (informally stated) theorem that holds for any finitary set functor T.

Theorem 1. *Let $(\mathbb{X}, x)$ be a pointed T-coalgebra that is behaviourally equivalent to a finite well-pointed (=minimal & reachable) T-coalgebra $(\mathbb{Y}, y)$. Let $\mathcal{L}$ be an expressive language for T-coalgebras. Our algorithm determines the well-pointed coalgebra $(\mathbb{Y}, y)$ using queries and counter-examples from $\mathcal{L}$.*

A key observation that led to the algorithm is that Angluin's algorithm essentially learns modal filtrations. These filtrations can be defined relative to any coalgebraic logic. In my talk I will discuss the above theorem and report on ongoing work on fitting filtrations and thus learning into the dual adjunction framework of coalgebraic logic.

Possible Application: Iterated Games. In our recent paper [45] we use the framework of open games [46] to represent an infinitely iterated strategic game (such as the well-known Prisoner's Dilemma) as a final coalgebra. As a byproduct, this work allows to characterise subgame perfect equilibiria using the (standard) coalgebraic μ-calculus. To give the reader a rough idea, let me spell out some of the details. Consider a simple game (think of the Prisoner's Dilemma) with set of moves Y where an element of Y represents the moves played by all players simultaneously. A play of the infinitely iterated game is an infinite sequence of moves $\rho \in Y^\omega$, strategies in this game are pointed coalgebras of the form

$$\langle \mathrm{now}, \mathrm{ltr} \rangle : X \to Y \times X^Y$$

where at each state $x \in X$ the coalgebra map determines the next move $\mathrm{now}(x)$ and moves to the next state $\mathrm{ltr}(x)(y')$ depending on which move $y' \in Y$ has been actually carried out in one round of the game. Pay-off functions for the infinitely iterated game are of type $k : Y^\omega \to R$ (where R is typically of the form $\mathbb{R}^n$ for an n-player game). This leads us to define a coalgebra

$$\langle \overline{\mathrm{now}}, \overline{\mathrm{ltr}} \rangle : (X \times R^{Y^\omega}) \longrightarrow Y \times (X \times R^{Y^\omega})^Y$$

by putting

$$\overline{\mathrm{now}}(x, k) := \mathrm{now}(x)$$
$$\overline{\mathrm{ltr}}(x, k) := \lambda y.\langle \mathrm{ltr}(x)(y), k_y \rangle$$

where $k_y(\rho) := k(y\rho)$ for $y \in Y$, $\rho \in Y^\omega$. The intuition behind this definition is to record the current strategy of the players and the payoff function - both based on the history of the game played thus far. With these definitions in place it is not difficult to see that subgame perfect equilibria in the infinitely repeated game can be characterised via a μ-calculus formula $\psi = \nu X.P \wedge \Box X$ for a suitable predicate P that is defined using the equilibrium of the stage game. The obtained characterisation has the following form:

$$(x, k) \models \psi \quad \text{iff } x \text{ represents an s.p.equilibrium of the game with payoff } k.$$

While the coalgebra $\langle \overline{\text{now}}, \overline{\text{ltr}} \rangle$ will in general be infinite, assumptions on the pay-off function (such as discounted sum) will allow us to obtain a finite equivalent coalgebra. In my talk I will provide the details of this construction and I will explain how this observation connects the afore mentioned automata and learning techniques to game theory.

Acknowledgements. The overview of current research is based on joint work with Simone Barlocco, Nick Bezhanisvili, Neil Ghani, Helle Hvid Hansen, Alasdair Lambert, Johannes Marti, Fredrik Nordvall Forsberg, Jurriaan Rot and Yde Venema.

References

1. Cîrstea, C., Kurz, A., Pattinson, D., Schröder, L., Venema, Y.: Modal logics are coalgebraic. Comput. J. **54**, 31–41 (2011)
2. Kupke, C., Pattinson, D.: Coalgebraic semantics of modal logics: an overview. TCS **412**(38), 5070–5094 (2011)
3. Rutten, J.: Universal coalgebra: a theory of systems. Theor. Comput. Sci. **249**, 3–80 (2000)
4. Gumm, H.P., Schröder, T.: Covarieties and complete covarieties. Theor. Comput. Sci. **260**(1–2), 71–86 (2001)
5. Kurz, A.: A co-variety-theorem for modal logic. In: Advances in Modal Logic, vol. 2, CSLI (2001). Selected Papers from AiML 2, Uppsala, 1998
6. Kurz, A.: Logics for Coalgebras and Applications to Computer Science. Ph.D. thesis, Ludwig-Maximilians-Universität (2000)
7. Moss, L.S.: Coalgebraic logic. Ann. Pure Appl. Logic **96**(1–3), 277–317 (1999)
8. Moss, L.S.: Erratum to "coalgebraic logic": Ann. pure appl. logic 96 (1999) 277–317. Ann. Pure Appl. Logic **99**(1–3) (1999) 241–259
9. Pattinson, D.: Coalgebraic modal logic: soundness, completeness and decidability of local consequence. Theor. Comput. Sci. **309**(1–3), 177–193 (2003)
10. Jacobs, B.: Many-sorted coalgebraic modal logic: a model-theoretic study. ITA **35**(1), 31–59 (2001)
11. Rößiger, M.: Coalgebras and modal logic. Electr. Notes Theor. Comput. Sci. **33**, 294–315 (2000)
12. Goldblatt, R.: Equational logic of polynomial coalgebras. In: Balbiani, P., Suzuki, N., Wolter, F., Zakharyaschev, M. (eds.) AIML 2002, pp. 149–184 (2002)
13. Cîrstea, C.: A compositional approach to defining logics for coalgebras. Theor. Comput. Sci. **327**(1–2), 45–69 (2004)
14. Cîrstea, C., Pattinson, D.: Modular construction of complete coalgebraic logics. Theor. Comput. Sci. **388**(1–3), 83–108 (2007)

15. Goldblatt, R.: Final coalgebras and the hennessy-milner property. Ann. Pure Appl. Logic **138**(1–3), 77–93 (2006)
16. Kupke, C., Leal, R.A.: Characterising behavioural equivalence: three sides of one coin. In: Kurz, A., Lenisa, M., Tarlecki, A. (eds.) CALCO 2009. LNCS, vol. 5728, pp. 97–112. Springer, Heidelberg (2009). https://doi.org/10.1007/978-3-642-03741-2_8
17. Levy, P.B.: Final coalgebras from corecursive algebras. In: Moss, L.S., Sobocinski, P. (eds.) 6th Conference on Algebra and Coalgebra in Computer Science (CALCO 2015). LIPIcs, vol. 35, pp. 221–237. Schloss Dagstuhl (2015)
18. Schröder, L.: Expressivity of coalgebraic modal logic: the limits and beyond. Theor. Comput. Sci. **390**(2), 230–247 (2008)
19. Kupke, C., Kurz, A., Venema, Y.: Stone coalgebras. Theor. Comput. Sci. **327**, 109–134 (2004)
20. Moss, L.S., Viglizzo, I.D.: Final coalgebras for functors on measurable spaces. Inf. Comput. **204**(4), 610–636 (2006)
21. Kupke, C., Kurz, A., Pattinson, D.: Algebraic semantics for coalgebraic modal logic. In: Adámek, J. (ed.) Proceedings of the Workshop on Coalgebraic Methods in Computer Science (CMCS). Electronic Notes in Theoretical Computer Science, vol. 106 (2004)
22. Klin, B.: Coalgebraic modal logic beyond sets. ENTCS **173**, 177–201 (2007)
23. Jacobs, B., Sokolova, A.: Exemplaric expressivity of modal logics. J. Log. Comput. **20**(5), 1041–1068 (2010)
24. Dahlqvist, F., Kurz, A.: The positivication of coalgebraic logics. In: Bonchi, F., König, B. (eds.) CALCO 2017. LIPIcs, vol. 72, pp. 9:1–9:15 (2017)
25. Klin, B.: Bialgebraic methods and modal logic in structural operational semantics. Inf. Comput. **207**(2), 237–257 (2009)
26. Klin, B., Rot, J.: Coalgebraic trace semantics via forgetful logics. Log. Methods Comput. Sci. **12**(4) (2016)
27. Jacobs, B.: Trace semantics for coalgebras. ENTCS **106**, 167–184 (2004)
28. Rutten, J.: Relators and metric bisimulation (extended abstract). Electron. Notes Theor. Comput. Sci. **11**, 1–7 (1998)
29. Kupke, C., Venema, Y.: Coalgebraic automata theory: basic results. Log. Methods Comput. Sci. **4**(4) (2008)
30. Kupke, C., Kurz, A., Venema, Y.: Completeness for the coalgebraic cover modality. Log. Methods Comput. Sci. **8**(3) (2012)
31. Bílková, M., Palmigiano, A., Venema, Y.: Proof systems for Moss' coalgebraic logic. Theor. Comput. Sci. **549**, 36–60 (2014)
32. Santocanale, L.: Completions of μ-algebras. Ann. Pure Appl. Log. **154**(1), 27–50 (2008)
33. Venema, Y.: Automata and fixpoint logics: a coalgebraic perspective. Inf. Comp. **204**, 637–678 (2006)
34. Fontaine, G., Leal, R., Venema, Y.: Automata for coalgebras: an approach using predicate liftings. In: Abramsky, S., Gavoille, C., Kirchner, C., Meyer auf der Heide, F., Spirakis, P.G. (eds.) ICALP 2010. LNCS, vol. 6199, pp. 381–392. Springer, Heidelberg (2010). https://doi.org/10.1007/978-3-642-14162-1_32
35. Cîrstea, C., Kupke, C., Pattinson, D.: EXPTIME tableaux for the coalgebraic μ-calculus. Log. Methods Comput. Sci. **7**(3) (2011)
36. Schröder, L., Pattinson, D.: PSPACE bounds for rank-1 modal logics. ACM Trans. Comput. Logic **10**(2), 13:1–13:33 (2009)
37. Santocanale, L., Venema, Y.: Completeness for flat modal fixpoint logics. Ann. Pure Appl. Logic **162**(1), 55–82 (2010)

38. Schröder, L., Venema, Y.: Flat coalgebraic fixed point logics. In: Gastin, P., Laroussinie, F. (eds.) CONCUR 2010. LNCS, vol. 6269, pp. 524–538. Springer, Heidelberg (2010). https://doi.org/10.1007/978-3-642-15375-4_36
39. Hansen, H.H., Kupke, C.: Weak completeness of coalgebraic dynamic logics. In: Matthes, R., Mio, M. (eds.) FICS 2015, pp. 90–104 (2015)
40. Enqvist, S., Seifan, F., Venema, Y.: Completeness for coalgebraic fixpoint logic. In: CSL 2016, pp. 7:1–7:19 (2016)
41. Hansen, H.H., Kupke, C., Marti, J., Venema, Y.: Parity games and automata for game logic. In: Madeira, A., Benevides, M. (eds.) DALI 2017. LNCS, vol. 10669, pp. 115–132. Springer, Cham (2018). https://doi.org/10.1007/978-3-319-73579-5_8
42. Bezhanishvili, N., Kupke, C.: Games for topological fixpoint logic. In: Cantone, D., Delzanno, G. (eds.) GandALF 2016. EPTCS, vol. 226, pp. 46–60 (2016)
43. Barlocco, S., Kupke, C.: Angluin learning via logic. In: Artemov, S., Nerode, A. (eds.) LFCS 2018. LNCS, vol. 10703, pp. 72–90. Springer, Cham (2018). https://doi.org/10.1007/978-3-319-72056-2_5
44. Angluin, D.: Learning regular sets from queries and counterexamples. Inf. Comput. **75**(2), 87–106 (1987)
45. Ghani, N., Kupke, C., Lambert, A., Forsberg, F.N.: A compositional treatment of iterated open games. CoRR abs/1711.07968 (2017)
46. Ghani, N., Hedges, J.: A compositional approach to economic game theory. CoRR abs/1603.04641 (2016)

Intuitionistic Podelski-Rybalchenko Theorem and Equivalence Between Inductive Definitions and Cyclic Proofs

Stefano Berardi[1] and Makoto Tatsuta[2]([⊠])

[1] Università di Torino, Torino, Italy
stefano@di.unito.it
[2] National Institute of Informatics/Sokendai, Tokyo, Japan
tatsuta@nii.ac.jp

Abstract. A cyclic proof system gives us another way of representing inductive and coinductive definitions and efficient proof search. Podelski-Rybalchenko termination theorem is important for program termination analysis. This paper first shows that Heyting arithmetic HA proves Kleene-Brouwer theorem for induction and Podelski-Rybalchenko theorem for induction. Then by using this theorem this paper proves the equivalence between the provability of the intuitionistic cyclic proof system and that of the intuitionistic system of Martin-Lof's inductive definitions when both systems contain HA.

1 Introduction

This paper studies two subjects: intuitionistic Podelski-Rybalchenko theorem for induction, and equivalence between intuitionistic system of Martin-Löf's inductive definitions and an intuitionistic cyclic proof system.

Podelski-Rybalchenko theorem [18] states that if a transition invariant is a finite union of well-founded relations then the transition invariant is also well-founded. This gives us good sufficient conditions for analysis of program termination [18]. Intuitionistic provability of this theorem is also interesting; if we can show this theorem is provable in some intuitionistic logical system, the theorem also gives us not only termination but also an upper bound of computation steps of a given program. For this purpose, we have to replace well-foundedness in the theorem by induction principle, since well-foundedness is a property of negation of existence and induction principle can show a property of existence. We say *Podelski-Rybalchenko theorem for induction* when we replace well-foundedness by induction principle in Podelski-Rybalchenko theorem. [3] shows Podelski-Rybalchenko theorem for induction is provable in intuitionistic second-order logic. [5] shows that this theorem for induction is provable in Peano arithmetic, by using the fact that Peano arithmetic can formalize Ramsey theorem. However until now it was not known whether Podelski-Rybalchenko theorem for induction is provable in some intuitionistic first-order logic. This paper will show this theorem for induction is provable in Heyting arithmetic and answer this question.

C. Cîrstea (Ed.): CMCS 2018, LNCS 11202, pp. 13–33, 2018.
https://doi.org/10.1007/978-3-030-00389-0_3

An inductive/coinductive definition is a way to define a predicate by an expression which may contain the predicate itself. The predicate is interpreted by the least/greatest fixed point of the defining equation. Inductive/coinductive definitions are important in computer science, since they can define useful recursive data structures such as lists, trees, and streams, and useful notions such as bisimulations. Inductive definitions are important also in mathematical logic, since they increase the proof theoretic strength. Martin-Löf's system of inductive definitions given in [16] is one of the most popular systems of inductive definitions. This system has production rules for an inductive predicate, and the production rule determines the introduction rules and the elimination rules for the predicate.

[8, 11] proposed an alternative formalization of inductive definitions, called a cyclic proof system. A proof, called *a cyclic proof*, is defined by proof search, going upwardly in a proof figure. If we encounter the same sequent (called a *bud*) as some sequent we already passed (called a *companion*) we can stop. The induction rule is replaced by a case rule, for this purpose. The soundness is guaranteed by some additional condition, called a *global trace condition*, which can show the case rule decreases some measure of a bud from that of the companion. In general, for proof search, a cyclic proof system can find an induction formula in a more efficient way than Martin-Löf's system, since a cyclic proof system does not have to choose fixed induction formulas in advance. A cyclic proof system enables us to get efficient implementation of theorem provers with inductive definitions [7, 9, 10, 12]. A cyclic proof system can also give us another logical system for coinductive predicates, since a coinductive predicate is a dual of an inductive predicate, and sequent calculus is symmetric for this dual.

[8, 11] investigated Martin-Löf's system **LKID** of inductive definitions in classical logic for the first-order language, and the cyclic proof system **CLKID**$^\omega$ for the same language, showed the provability of **CLKID**$^\omega$ includes that of **LKID**, and conjectured the equivalence.

As the second subject, this paper studies the equivalence for intuitionistic logic, namely, the provability of the intuitionistic cyclic proof system, called **CLJID**$^\omega$, is the same as that of the intuitionistic system of Martin-Lof's inductive definitions, called **LJID**. This question is theoretically interesting, and answers will potentially give new techniques of theorem proving by cyclic proofs to type theories with inductive/coinductive types and program extraction by constructive proofs.

This paper first points out that the countermodel of [4] also shows the equivalence is false in general. Then this paper shows the equivalence is true under arithmetic, namely, the provability of **CLJID**$^\omega$ is the same as that of **LJID**, when both systems contain Heyting arithmetic **HA**.

There are not papers that study the equivalence for intuitionistic logic or Kleene-Brouwer theorem for induction in intuitionistic first-order logic. For Podelski-Rybalchenko theorem for induction, [3] intuitionistically showed it but the paper used second-order logic.

Section 2 proves Kleene-Brouwer theorem for induction and Podelski-Rybalchenko theorem for induction. Section 3 defines **LJID** and **CLJID**$^\omega$ and discuss a cyclic proof system for streams. Section 4 discusses the countermodel, defines **CLJID**$^\omega$ + **HA** and **LJID** + **HA**, states the equivalence theorem, and explains ideas of the equivalence proof. Section 5 discusses proof transformation and proves the equivalence theorem. Section 6 discusses related work. We conclude in Sect. 7.

2   HA-Provable Podelski-Rybalchenko Theorem for Induction

This section will prove Podelski-Rybalchenko theorem for induction, inside Heyting arithmetic **HA**. First we will prove Kleene-Brouwer theorem for induction, inside **HA**. This is done by carefully using some double induction. This theorem is new. Next we will show induction for the set MS of monotonically-colored subsequences. Monotonically-colored subsequences are used in ordinary proof of Ramsey theorem and we will show some intuitionistic property of them. Then by applying Kleene-Brouwer theorem to a part of MS and some orders $>_{u,\text{Left}}$ and $>_{u,\text{Right}}$, we will obtain two Kleene-Brouwer relations $>_{\text{KB}1,r}$ and $>_{\text{KB}2,r}$ and show their induction principle. These two relations are simple but necessary preparation for the next relation. Then by applying Kleene-Brouwer theorem to some lifted tree determined by $>_{\text{KB}2,r}$ and the relation $>_{\text{KB}1,r}$, we will obtain a Kleene-Brouwer relation $>_{\text{KB},r}$ and show its induction principle. This relation is a key of the proof. Then we will show that induction for decreasing transitive sequences is reduced to induction for Erdös trees with the relation $>_{\text{KB},r}$. An Erdös tree is some set of monotonically-colored sequences and implicitly used in ordinary proof of Ramsey theorem. Since Erdös trees are in the lifted tree, by combining them, finally we will prove Podelski-Rybalchenko theorem for induction.

2.1   Kleene-Brouwer Theorem

We will show Kleene-Brouwer theorem for induction, which states that if we have both induction principle for a lifted tree (namely $\langle u \rangle * T$ for some tree T) with respect to the one-step extension relation and induction principle for relations on children, then we have induction principle for the Kleene-Brouwer relation. We can prove it by refining an ordinary proof of Kleene-Brouwer theorem for orders.

We assume Heyting arithmetic **HA** is defined in an ordinary way with constants and function symbols $0, s, +, \times$. We define $x < y$ by $\exists z.x + sz = y$ and $x \leq y$ by $x = y \lor x < y$. We can assume some coding of a sequence of numbers by a number in Heyting arithmetic, because the definitions on pages 115–117 of [19] work also in **HA**. We write $\langle t_0, \ldots, t_n \rangle$ for the sequence of $t_0, \ldots, t_n$. We also write $|t|$, and $(t)_u$ for the length of the sequence t, and the u-th element of the sequence t respectively. We write $*$ for the concatenation operation of sequences.

We write $>_R$ or $>$ for a binary relation. We write $<_R$ for the binary relation of the inverse of $>_R$. For notational simplicity, we say X is a set in order to say there is some first-order formula Fx such that $x \in X \leftrightarrow Fx$. Then we also say $t \in X$ in order to say Ft. We write $y <_R x \in X$ for $y <_R x \wedge y \in X$. We write $x \in \sigma$ when x is an element of the sequence σ. We write $U^{<\omega}$ for the set of finite sequences of elements in U. For a set S of sequences, we write $\langle u \rangle * S$ for $\{\langle u \rangle * \sigma \mid \sigma \in S\}$. For a set U and a binary relation $>_R$ for U, the *induction principle* for $(U, >_R)$ is defined as

$$\mathrm{Ind}(U, >_R, F) \equiv \forall x \in U((\forall y <_R x \in U.Fy) \rightarrow Fx) \rightarrow \forall x \in U.Fx,$$
$$\mathrm{Ind}(U, >_R) \equiv \mathrm{Ind}(U, >_R, F) \text{ (for every formula } Fx).$$

For a set U a set T is called a *tree* of U if $T \subseteq U^{<\omega}$ and T is nonempty and closed under prefix operations. Note that the empty sequence is a prefix of any sequence. As a graph, the set of nodes is T and the set of edges is $\{(x, y) \in T^2 \mid y = x * \langle u \rangle\}$. We call a set $T \subseteq U^{<\omega}$ a *lifted tree* of U when there are a tree $T' \subseteq U^{<\omega}$ and $r \in U$ such that $T = \langle r \rangle * T'$. We define $\mathrm{LiftedTree}(T, U)$ as a first-order formula that means T is a lifted tree of U.

For $x, y \in U^{<\omega}$ we define the one-step extension relation $x >_{\mathrm{ext}} y$ if $y = x * \langle u \rangle$ for some u. For a set $T \subseteq U^{<\omega}$ and $\sigma \in U^{<\omega}$, we define T_σ as $\{\rho \in T \mid \rho = \sigma * \rho'\}$. Note that T_σ is a subset of T. For a nonempty sequence σ, we define $\mathrm{first}(\sigma)$ and $\mathrm{last}(\sigma)$ as the first and the last element of σ respectively.

The next lemma shows induction implies $x \not> x$. The proof is in [6].

Lemma 2.1. *If* **HA** $\vdash \mathrm{Ind}(U, >)$, *then* **HA** $\vdash \forall x, y \in U(y < x \rightarrow y \neq x)$.

Definition 2.2 (Kleene-Brouwer Relation). For a set U, a lifted tree T of U, and a set of binary relations $>_u$ on U for every $u \in U$, we define the *Kleene-Brouwer relation* $>_{\mathrm{KB}}$ for T and $\{(>_u) \mid u \in U\}$ as follows: for $x, y \in T$, $x >_{\mathrm{KB}} y$ if (1) $x = z * \langle u, u_1 \rangle * w_1$, $y = z * \langle u, u_2 \rangle * w_2$, and $u_1 >_u u_2$ for some z, u, u_1, w_1, u_2, w_2, or (2) $y = x * z$ for some $z \neq \langle \rangle$.

When $(>_u)$ is some fixed $(>)$ for all u, for simplicity we call the relation $(>_{\mathrm{KB}})$ the Kleene-Brouwer relation for T and $>$.

Note that $(>_{\mathrm{KB}})$ is a relation on T. This Kleene-Brouwer relation is slightly different from ordinary Kleene-Brouwer order for the following points: it creates a relation instead of an order, it uses a set of relations indexed by an element, and it is defined for a lifted tree instead of a tree (in order to use indexed relations).

The next theorem shows induction principle for the Kleene-Brouwer relation.

Theorem 2.3 (Kleene-Brouwer Theorem for Induction). *If* **HA** $\vdash$ $\mathrm{LiftedTree}(T, U)$, **HA** $\vdash$ $\mathrm{Ind}(T, >_{\mathrm{ext}})$ *and* **HA** $\vdash$ $\forall u \in U.\mathrm{Ind}(U, >_u)$, *then* **HA** $\vdash \mathrm{Ind}(T, >_{\mathrm{KB}})$.

Proof. By induction on $(T, >_{\mathrm{ext}})$ with the induction principle $\mathrm{Ind}(T, >_{\mathrm{ext}})$, we will show $\forall \sigma \in T.\mathrm{Ind}(T_\sigma, >_{\mathrm{KB}})$. After we prove it, we can take σ to be $\langle \rangle$ to show the theorem, since $T_{\langle \rangle} = T$.

Fix $\sigma \in T$ in order to show $\mathrm{Ind}(T_\sigma, >_{\mathrm{KB}})$. Note that we can use induction hypothesis for every $\sigma * \langle u \rangle \in T$:

$$\mathrm{Ind}(T_{\sigma*\langle u \rangle}, >_{\mathrm{KB}}). \tag{1}$$

Assume

$$\forall x \in T_\sigma((\forall y <_{\mathrm{KB}} x \in T_\sigma.Fy) \to Fx) \tag{2}$$

in order to show $\forall x \in T_\sigma.Fx$. For simplicity we write $F(X)$ for $\forall x \in X.Fx$. Let $Gu \equiv F(T_{\sigma*\langle u \rangle})$. By $\mathrm{Ind}(U, >_{\mathrm{last}(\sigma)})$ we will show the following claim.

Claim: $\forall u \in U.Gu$.

Fix $u \in U$ in order to show Gu.

By IH for v with $>_{\mathrm{last}(\sigma)}$ we have

$$v <_{\mathrm{last}(\sigma)} u \to F(T_{\sigma*\langle v \rangle}). \tag{3}$$

We can show

$$\forall x \in T_{\sigma*\langle u \rangle}((\forall y <_{\mathrm{KB}} x \in T_{\sigma*\langle u \rangle}.Fy) \to (\forall y <_{\mathrm{KB}} x \in T_\sigma.Fy)) \tag{4}$$

as follows. Fix $x \in T_{\sigma*\langle u \rangle}$, assume

$$\forall y <_{\mathrm{KB}} x \in T_{\sigma*\langle u \rangle}.Fy \tag{5}$$

and assume $y <_{\mathrm{KB}} x \in T_\sigma$ in order to show Fy. By definition of $>_{\mathrm{KB}}$, we have $y \in T_{\sigma*\langle v \rangle}$ for some $v <_{\mathrm{last}(\sigma)} u$, or $y \in T_{\sigma*\langle u \rangle}$. In the first case, Fy by (3). In the second case, Fy by (5). Hence we have shown (4).

Combining (4) with (2), we have

$$\forall x \in T_{\sigma*\langle u \rangle}((\forall y <_{\mathrm{KB}} x \in T_{\sigma*\langle u \rangle}.Fy) \to F(x)). \tag{6}$$

By IH (1) for $\sigma * \langle u \rangle$, we have $\mathrm{Ind}(T_{\sigma*\langle u \rangle}, >_{\mathrm{KB}})$, namely,

$$\forall x \in T_{\sigma*\langle u \rangle}((\forall y <_{\mathrm{KB}} x \in T_{\sigma*\langle u \rangle}.Fy) \to Fx) \to \forall x \in T_{\sigma*\langle u \rangle}.Fx. \tag{7}$$

By (6), (7), $F(T_{\sigma*\langle u \rangle})$. Hence we have shown the claim.

If $y <_{\mathrm{KB}} \sigma \in T_\sigma$, we have $y \in T_{\sigma*\langle u \rangle}$ for some u, since $y <_{\mathrm{KB}} \sigma$ implies $y \neq \sigma$ by definition of KB and Lemma 2.1 for $>_u$. By the claim, Fy. Hence

$$\forall y <_{\mathrm{KB}} \sigma \in T_\sigma.Fy. \tag{8}$$

By letting $x := \sigma$ in (2), we have $(\forall y <_{\mathrm{KB}} \sigma \in T_\sigma.Fy) \to F\sigma$. By (8), $F\sigma$. Combining it with the claim, $\forall x \in T_\sigma.Fx$. $\qquad\square$

2.2 Proof Ideas for Podelski-Rybalchenko Theorem for Induction

In this subsection we will explain proof ideas of Theorem 2.15.

A sequence $u_1 >_R u_2 >_R u_3 >_R \dots$ is called *transitive* when $u_i >_R u_j$ for any $i < j$. We say the edge from u to v is of color R when $u >_R v$. A sequence is called *monotonically-colored* when for any element there is a color such that the edge from the element to any element after it in the sequence has the same color.

Definition 2.4. For a set U and a relation $>$ for U, we define the set $\mathrm{DS}(U, >)$ of decreasing sequences as $\{\langle x_0, \ldots, x_{n-1}\rangle \mid n \geq 0, x_i \in U, \forall i < n-1.(x_i > x_{i+1})\}$.

We define the set $\mathrm{DT}(U, >)$ of decreasing transitive sequences by $\{\langle x_0, \ldots, x_{n-1}\rangle \mid n \geq 0, x_i \in U, \forall i(\forall j \leq n-1.(i < j \to x_i > x_j))\}$.

We define $>_{R_1 \cup \ldots \cup R_k}$ as the union of $>_{R_i}$ for all $1 \leq i \leq k$. We define $>_{R_1 + \ldots + R_k}$ as the disjoint union of $>_{R_i}$ for all $1 \leq i \leq k$. (Whenever we use it, we implicitly assume the disjointness is provable in **HA**.)

We define $\mathrm{Monoseq}_{R_1, \ldots, R_k}(x)$ to hold when $x = \langle x_0, \ldots, x_{n-1}\rangle \in \mathrm{DT}(U, >_{R_1 + \ldots + R_k})$ and $\forall i < n-1.(\forall j \leq n-1.(i < j \to \bigwedge_{1 \leq l \leq k} (x_i >_{R_l}$ $x_{i+1} \to x_i >_{R_l} x_j)))$. Note that n may be 0.

We define MS as $\{x \in \mathrm{DT}(U, >_{R_1 + \ldots + R_k}) \mid \mathrm{Monoseq}_{R_1, \ldots, R_k}(x)\}$.

MS is the set of monotonically-colored finite sequences. Note that $\mathrm{MS}_{\langle r \rangle}$ is a subset of MS (by taking T and σ to be MS and $\langle r \rangle$ in our notation T_σ) and a lifted tree of U for any $r \in U$.

We will show Podelski-Rybalchenko theorem for induction stating that if a transition invariant $>_\Pi$ is a finite union of relations $>_\pi$ such that each $\mathrm{Ind}(>_\pi^n)$ is provable for some n, and each $(>_\pi)$ is decidable, then $\mathrm{Ind}(>_\Pi)$ is provable.

First each $\mathrm{Ind}(>_\pi)$ is obtained by $\mathrm{Ind}(>_\pi^n)$. Next by the decidability of each $(>_\pi)$, we can assume all of $(>_\pi)$ are disjoint to each other. For simplicity, we explain the idea of our proof for well-foundedness instead of induction principle.

Assume the relation $>_\Pi$ has some infinite decreasing transitive sequence

$$u_1 >_\Pi u_2 >_\Pi u_3 >_\Pi \cdots$$

in order to show contradiction.

The set MS will be shown to be well-founded with the one-step extension relation. For a decreasing transitive sequence x of U, a lifted tree $T \in U^{<\omega}$ is called an *Erdös tree* of x when the elements of x are the same as elements of elements of T, every element of T is monotonically-colored, and the edges from a parent to its children have different colors. Let ET be a function that returns an Erdös tree of a given decreasing transitive sequence. Then we consider

$$\mathrm{ET}(\langle u_1 \rangle), \mathrm{ET}(\langle u_1, u_2 \rangle), \mathrm{ET}(\langle u_1, u_2, u_3 \rangle), \ldots.$$

Define $\mathrm{MS}_{\langle r \rangle}$ as the set of sequences beginning with r in MS. Define $>_{\mathrm{KB1}, r}$ as the Kleene-Brouwer relation for the lifted tree $\mathrm{MS}_{\langle r \rangle}$ and some left-to-right-decreasing relation on children of the lifted tree. Define $>_{\mathrm{KB2}, r}$ as the Kleene-Brouwer relation for the lifted tree $\mathrm{MS}_{\langle r \rangle}$ and some right-to-left-decreasing relation on children of the lifted tree. By Kleene-Brouwer theorem, $(>_{\mathrm{KB1}, r})$ and $(>_{\mathrm{KB2}, r})$ are well-founded. Define $\mathrm{ET2}(\langle u_1, \ldots, u_n \rangle)$ as the $(>_{\mathrm{KB2}, u_1})$-sorted sequence of elements in $\mathrm{ET}(\langle u_1, \ldots, u_n \rangle)$. Then consider

$$\mathrm{ET2}(\langle u_1 \rangle), \mathrm{ET2}(\langle u_1, u_2 \rangle), \mathrm{ET2}(\langle u_1, u_2, u_3 \rangle), \ldots.$$

Define $>_{\mathrm{KB}, r}$ as the Kleene-Brouwer relation for $>_{\mathrm{KB1}, r}$ and the set of $(>_{\mathrm{KB2}, r})$-sorted finite sequences of elements in $\mathrm{MS}_{\langle r \rangle}$. This definition is a key idea. By this definition, we can show the most difficult step in this proof:

$$\text{ET2}(\langle u_1 \rangle) >_{\text{KB},u_1} \text{ET2}(\langle u_1, u_2 \rangle) >_{\text{KB},u_1} \text{ET2}(\langle u_1, u_2, u_3 \rangle) >_{\text{KB},u_1} \ldots .$$

Since $(>_{\text{KB},u_1})$ is well-founded by Kleene-Brouwer theorem, we have contradiction. Hence we have shown $u_1 >_{\Pi} u_2 >_{\Pi} u_3 >_{\Pi} \ldots$ terminates.

In general we need classical logic to derive induction principle from well-foundedness, but the idea we have explained will work well for showing induction principle in intuitionistic logic.

2.3 Proof of Podelski-Rybalchenko Theorem for Induction

This subsection gives a proof of Podelski-Rybalchenko Theorem for Induction.

The next lemma shows that induction principle for each relation implies induction principle for monotonically-colored sequences. This lemma can be proved by refining Lemma 6.4 (1) of [3] from second-order logic to first-order logic. The proof is given in [6].

Lemma 2.5. *If* **HA** $\vdash \text{Ind}(\text{DT}(U, >_{R_i}), >_{\text{ext}})$ *for all* $1 \le i \le k$, *then* **HA** $\vdash \forall r \in U.\text{Ind}(\text{MS}_{\langle r \rangle}, >_{\text{ext}})$.

Next we create Kleene-Brouwer relations $>_{\text{KB1},r}$ and $>_{\text{KB2},r}$ for monotonically-colored sequences beginning with r. Then we consider the set of $(>_{\text{KB2},r})$-sorted finite sequences of monotonically-colored finite sequences beginning with r. It is a lifted tree. Then, by induction principle for MS, the lifted tree is well-founded with the one-step extension relation. The Kleene-Brouwer relation for the lifted tree and $>_{\text{KB1},r}$ gives us $>_{\text{KB},r}$ for the lifted tree. Since an Erdös tree is in the lifted tree, this will later show induction principle for Erdös trees.

Definition 2.6. For $u \in U$, we define $>_{u,\text{Left}}$ for U by: $u_1 >_{u,\text{Left}} u_2$ if $u >_{R_j} u_1$, $u >_{R_l} u_2$, and $j < l$ for some j, l.

We define $>_{\text{KB1},r}$ for $\text{MS}_{\langle r \rangle}$ as the KB relation for $\text{MS}_{\langle r \rangle} \subseteq U^{<\omega}$ and $(>_{u,\text{Left}}) \subseteq U^2$ for all $u \in U$.

For $u \in U$, we define $>_{u,\text{Right}}$ for U by: $u_1 >_{u,\text{Right}} u_2$ if $u_1 <_{u,\text{Left}} u_2$.

We define $>_{\text{KB2},r}$ for $\text{MS}_{\langle r \rangle}$ as the KB relation for $\text{MS}_{\langle r \rangle} \subseteq U^{<\omega}$ and $(>_{u,\text{Right}}) \subseteq U^2$ for all $u \in U$.

We define $>_{\text{KB},r}$ for $\text{DS}(\text{MS}_{\langle r \rangle}, >_{\text{KB2},r})_{\langle\langle r \rangle\rangle}$ as the KB relation for $\text{DS}(\text{MS}_{\langle r \rangle}, >_{\text{KB2},r})_{\langle\langle r \rangle\rangle} \subseteq \text{MS}_{\langle r \rangle}^{<\omega}$ and $>_{\text{KB1},r}$.

$>_{u,\text{Left}}$ is the left-to-right-decreasing order of children of u in some ordered tree of U in which the edge label R_i is put to an edge (x, y) such that $x >_{R_i} y$, each parent has at most one child of the same edge label, and children are ordered by their edge labels with $R_1 < \ldots < R_k$. Similarly $>_{u,\text{Right}}$ is the right-to-left-decreasing order of children of u in the ordered tree.

Definition 2.7. For $u \in U \subseteq N$, finite $T \subseteq$ MS such that $\forall \rho \in T.\forall v \in \rho.(v >_{R_1+\ldots+R_k} u)$, and for $\sigma \in T$, we define the function insert by:

insert$(u, T, \sigma) =$

   insert$(u, T, \sigma * \langle v \rangle)$ if last$(\sigma) >_{R_i} u, v = \mu v.(\sigma * \langle v \rangle \in T \wedge$ last$(\sigma) >_{R_i} v)$,

   $T \cup \{\sigma * \langle u \rangle\}$ otherwise,

where $\mu v.F(v)$ denotes the least element v with the natural number order such that $F(v)$. Formally insert$(u, T, \sigma) = T'$ is an abbreviation of some **HA**-formula $G(u, T, \sigma, T')$. It is the same for ET below.

For $x \in DT(U, >_{R_1+\ldots+R_k}) - \{\langle \ \rangle\}$, we define ET$(x) \subseteq$ MS by

$$ET(\langle u \rangle) = \{\langle u \rangle\},$$
$$ET(x * \langle u \rangle) = \text{insert}(u, ET(x), \langle \text{first}(x) \rangle) \text{ if } x \neq \langle \ \rangle.$$

Note that insert(u, T, σ) adds a new element u to the set T at some position below σ to obtain a new set. ET(x) is an Erdös tree obtained from the decreasing transitive sequence x.

The next lemma (1) states a new element is inserted at a leaf. It is proved by induction on the number of elements in T. The claim (2) states that edges from a parent to its children have different colors. It is proved by induction on the length of x.

Lemma 2.8. *(1) For $u \in U$, $T \subseteq$ MS, and $\sigma \in T$, if $u \notin \rho$ for all $\rho \in T$, $\sigma = \langle x_0, \ldots, x_{n-1} \rangle$, $x_i >_{R_j} x_{i+1}$ implies $x_i >_{R_j} u$ for all $i < n - 1$, and insert$(u, T, \sigma) = T'$, then there is some $\rho \in T_\sigma$ such that $\rho * \langle u \rangle \in$ MS, $T' = T + \{\rho * \langle u \rangle\}$, and $\rho * \langle u \rangle$ is a maximal sequence in T'.*

(2) If $\sigma * \langle u, u_1 \rangle * \rho_1, \sigma * \langle u, u_2 \rangle * \rho_2 \in ET(x)$, $u >_{R_i} u_1$, and $u >_{R_i} u_2$, then $u_1 = u_2$.

Definition 2.9. For $x \in DT(U, >_{R_1+\ldots+R_k}) - \{\langle \ \rangle\}$, we define

$$ET2(x) \equiv \langle x_0, \ldots, x_{n-1} \rangle$$

where $\{x_0, \ldots, x_{n-1}\} = ET(x)$ and $\forall i < n - 1.(x_i >_{\text{KB2,first}(x)} x_{i+1})$.

Note that $>_{\text{KB2,first}(x)}$ is a total order on ET(x) by Lemma 2.8 (2). ET2(x) is the decreasing sequence of all nodes in the Erdös tree ET(x) ordered by $>_{\text{KB2,first}(x)}$.

The next lemma shows ET2 is monotone. It is the key property of reduction in Lemma 2.11.

Lemma 2.10. HA $\vdash \forall r \in U.\forall x, y \in DT(U, >_{R_1+\ldots+R_k})_{\langle r \rangle}.(x >_{\text{ext}} y \rightarrow ET2(x) >_{\text{KB},r} ET2(y))$.

Proof. Fix $r \in U$ and $x, y \in DT(U, >_{R_1+\ldots+R_k})_{\langle r \rangle}$ and assume $x >_{\text{ext}} y$. Let $y = x * \langle u \rangle$. Then ET$(y) = $ insert$(u, ET(x), \langle r \rangle)$. By Lemma 2.8 (1), we have σ such that ET$(y) = ET(x) + \{\sigma * \langle u \rangle\}$. Then we have two cases:

   Case 1. last(ET2(x)) $>_{\text{KB2,}r} \sigma * \langle u \rangle$.

Then $ET2(y) = ET2(x) * \langle \sigma * \langle u \rangle \rangle$. By definition, $ET2(x) >_{KB,r} ET2(y)$.

Case 2. $\sigma * \langle u \rangle >_{KB2,r} \tau$ for some $\tau \in ET2(x)$.

Let ρ be the next element of $\sigma * \langle u \rangle$ in $ET2(y)$. Then $ET2(x) = \alpha * \langle \rho \rangle * \beta$ and $ET2(y) = \alpha * \langle \sigma * \langle u \rangle, \rho \rangle * \beta$. By definition of ET2, $\sigma * \langle u \rangle >_{KB2,r} \rho$. Since $\sigma * \langle u \rangle$ is maximal in $ET(y)$ by Lemma 2.8 (1), there is no $\alpha \neq \langle \ \rangle$ such that $\sigma * \langle u \rangle * \alpha = \rho$. Hence $\sigma * \langle u \rangle <_{KB1,r} \rho$. Hence $ET2(x) >_{KB,r} ET2(y)$.   □

The next lemma shows that induction for decreasing transitive sequences is reduced to induction for Erdös trees with $>_{KB,r}$.

Lemma 2.11. HA $\vdash \forall r \in U.\mathrm{Ind}(ET2(DT(U, >_{R_1+...+R_k})\langle r \rangle), >_{KB,r})$ *implies* **HA** $\vdash \mathrm{Ind}(DT(U, >_{R_1+...+R_k}), >_{ext})$.

Proof sketch. In order to show $\mathrm{Ind}(DT(U, >_{R_1+...+R_k}), >_{ext})$ for F, define $Gy \equiv \forall z \in DT(z \neq \langle \ \rangle \to ET2(z) = y \to Fz)$ and use $\mathrm{Ind}(ET2(DT(U, >_{R_1+...+R_k})\langle r \rangle), >_{KB,r})$ for G and Lemma 2.10. The proof is in [6].   □

The next lemma shows induction holds when we restrict the universe. The proof is in [6].

Lemma 2.12. HA $\vdash \mathrm{Ind}(U, >)$ *and* **HA** $\vdash V \subseteq U$ *imply* **HA** $\vdash \mathrm{Ind}(V, >)$.

The next lemma shows induction is implied from induction for decreasing sequences. The proof is in [6].

Lemma 2.13. HA $\vdash \mathrm{Ind}(DS(U, >), >_{ext})$ *implies* **HA** $\vdash \mathrm{Ind}(U, >)$.

The next lemma shows induction for a power of a relation implies induction for the relation. The proof is in [6].

Lemma 2.14. HA $\vdash \mathrm{Ind}(U, >^n)$ *implies* **HA** $\vdash \mathrm{Ind}(U, >)$.

Define

$$\mathrm{Trans}(U, >_R) \equiv \forall xyz \in U(x >_R y \wedge y >_R z \to x >_R z),$$
$$\mathrm{Decide}(U, >_R) \equiv \forall xy \in U(x >_R y \vee \neg(x >_R y)).$$

The next theorem states that if some powers of relations $>_{R_i}$ have induction principle, $>_{R_i}$ are decidable and their union is transitive, then the union has induction principle. This theorem is the same as Theorem 6.1 in [5] except **HA** and the decidability condition $\mathrm{Decide}(U, >_{R_i})$.

Theorem 2.15 (Podelski-Rybalchenko Theorem for Induction). *If* **HA** $\vdash \mathrm{Ind}(U, >_{R_1}^{n_1})$, **HA** $\vdash \mathrm{Decide}(U, >_{R_1})$, ..., **HA** $\vdash \mathrm{Ind}(U, >_{R_k}^{n_k})$, **HA** $\vdash \mathrm{Decide}(U, >_{R_k})$, *and* **HA** $\vdash \mathrm{Trans}(U, >_{R_1+...+R_k})$, *then* $\mathrm{Ind}(U, >_{R_1+...+R_k})$.

Proof. We will discuss in **HA**.

By Lemma 2.14, we can replace n_i by 1 and obtain $\mathrm{Ind}(U, >_{R_i})$. In order to obtain disjoint relations, we define $>_{R'_1}$ as $>_{R_1}$ and $>_{R'_{i+1}}$ as $(>_{R_{i+1}}) - (>_{R'_1})$ $- ... - (>_{R'_i})$. Then $(>_{R'_1})$, ..., $(>_{R'_k})$ are disjoint and $\forall xy \in U(x >_{R_1 \cup ... \cup R_k}$

$y \to x >_{R'_1 + \ldots + R'_k} y$) by Decide($U, >_{R_i}$) for $1 \le i \le k$. Since $(>_{R'_i}) \subseteq (>_{R_i})$, Ind($U, >_{R'_i}$). For simplicity, from now on we write $>_{R_i}$ for $>_{R'_i}$ in this proof. We will show Ind($U, >_{R_1 + \ldots + R_k}$).

From Ind($U, >_{R_i}$), by replacing induction on elements by induction on sequences, we have Ind(DT($U, >_{R_i}$), $>_{\text{ext}}$) for $1 \le i \le k$. By Lemma 2.5, we have $\forall r \in U$.Ind(MS$_{\langle r \rangle}$, $>_{\text{ext}}$). Apparently $\forall u \in U$.Ind($U, >_{u, \text{Left}}$). By taking U to be U, T to be MS$_{\langle r \rangle}$, and $>_u$ to be $>_{u, \text{Left}}$ in Theorem 2.3 for $>_{\text{KB1},r}$, we have $\forall r \in U$.Ind(MS$_{\langle r \rangle}$, $>_{\text{KB1},r}$). By Theorem 2.3 for $>_{\text{KB2},r}$, we have $\forall r \in U$.Ind(MS$_{\langle r \rangle}$, $>_{\text{KB2},r}$) similarly. By replacing induction on elements by induction on sequences, we have $\forall r \in U$.Ind(DS(MS$_{\langle r \rangle}$, $>_{\text{KB2},r}$), $>_{\text{ext}}$). Since DS(MS$_{\langle r \rangle}$, $>_{\text{KB2},r}$)$_{\langle\langle r \rangle\rangle}$ is a subset of DS(MS$_{\langle r \rangle}$, $>_{\text{KB2},r}$), from Lemma 2.12, we have $\forall r \in U$.Ind(DS(MS$_{\langle r \rangle}$, $>_{\text{KB2},r}$)$_{\langle\langle r \rangle\rangle}$, $>_{\text{ext}}$). By taking T to be DS(MS$_{\langle r \rangle}$, $>_{\text{KB2},r}$)$_{\langle\langle r \rangle\rangle}$, U to be MS$_{\langle r \rangle}$, and $(>_u)$ to be $(>_{\text{KB1},r})$ in Theorem 2.3 for $>_{\text{KB},r}$, we have $\forall r \in U$.Ind(DS(MS$_{\langle r \rangle}$, $>_{\text{KB2},r}$)$_{\langle\langle r \rangle\rangle}$, $>_{\text{KB},r}$). This is a key step of this proof. Since ET2(DT($U, >_{R_1 + \ldots + R_k}$)$_{\langle r \rangle}$) $\subseteq$ DS(MS$_{\langle r \rangle}$, $>_{\text{KB2},r}$)$_{\langle\langle r \rangle\rangle}$, by Lemma 2.12, we have $\forall r \in U$.Ind(ET2(DT($U, >_{R_1 + \ldots + R_k}$)$_{\langle r \rangle}$), $>_{\text{KB},r}$). By Lemma 2.11, Ind(DT($U, >_{R_1 + \ldots + R_k}$), $>_{\text{ext}}$). By Trans(U, $>_{R_1 + \ldots + R_k}$), DT($U, >_{R_1 + \ldots + R_k}$) is DS($U, >_{R_1 + \ldots + R_k}$). Hence we have Ind(DS($U$, $>_{R_1 + \ldots + R_k}$), $>_{\text{ext}}$). From Lemma 2.13, by replacing induction on sequences by induction on elements, we have Ind($U, >_{R_1 + \ldots + R_k}$).   □

3 Cyclic Proofs

3.1 Intuitionistic Martin-Löf's Inductive Definition System LJID

We define an intuitionistic Martin-Löf's inductive definition system, called **LJID**.

The language of **LJID** is determined by a first-order language with inductive predicate symbols. The logical system **LJID** is determined by production rules for inductive predicate symbols. These production rules mean that the inductive predicate denotes the least fixed point defined by these production rules.

We assume the first order terms $t, u, \ldots$. We assume $\forall x$ and $\exists y$ are less tightly connected than other logical connectives. To save space, we sometimes write Pxy and Fxy for $P(x, y)$ and $F(x, y)$.

For example, the production rules of the inductive predicate symbol N are

$$\frac{}{N0} \qquad \frac{Nx}{Nsx}$$

These production rules mean that N denotes the smallest set closed under 0 and s, namely the set of natural numbers.

The inference rules of **LJID** contains the introduction rules and the elimination rules for inductive predicates, determined by the production rules. These rules describe that the predicate actually denotes the least fixed point. In particular, the elimination rule describes the induction principle.

For example, the above production rules give the introduction rules

$$\frac{}{\Gamma \vdash N0} \qquad \frac{\Gamma \vdash Nx}{\Gamma \vdash Nsx}$$

and the elimination rule

$$\frac{\Gamma \vdash F0 \quad \Gamma, Fx \vdash Fsx}{\Gamma, Nt \vdash Ft}$$

This elimination rule describes mathematical induction principle.

$$\frac{}{\Gamma, A \vdash A} \text{ (Axiom)} \qquad \frac{\Gamma' \vdash \Delta'}{\Gamma \vdash \Delta} \text{ (Wk)}_{(\Gamma' \subseteq \Gamma, \Delta' \subseteq \Delta)} \qquad \frac{\Gamma \vdash F \quad \Gamma, F \vdash \Delta}{\Gamma \vdash \Delta} \text{ (Cut)}$$

$$\frac{\Gamma \vdash \Delta}{\Gamma\theta \vdash \Delta\theta} \text{ (Subst)} \qquad \frac{\Gamma \vdash F}{\Gamma, \neg F \vdash} \text{ (}\neg L\text{)} \qquad \frac{\Gamma, F \vdash}{\Gamma \vdash \neg F} \text{ (}\neg R\text{)} \qquad \frac{\Gamma, F \vdash \Delta \quad \Gamma, G \vdash \Delta}{\Gamma, F \vee G \vdash \Delta} \text{ (}\vee L\text{)}$$

$$\frac{\Gamma \vdash F}{\Gamma \vdash F \vee G} \text{ (}\vee R_l\text{)} \qquad \frac{\Gamma \vdash G}{\Gamma \vdash F \vee G} \text{ (}\vee R_r\text{)} \qquad \frac{\Gamma, F, G \vdash \Delta}{\Gamma, F \wedge G \vdash \Delta} \text{ (}\wedge L\text{)} \qquad \frac{\Gamma \vdash F \quad \Gamma \vdash G}{\Gamma \vdash F \wedge G} \text{ (}\wedge R\text{)}$$

$$\frac{\Gamma \vdash F \quad \Gamma, G \vdash \Delta}{\Gamma, F \to G \vdash \Delta} \text{ (}\to L\text{)} \qquad \frac{\Gamma, F \vdash G}{\Gamma \vdash F \to G} \text{ (}\to R\text{)} \qquad \frac{\Gamma, F[x := t] \vdash \Delta}{\Gamma, \forall x F \vdash \Delta} \text{ (}\forall L\text{)}$$

$$\frac{\Gamma \vdash F}{\Gamma \vdash \forall x F} \text{ (}\forall R\text{)}_{(x \notin \text{FV}(\Gamma))} \qquad \frac{\Gamma, F \vdash \Delta}{\Gamma, \exists x F \vdash \Delta} \text{ (}\exists L\text{)}_{(x \notin \text{FV}(\Gamma, \Delta))} \qquad \frac{\Gamma \vdash F[x := t]}{\Gamma \vdash \exists x F} \text{ (}\exists R\text{)}$$

$$\frac{\Gamma[x := t] \vdash \Delta[x := t]}{\Gamma[x := u], t = u \vdash \Delta[x := u]} \text{ (}= L\text{)} \qquad \frac{}{\Gamma \vdash t = t} \text{ (}= R\text{)} \qquad \frac{\text{minor premises}}{\Gamma, P_j \overrightarrow{u} \vdash F_j \overrightarrow{u}} \text{ (Ind } P_j\text{)}$$

$$\frac{\Gamma \vdash Q_1 \overrightarrow{u}_1 \quad \cdots \quad \Gamma \vdash Q_n \overrightarrow{u}_n \quad \Gamma \vdash P_1 \overrightarrow{t}_1 \quad \cdots \quad \Gamma \vdash P_m \overrightarrow{t}_m}{\Gamma \vdash P \overrightarrow{t}} \text{ (}P\ R\text{)}$$

Fig. 1. Inference rules

The inference rules are given in Fig. 1 where for $(P_i\ R)$ we assume the production rule

$$\frac{Q_1 \overrightarrow{u}_1 \quad \cdots \quad Q_n \overrightarrow{u}_n \quad P_1 \overrightarrow{t}_1 \quad \cdots \quad P_m \overrightarrow{t}_m}{P \overrightarrow{t}}$$

and for $(\text{Ind } P_j)$ we assume a predicate F_i for each P_i and the minor premises are defined as

$$\Gamma, Q_{i1} \overrightarrow{u}_{i1}, \ldots, Q_{in_i} \overrightarrow{u}_{in_i}, F_1 \overrightarrow{t}_{i1}, \ldots, F_{im_i} \overrightarrow{t}_{im_i} \vdash F_i \overrightarrow{t}_i$$

for each production rule

$$\frac{Q_{i1} \overrightarrow{u}_{i1} \quad \cdots \quad Q_{in_i} \overrightarrow{u}_{in_i} \quad P_1 \overrightarrow{t}_{i1} \quad \cdots \quad P_{im_i} \overrightarrow{t}_{im_i}}{P_i \overrightarrow{t}_i}$$

Note that the antecedents and the succedents are sets and the succedent is empty or a formula.

The system **LJID** is the same as the system obtained from classical Martin-Löf's inductive definition system **LKID** defined in [11] by restricting every sequent to intuitionistic sequents and replacing $(\to L)$, $(\vee R)$, and $(\text{Ind } P_j)$ accordingly. The provability of the system **LJID** is the same as that of the natural deduction system given in [16].

3.2 Cyclic Proof System CLJID$^\omega$

An intuitionistic cyclic proof system, called **CLJID**$^\omega$, is defined as the system obtained from classical cyclic proof system **CLKID**$^\omega$ defined in [11] by restricting every sequent to intuitionistic sequents and replacing $(\to L)$ and $(\vee R)$ in the same way as **LJID**. Note that the *global trace condition* in **CLJID**$^\omega$ is the same as that in **CLKID**$^\omega$ (Definition 5.5 of [11]).

Namely, the inference rules of **CLJID**$^\omega$ are obtained from **LJID** by replacing $(\text{Ind} P_j)$ by

$$\frac{\text{case distinctions}}{\Gamma, P\vec{u} \vdash \Delta} \ (\text{Case} P)$$

where the case distinctions are

$$\Gamma, \vec{u} = \vec{t}, Q_1 \vec{u}_1, \ldots, Q_n \vec{u}_n, P_1 \vec{t}_1, \ldots, P_m \vec{t}_m \vdash \Delta$$

for each production rule

$$\frac{Q_1 \vec{u}_1 \quad \cdots \quad Q_n \vec{u}_n \quad P_1 \vec{t}_1 \quad \cdots \quad P_m \vec{t}_m}{P\vec{t}}$$

A cyclic proof in **CLJID**$^\omega$ is defined by (1) allowing a bud as an open assumption and requiring a companion for each bud, (2) requiring the global trace condition.

The *global trace condition* [8,10] is the condition that for every infinite path in the infinite unfolding of a given cyclic proof, there is a trace that passes main formulas of case rules infinitely many times. The global trace condition ensures that when we think some measure by counting case rules, the measure of a bud is smaller than that of the companion. For example, in the next example the companion (a) uses $Px0y$, but the bud (a) uses $Px0y$ where x is x' and $x' < x$, so their actual meanings are different even though they are of the same form. The global trace condition guarantees the soundness of a cyclic proof system.

An example of a cyclic proof (trivial steps are omitted) is as follows:

$$\frac{\displaystyle \frac{\displaystyle \frac{(a)Px0y \vdash x = y}{Px'0y' \vdash x' = y'} \ (\text{Subst})}{\displaystyle \frac{Px'0y', x = sx', y = sy' \vdash x' = y'}{Px'0y', x = sx', y = sy' \vdash sx' = sy'} \ (\text{Wk})}}{\displaystyle \frac{\displaystyle \frac{\vdash 0 = 0}{x = 0, y = 0, Px0y \vdash x = y} \qquad Px'0y', x = sx', y = sy' \vdash x = y}{(a)Px0y \vdash x = y}} \ (\text{Case } P)$$

where the mark (a) denotes the bud-companion relation, and the production rules are

$$\frac{}{P0yy} \qquad \frac{Pxyz}{P(sx)y(sz)}$$

Note that the predicate P is addition on natural numbers and the proof is, essentially, deriving the arithmetic identity $x + 0 = x$.

We call an atomic formula an *inductive atomic formula* when its predicate symbol is an inductive predicate symbol.

3.3   Cyclic Proofs for Coinductive Predicates

This subsection shows how we can use a cyclic proof system to formalize coinductive predicates. Since a coinductive predicate is a dual of an inductive predicate, and sequent calculus is symmetric for this dual, we can construct a cyclic proof system for coinductive predicates. For example, for stream predicates we can define a cyclic proof system $\mu\nu$LK from **CLKID**$^{\omega}$ as follows:

(1) Add function symbols head, tail, and the pair $\langle \, , \, \rangle$ with the axioms $\langle x, y \rangle = \langle x', y' \rangle \rightarrow x = x' \wedge y = y'$ and $x = \langle \text{head } x, \text{tail } x \rangle$, and a coinductive predicate symbol P with its coproduction rule

$$\frac{Qyx \quad Px}{P\langle y, x \rangle} \text{ co}$$

which means P is defined coinductively by this rule. Note that P represents the set of streams $\langle x_0, \langle x_1, \langle x_2, \langle \ldots \rangle \rangle \rangle \rangle$ such that $Q(x_i, \langle x_{i+1}, \langle x_{i+2}, \langle \ldots \rangle \rangle \rangle)$ for all i.

(2) Add the inference rules $(P\ R)$ and (Case P) in the same way as **CLKID**$^{\omega}$, namely,

$$\frac{\Gamma, t = \langle y, x \rangle, Qyx, Px \vdash \Delta}{\Gamma, Pt \vdash \Delta} \text{ (Case } P) \qquad \frac{\Gamma \vdash Qyx, \Delta \quad \Gamma \vdash Px, \Delta}{\Gamma \vdash P\langle y, x \rangle, \Delta} \text{ (}P\ R\text{)}$$

(3) We call an atomic formula a *coinductive atomic formula* when its predicate symbol is a coinductive predicate symbol. We define a *cotrace* as a sequence of coinductive atomic formulas in the succedents of a path such that two atomic formulas are related by an inference rule in a similar way to a trace defined in [11]. The *global trace and cotrace condition* is the condition that for every infinite path in the infinite unfolding of a given cyclic proof, the path contains either a trace that passes main formulas of case rules infinitely many times, or a cotrace that passes main formulas of rules $(P\ R)$ infinitely many times.

(4) A cyclic proof is a preproof that satisfies the global trace and cotrace condition.

Example. We define the bit stream predicate BS by the following coproduction rules:

$$\frac{\text{Bit } y \quad \text{BS } x}{\text{BS}\langle y, x\rangle} \text{ co}$$

where Bit is an ordinary predicate symbol with the axiom Bit $x \leftrightarrow x = 0 \lor x = 1$. The inference rules from this production rule are:

$$\frac{\Gamma \vdash \text{Bit } y, \Delta \quad \Gamma \vdash \text{BS } x, \Delta}{\Gamma \vdash \text{BS}\langle y, x\rangle, \Delta} \text{ (BS R)} \qquad \frac{\Gamma, t = \langle y, x\rangle, \text{Bit } y, \text{BS } x \vdash \Delta}{\Gamma, \text{BS } t \vdash \Delta} \text{ (Case BS)}$$

Then we can show $x = \langle 0, x\rangle \vdash \text{BS } x$, namely, the zero stream is a bit stream, as follows (trivial steps are omitted):

$$\frac{\dfrac{\overline{x = \langle 0, x\rangle \vdash \text{Bit } 0} \quad (a) \; x = \langle 0, x\rangle \vdash \text{BS } x}{x = \langle 0, x\rangle \vdash \text{BS}\langle 0, x\rangle} \text{ (BS R)}}{(a) \; x = \langle 0, x\rangle \vdash \text{BS } x}$$

where (a) denotes the bud-companion relation.

The cyclic proof system $\mu\nu$LK is sound for the standard model.

Theorem 3.1. *If a sequent is provable in $\mu\nu$LK, then it is true in the standard model where a coinductive predicate is interpreted as the greatest fixed point that satisfies the coproduction rules.*

Proof sketch. We add an ordinary predicate symbol $\tilde{Q}$ with the axiom $\tilde{Q}yx \leftrightarrow \neg Qyx$ and add an inductive predicate symbols $\tilde{P}$ with the production rules

$$\frac{\tilde{Q}(\text{head } x)(\text{tail } x)}{\tilde{P}x} \qquad \frac{\tilde{P}(\text{tail } x)}{\tilde{P}x}$$

In the standard model, P is the greatest solution of the equation

$$Px \leftrightarrow \exists yx'(x = \langle y, x'\rangle \land Qyx' \land Px')$$

and $\tilde{P}$ is the least solution of the equation

$$\tilde{P}x \leftrightarrow \tilde{Q}(\text{head } x)(\text{tail } x) \lor \tilde{P}(\text{tail } x).$$

By putting $\neg$ on both sides of the equation for P and taking y and x' to be head x and tail x, we can show $\neg P$ is a solution of the equation for $\tilde{P}$. Hence $\tilde{P}x \to \neg Px$ is true. In the same say by putting $\neg$ on both sides of the equation for $\tilde{P}$, and using $x = \langle \text{head}x, \text{tail}x\rangle$, we can show $\neg\tilde{P}$ is a solution of the equation for P. Hence $\neg\tilde{P}x \to Px$ is true. Therefore $\tilde{P}x \leftrightarrow \neg Px$ is true.

We define a transformation $(\ )^-$ for a sequent and a proof, in order to replace P by $\tilde{P}$. For a sequent J, we define J^- by replacing P by $\neg\tilde{P}$ and then moving an atomic formula $\neg\tilde{P}t$ of the antecedent to $\tilde{P}t$ of the succedent and moving an atomic formula $\neg\tilde{P}t$ of the succedent to $\tilde{P}t$ of the antecedent.

Given a cyclic proof π, we define π^- by replacing each sequent J by J^- and then replacing $(P\ R)$ by

$$\frac{\dfrac{\Gamma^- \vdash Qyx', \Delta^-}{\Gamma^-, \tilde{Q}yx' \vdash \Delta^-} \quad \Gamma^-, \tilde{P}x' \vdash \Delta^-}{\Gamma^-, \tilde{P}\langle y, x'\rangle \vdash \Delta^-} \ (\text{Case } \tilde{P})$$

(trivial steps are omitted)

and replacing $(\text{Case}P)$ by

$$\frac{\dfrac{\dfrac{\Gamma^-, x = \langle y, x'\rangle, Qyx' \vdash \tilde{P}x', \Delta^-}{\Gamma^-, x = \langle y, x'\rangle \vdash \tilde{Q}yx', \tilde{P}x', \Delta^-} \ (\tilde{P}\ R)}{\Gamma^-, x = \langle y, x'\rangle \vdash \tilde{Q}yx', \tilde{P}x, \Delta^-} \ (\tilde{P}\ R)}{\dfrac{\Gamma^-, x = \langle y, x'\rangle \vdash \tilde{P}x, \Delta^-}{\Gamma^- \vdash \tilde{P}x, \Delta^-}}$$

(trivial steps are omitted)

Then a cotrace in π corresponds to a trace in π^-. Hence π^- is a cyclic proof of J^- in **CLKID**$^\omega$ when π is a cyclic proof of J in $\mu\nu$LK. By the soundness of **CLKID**$^\omega$, J^- is true in the standard model. Since $\tilde{P}x \leftrightarrow \neg Px$ is true, J is true in the standard model where P is interpreted as the greatest fixed point.   □

4 Equivalence Between LJID and CLJID$^\omega$

This section studies the equivalence between **CLJID**$^\omega$ and **LJID**.

4.1 Countermodel and Addition of Heyting Arithmetic

This subsection gives a countermodel and adds arithmetic to the logical systems.

The counterexample given in [4] also shows that the equivalence between **CLJID**$^\omega$ and **LJID** does not hold in general, because the proof of the statement H in [4] is actually in **CLJID**$^\omega$, and **LJID** does not prove H since **LKID** does not prove H. This gives us the following theorem (it is not new in the sense [4] immediately implies it).

Theorem 4.1. *There are some signature and some set of production rules for which the provability of* **CLJID**$^\omega$ *is not the same as that of* **LJID**.

There is a possibility of the equivalence under some conditions. We will show the equivalence holds by adding arithmetic to both systems.

We add arithmetic to both **LJID** and **CLJID**$^\omega$.

Definition 4.2. **CLJID**$^\omega$ + **HA** and **LJID** + **HA** are defined to be obtained from **CLJID**$^\omega$ and **LJID** by adding Heyting arithmetic. Namely, we add constants and function symbols $0, s, +, \times$, the inductive predicate symbol N, the productions for N, and Heyting axioms:

$$\overline{N0} \qquad \frac{Nx}{Nsx} \qquad \vdash Nx \to sx \neq 0, \qquad \vdash Nx \wedge Ny \to sx = sy \to x = y,$$

$$\vdash Nx \to x + 0 = x, \qquad \vdash Nx \wedge Ny \to x + sy = s(x + y),$$

$$\vdash Nx \to x \times 0 = 0, \qquad \vdash Nx \wedge Ny \to x \times sy = x \times y + x.$$

4.2   Equivalence Theorem

In this subsection we state the equivalence theorem and explains proof ideas.

First we assume a new inductive predicate symbol P' for each inductive predicate symbol P and define the production rules of P' in the same way as [5].

Definition 4.3. We define the production rule of P'

$$\frac{Q_1 \vec{u}_1 \quad \cdots \quad Q_n \vec{u}_n \quad v > v_1 \quad P_1' \vec{t}_1 v_1 \quad \cdots \quad v > v_m \quad P_m' \vec{t}_m v_m \quad Nv}{P' \vec{t} v}$$

for each production rule of P

$$\frac{Q_1 \vec{u}_1 \quad \cdots \quad Q_n \vec{u}_n \quad P_1 \vec{t}_1 \quad \cdots \quad P_m \vec{t}_m}{P \vec{t}}$$

where $v, v_1, \ldots, v_m$ are fresh variables.

We write $\mathbf{LJID} + \mathbf{HA} + (\Sigma, \Phi)$ for the system $\mathbf{LJID} + \mathbf{HA}$ with the signature Σ and the set Φ of production rules. Similarly we write $\mathbf{CLJID}^\omega + \mathbf{HA} + (\Sigma, \Phi)$. For simplicity, in Φ we write only P for the set of production rules for P. We define $\Sigma_N = \{0, s, +, \times, <, N\}$ and $\Phi_N = \{N\}$. We write P'' for $(P')'$.

The next theorem shows the equivalence of $\mathbf{LJID} + \mathbf{HA}$ and $\mathbf{CLJID}^\omega + \mathbf{HA}$ with signatures.

Theorem 4.4 (Equivalence of $\mathbf{LJID} + \mathbf{HA}$ and $\mathbf{CLJID}^\omega + \mathbf{HA}$). *Let* $\Sigma = \Sigma_N \cup \{\vec{Q}, \vec{P}, \vec{P}'\}$ *and* $\Phi = \Phi_N \cup \{\vec{P}, \vec{P}'\}$. *Then the provability of* $\mathbf{CLJID}^\omega + \mathbf{HA} + (\Sigma, \Phi)$ *is the same as that of* $\mathbf{LJID} + \mathbf{HA} + (\Sigma, \Phi)$.

We explain our ideas of proofs of this theorem. [5] shows the equivalence between classical systems by using classical Podelski-Rybalchenko theorem for induction. This proof goes well even if we replace classical systems by intuitionistic systems except that we have to replace classical Podelski-Rybalchenko theorem for induction by intuitionistic Podelski-Rybalchenko theorem for induction. Since we proved intuitionistic Podelski-Rybalchenko theorem for induction in Theorem 2.15, by combining them, we can show the equivalence between $\mathbf{LJID}$ and $\mathbf{CLJID}^\omega$.

5   Proof Transformation

This section gives the proof of the equivalence. More detailed discussions are given in [6]. We define proof transformation from $\mathbf{CLJID}^\omega + \mathbf{HA}$ to $\mathbf{LJID} + \mathbf{HA}$. First we will define stage numbers and path relations, and then define proof transformation using them.

For notational convenience, we assume a cyclic proof Π in this section. Let the buds in Π be J_{1i} $(i \in I)$ and the companions be J_{2j} $(j \in K)$. Assume $f : I \to K$ such that the companion of a bud J_{1i} is $J_{2,f(i)}$.

5.1 Stage Numbers for Inductive Definitions

In this subsection, we define and discuss stage transformation.

We introduce a stage number to each inductive atomic formula so that the argument of the formula comes into the inductive predicate at the stage of the stage number. This stage number will decrease by a progressing trace. A proof in **LJID + HA** will be constructed by using the induction on stage numbers.

First we give stage transformation of an inductive atomic formula. We assume a fresh inductive predicate symbol P' for each inductive predicate symbol P and we call it a *stage-number inductive predicate symbol*. $P'(\overrightarrow{t}, v)$ means that the element $\overrightarrow{t}$ comes into P at the stage v. We transform $P(\overrightarrow{t})$ into $\exists v P'(\overrightarrow{t}, v)$. We call a variable v a *stage number* of $\overrightarrow{t}$ when $P'(\overrightarrow{t}, v)$. $P(\overrightarrow{t})$ and $\exists v P'(\overrightarrow{t}, v)$ will become equivalent by inference rules introduced by the transformation of production rules. We call $P'(\overrightarrow{t}, v)$ a *stage-number inductive atomic formula*.

Secondly we give stage transformation of a production rule. We transform the production of P into the production of P' given in Definition 4.3.

Next we give the stage transformation of a sequent. For given fresh variables $\overrightarrow{v}$, we transform a sequent J into $J^{\circ}_{\overrightarrow{v}}$ defined as follows. We define $\Gamma^{\bullet}$ as the set obtained from Γ by replacing $P(\overrightarrow{t})$ by $\exists v P'(\overrightarrow{t}, v)$. For fresh variables $\overrightarrow{v}$, we define $(\Gamma)^{\circ}_{\overrightarrow{v}}$ as the sequent obtained from $\Gamma^{\bullet}$ by replacing the i-th element of the form $\exists v P'(\overrightarrow{t}, v)$ in the sequent $\Gamma^{\bullet}$ by $P'(\overrightarrow{t}, v_i)$. We define $(\Gamma \vdash \Delta)^{\bullet}$ by $\Gamma^{\bullet} \vdash \Delta^{\bullet}$, and define $(\Gamma \vdash \Delta)^{\circ}_{\overrightarrow{v}}$ by $(\Gamma)^{\circ}_{\overrightarrow{v}} \vdash \Delta^{\bullet}$.

We write $(a_i)_{i \in I}$ for the sequence of elements a_i where i varies in I. We extend the notion of proofs by allowing open assumptions. We write $\Gamma \vdash_{\mathbf{CLJID}^{\omega} + \mathbf{HA}} \Delta$ with assumptions $(J_i)_{i \in I}$ when there is a proof with assumptions $(J_i)_{i \in I}$ and the conclusion $\Gamma \vdash \Delta$ in $\mathbf{CLJID}^{\omega} + \mathbf{HA}$.

Definition 5.1. In a path π in a proof, we define $\mathrm{Ineq}(\pi)$ as the set of the forms $v > v'$ and $v = v'$ for any stage numbers v, v' eliminated by every case distinction in π.

The proof of the next proposition gives *stage transformation* of a proof into a proof of the stage transformation of the conclusion of the original proof. We write Π° for the stage transformation of Π.

Proposition 5.2 (Stage Transformation). *For any fresh variables $\overrightarrow{v}$, if $\Gamma \vdash_{\mathbf{CLJID}^{\omega} + \mathbf{HA}} \Delta$ with assumptions $(\Gamma_i \vdash \Delta_i)_{i \in I}$ without any buds, then for some fresh variables $(\overrightarrow{v}_i)_{i \in I}$ we have $(\Gamma)^{\circ}_{\overrightarrow{v}} \vdash_{\mathbf{CLJID}^{\omega} + \mathbf{HA}} \Delta^{\bullet}$ with assumptions $(\mathrm{Ineq}(\pi_i), (\Gamma_i)^{\circ}_{\overrightarrow{v}_i} \vdash \Delta_i^{\bullet})_{i \in I}$ without any buds, where π_i is the path from the conclusion to the assumption $(\Gamma_i)^{\circ}_{\overrightarrow{v}_i} \vdash \Delta_i^{\bullet}$.*

5.2 Path Relation

In this section, we will introduce path relations and discuss them.

We assume a subproof Π_j of Π such that it does not have buds, its conclusion is J_{2j} and its assumptions are J_{1i} ($i \in I_j$).

For J in Π_j°, we define $\tilde{J}$ as $\langle v_1, \ldots, v_k \rangle$ where J is $\Gamma_{v_1 \ldots v_k}^\circ \vdash \Delta^\bullet$.

For a path π from the conclusion to an assumption in Π_j°, we write $\check{\pi}$ for the corresponding path in Π. We extend this notation to a finite composition of π's. By the correspondence ($\check{\ }$), a stage-number inductive atomic formula in Π_j° corresponds to an inductive atomic formula in Π, and a path, a trace, and a progressing trace in Π_j° correspond to the same kind of objects in Π.

Definition 5.3. For a finite composition π of paths in $\{\Pi_j^\circ \mid j \in K\}$ such that $\check{\pi}$ is a path in the infinite unfolding of Π, we define the *path relation* $\tilde{>}_\pi$ by

$$x \tilde{>}_\pi y \equiv |x| = |\tilde{J}_2| \wedge |y| = |\tilde{J}_1| \wedge \bigwedge_{F(q_1, q_2)} (x)_{q_2} > (y)_{q_1} \wedge \bigwedge_{G(q_1, q_2)} (x)_{q_2} = (y)_{q_1}$$

where J_1 and J_2 are the top and bottom sequents of π respectively, $\check{J}_1$ and $\check{J}_2$ are those of the path $\check{\pi}$, $F(q_1, q_2)$ is that there is some progressing trace from the q_2-th atomic formula in $\check{J}_2$ to the q_1-th atomic formula in $\check{J}_1$, $G(q_1, q_2)$ is that there is some non-progressing trace from the q_2-th atomic formula in $\check{J}_2$ to the q_1-th atomic formula in $\check{J}_1$.

We define B_1 as the set of paths from conclusions to assumptions in Π_j° ($j \in K$). We define B as the set of finite compositions of elements in B_1 such that if $\pi \in B$ then $\check{\pi}$ is a path in the infinite unfolding of Π.

Definition 5.4. For $\pi \in B$, define $x >_\pi y$ by

$$x >_\pi y \equiv (x)_0 = j \wedge (y)_0 = f(i) \wedge (x)_1 \tilde{>}_\pi (y)_1,$$

where J_{1i} is the top sequent of $\check{\pi}$, and J_{2j} is the bottom sequent of $\check{\pi}$.

Note that $(\ )_0$ and $(\ )_1$ are operations for a number that represents a sequence of numbers defined in Sect. 3. The first element is a companion number.

Lemma 5.5. $\{>_\pi \mid \pi \in B\}$ *is finite.*

Proof. Define C_n as $\{>_{\pi_1 \ldots \pi_m} \mid m \leq n, \pi_i \in B_1\}$. Since $>_\pi$ is a relation on $N \times N^{\leq p}$ where p is the maximum number of inductive atomic formulas in the antecedents of Π, there is L such that $|C_n| \leq L$ for all n. Then we have the least n such that $C_{n+1} = C_n$. Then $|\{>_\pi \mid \pi \in B\}| = |C_n|$. $\square$

The next lemma is the only lemma that uses the global trace condition.

Lemma 5.6. *For all $\pi \in B$, there is $n > 0$ such that $\vdash_{\mathbf{HA}} \mathrm{Ind}(U, >_\pi^n)$.*

We define $>_\Pi$ as $\bigcup \{>_\pi \mid \pi \in B\}$. Note that $>_\Pi$ is transitive, since the top sequent of π_1 is the bottom sequent of π_2 by the first element, and $((>_{\pi_1}) \circ (>_{\pi_2})) \subseteq (>_{\pi_1 \pi_2})$.

5.3 Proof Transformation

This section gives proof transformation.

The next lemma shows we can replace (Case) rules of $\mathbf{CLJID}^\omega + \mathbf{HA}$ by (Ind) rules of $\mathbf{LJID} + \mathbf{HA}$.

Lemma 5.7. *If there is a proof with some assumptions and without any buds in* $\mathbf{CLJID}^\omega + \mathbf{HA}$, *then there is a proof of the same conclusions with the same assumptions in* $\mathbf{LJID} + \mathbf{HA}$.

The next is a key lemma and shows each bud in a cyclic proof is provable in $\mathbf{LJID} + \mathbf{HA}$, which is proved by using Theorem 2.15.

Lemma 5.8. *For every bud J of a proof in* $\mathbf{CLJID}^\omega + \mathbf{HA}$ *and fresh variables* $\overrightarrow{v}$, $(J)^\circ_{\overrightarrow{v}}$ *is provable in* $\mathbf{LJID} + \mathbf{HA}$.

The next is the main proposition stating that a cyclic proof is transformed into an ($\mathbf{LJID} + \mathbf{HA}$)-proof with stage-number inductive predicates.

Proposition 5.9. *If a sequent J is provable in* $\mathbf{CLJID}^\omega + \mathbf{HA} + (\Sigma_N \cup \{\overrightarrow{P}\}, \Phi_N \cup \{\overrightarrow{P}\})$, *then J is provable in* $\mathbf{LJID} + \mathbf{HA} + (\Sigma_N \cup \{N', \overrightarrow{P}, \overrightarrow{P'}\}, \Phi_N \cup \{\overrightarrow{P}, \overrightarrow{P'}\})$ *where $N', \overrightarrow{P'}$ are the stage-number inductive predicates of $N, \overrightarrow{P}$.*

The next shows conservativity for stage-number inductive predicates.

Proposition 5.10 (Conservativity of N' and P''). *Let $\Sigma = \Sigma_N \cup \{\overrightarrow{Q}, \overrightarrow{P}, \overrightarrow{P'}\}$, $\Phi = \Phi_N \cup \{\overrightarrow{P}, \overrightarrow{P'}\}$, $\Sigma' = \Sigma \cup \{N', \overrightarrow{P''}\}$, and $\Phi' = \Phi \cup \{N', \overrightarrow{P''}\}$. Then* $\mathbf{LJID} + \mathbf{HA} + (\Sigma', \Phi')$ *is conservative over* $\mathbf{LJID} + \mathbf{HA} + (\Sigma, \Phi)$.

Proof of Theorem 4.4. (1) $\mathbf{LJID} + \mathbf{HA} + (\Sigma, \Phi)$ to $\mathbf{CLJID}^\omega + \mathbf{HA} + (\Sigma, \Phi)$.

For this claim, we can obtain a proof from the proof of Lemma 7.5 in [11] by restricting every sequent into intuitionistic sequents and replacing $\mathbf{LKID} + (\Sigma, \Phi)$ and $\mathbf{CLKID}^\omega + (\Sigma, \Phi)$ by $\mathbf{LJID} + (\Sigma, \Phi)$ and $\mathbf{CLJID}^\omega + (\Sigma, \Phi)$ respectively.

(2) $\mathbf{CLJID}^\omega + \mathbf{HA} + (\Sigma, \Phi)$ to $\mathbf{LJID} + \mathbf{HA} + (\Sigma, \Phi)$.

Let $\Sigma' = \Sigma \cup \{N', \overrightarrow{P''}\}$ and $\Phi' = \Phi \cup \{N', \overrightarrow{P''}\}$. Assume J is provable in $\mathbf{CLJID}^\omega + \mathbf{HA} + (\Sigma, \Phi)$. By Proposition 5.9, J is provable in $\mathbf{LJID} + \mathbf{HA} + (\Sigma', \Phi')$. By Proposition 5.10, J is provable in $\mathbf{LJID} + \mathbf{HA} + (\Sigma, \Phi)$. $\square$

6 Related Work

The conjecture 7.7 in [11] (also in [8]) is that the provability of $\mathbf{LKID}$ is the same as that of $\mathbf{CLKID}^\omega$. In general, the equivalence was proved to be false in [4], by showing a counterexample. However, if we restrict both systems to only the natural number inductive predicate and add Peano arithmetic to both systems, the equivalence was proved to be true in [20], by internalizing a cyclic proof in ACA_0 and using some results in reverse mathematics. [5] proved that if we add Peano arithmetic to both systems, $\mathbf{CLKID}^\omega$ and $\mathbf{LKID}$ are equivalent, namely the equivalence is true under arithmetic, by showing arithmetical Ramsey theorem and Podelski-Rybalchenko theorem for induction.

This paper shows that similar results as shown in [5] hold for intuitionistic logic, namely, the provability of **LJID** is the same as that of **CLJID**$^\omega$ if we add Heyting arithmetic to both systems.

The results of this paper immediately give another proof to the equivalence under arithmetic for classical logic shown in [5] by using the fact $\Gamma \vdash_{\textbf{CLKID}^\omega + \textbf{PA}}$ Δ implies $E, \Gamma, \neg\Delta \vdash_{\textbf{CLJID}^\omega + \textbf{HA}}$ for some finite set E of excluded middles.

By taking $\overrightarrow{Q}$ and $\overrightarrow{P}$ to be empty in Theorem 4.4, we have conservativity of **CLJID**$^\omega$ + **HA** over **LJID** + **HA** with only the inductive predicate N, which answers the question (iv) in Sect. 7 of [20].

[15] presented the first logical system for inductive/coinductive predicates. [17] also gave a similar system. They are both based on a finite system with unfold and fold, and limited to propositional logic. [21] showed the completeness of the system by using a cyclic proof system but it is also limited to propositional logic. [2,14] investigated cyclic proof systems for inductive/coinductive predicates. [1, 13] also used cyclic proof systems for inductive/coinductive predicates to show some completeness results. But these systems are all limited to propositional logic.

7   Conclusion

We have first shown intuitionistic Podelski-Rybalchenko theorem for induction in HA, and we have secondly shown the provability of the intuitionistic cyclic proof system is the same as that of the intuitionistic system of Martin-Lof's inductive definitions when both systems contain HA. We have also constructed a cyclic proof system $\mu\nu$LK for stream predicates.

One future work would be to construct a cyclic proof system for coinductive predicates in a general way and show the equivalence between the cyclic proof system and other logical systems for coinductive predicates.

References

1. Afshari, B., Leigh, G.: Cut-free completeness for modal mu-calculus. In: Proceedings of LICS 2017, pp. 1–12 (2017)
2. Baelde, D., Doumane, A., Saurin, A.: Infinitary proof theory: the multiplicative additive case. In: Proceedings of 25th EACSL Annual Conference on Computer Science Logic (CSL 2016). LIPIcs, vol. 62, pp. 42:1–42:17 (2016)
3. Berardi, S., Steila, S.: An intuitionistic version of Ramsey's Theorem and its use in Program Termination. Ann. Pure Appl. Log. **166**, 1382–1406 (2015)
4. Berardi, S., Tatsuta, M.: Classical system of Martin-Löf's inductive definitions is not equivalent to cyclic proof system. In: Esparza, J., Murawski, A.S. (eds.) FoSSaCS 2017. LNCS, vol. 10203, pp. 301–317. Springer, Heidelberg (2017). https://doi.org/10.1007/978-3-662-54458-7_18

5. Berardi, S., Tatsuta, M.: Equivalence of inductive definitions and cyclic proofs under arithmetic. In: Proceedings of Thirty-Second Annual IEEE Symposium on Logic in Computer Science (LICS 2017), pp. 1–12 (2017)
6. Berardi, S., Tatsuta, M.: Equivalence of Intuitionistic Inductive Definitions and Intuitionistic Cyclic Proofs under Arithmetic, arXiv:1712.03502 (2017)
7. Brotherston, J.: Cyclic proofs for first-order logic with inductive definitions. In: Beckert, B. (ed.) TABLEAUX 2005. LNCS (LNAI), vol. 3702, pp. 78–92. Springer, Heidelberg (2005). https://doi.org/10.1007/11554554_8
8. Brotherston, J.: Sequent calculus proof systems for inductive definitions, PhD. thesis, University of Edinburgh (2006)
9. Brotherston, J., Bornat, R., Calcagno, C.: Cyclic proofs of program termination in separation logic. In: Proceedings of POPL 2008 (2008)
10. Brotherston, J., Distefano, D., Petersen, R.L.: Automated cyclic entailment proofs in separation logic. In: Bjørner, N., Sofronie-Stokkermans, V. (eds.) CADE 2011. LNCS (LNAI), vol. 6803, pp. 131–146. Springer, Heidelberg (2011). https://doi.org/10.1007/978-3-642-22438-6_12
11. Brotherston, J., Simpson, A.: Sequent calculi for induction and infinite descent. J. Log. Comput. **21**(6), 1177–1216 (2011)
12. Brotherston, J., Gorogiannis, N., Petersen, R.L.: A generic cyclic theorem prover. In: Jhala, R., Igarashi, A. (eds.) APLAS 2012. LNCS, vol. 7705, pp. 350–367. Springer, Heidelberg (2012). https://doi.org/10.1007/978-3-642-35182-2_25
13. Doumane, A.: Constructive completeness for the linear-time μ-calculus. In: Proceedings of LICS 2017, pp. 1–12 (2017)
14. Fortier, J., Santocanale, L.: Cuts for circular proofs: semantics and cut-elimination. In: Proceedings of Computer Science Logic 2013 (CSL 2013). LIPIcs, vol. 23, pp. 248–262 (2013)
15. Kozen, D.: Results on the propositional μ-calculus. Theor. Comput. Sci. **27**, 333–354 (1983)
16. Martin-Löf, P.: Haupstatz for the intuitionistic theory of iterated inductive definitions. In: Proceedings of the Second Scandinavian Logic Symposium, North-Holland, pp. 179–216 (1971)
17. Niwiński, D., Walukiewicz, I.: Games for the μ-calculus. Theor. Comput. Sci. **163**, 99–116 (1997)
18. Podelski, A., Rybalchenko, A.: Transition invariants. In: Proceedings of 19th IEEE Symposium on Logic in Computer Science (LICS 2004), pp. 32–41 (2004)
19. Shoenfield, J.R.: Mathematical Logic. Addison-Wesley, Boston (1967)
20. Simpson, A.: Cyclic arithmetic is equivalent to Peano arithmetic. In: Esparza, J., Murawski, A.S. (eds.) FoSSaCS 2017. LNCS, vol. 10203, pp. 283–300. Springer, Heidelberg (2017). https://doi.org/10.1007/978-3-662-54458-7_17
21. Walukiewicz, I.: Completeness of Kozen's axiomatisation of the propositional μ-calculus. Inf. Comput. **157**, 142–182 (2000)

Undecidability of Equality for Codata Types

Ulrich Berger and Anton Setzer[✉]

Department of Computer Science, Swansea University, Swansea SA2 8PP, UK
{u.berger,a.g.setzer}@swan.ac.uk
http://www.cs.swan.ac.uk/~csetzer/, http://www.cs.swan.ac.uk/~csulrich/

Abstract. Decidability of type checking for dependently typed languages usually requires a decidable equality on types. Since bisimilarity on (weakly final) coalgebras such as streams is undecidable, one cannot use it as the equality in type checking. Instead, languages based on dependent types with decidable type checking such as Coq or Agda use intensional equality for type checking. Two streams are definitionally equal if the underlying terms reduce to the same normal form, i.e. if the underlying programs are syntactically equivalent. For reasoning about equality of streams one introduces bisimilarity as a propositional rather than judgemental equality.

In this paper we show that it is not possible to strengthen intensional equality in a decidable way while having the property that equality respects one step expansion, which means that a stream with head n and tail s is equal to cons(n, s). This property, which would be very useful in type checking, would not necessarily imply that bisimilar streams are equal, and we prove that there exist equalities with this properties which do not coincide with bisimilarity. Whereas a proof that bisimilarity on streams is undecidable is straightforward, proving that respecting one step expansion makes equality undecidable is much more involved and relies on an inseparability result for sets of codes for Turing machines. We prove this theorem both for streams with primitive corecursion and with coiteration as introduction rule.

Therefore, pattern matching on streams is, understood literally, not a valid principle, since it assumes that every stream is equal to a stream of the form cons(n, s). We relate this problem to the subject reduction problem found when adding pattern matching on coalgebras to Coq and Agda. We discuss how this was solved in Agda by defining coalgebras by their elimination rule and replacing pattern matching on coalgebras by copattern matching, and how this relates to the approach in Agda which uses the type of delayed computations, i.e. the so called "musical notation" for codata types.

Keywords: Coalgebra · Weakly final coalgebras · Codata
Decidable type checking · Martin-Löf type theory
Intensional equality · Intensional type theory
Dependent type theory · Undecidability results · Inseparability
Pattern matching · Copattern matching

© IFIP International Federation for Information Processing 2018
Published by Springer Nature Switzerland AG 2018. All Rights Reserved
C. Cîrstea (Ed.): CMCS 2018, LNCS 11202, pp. 34–55, 2018.
https://doi.org/10.1007/978-3-030-00389-0_4

1   Introduction

Many programs in computing are interactive in nature. We use user interfaces, text editors, data bases, interact with sensors and actuators, and communicate with other devices such as mobile phones or servers. Such programs potentially run forever – a text editor will never terminate, unless we terminate it explicitly or by accident, communication with a data base will never stop unless the server is shut down, etc. In a series of articles [HS05, HS00, HS04] Peter Hancock and the second author have shown how to represent interactive programs as non-well-founded trees. Such a connection has been observed in other contexts as well such as in order to give semantics to process algebras or describe interactive programs in functional programs using streams or monads. Because of this, non-well-founded data types play an important rôle in computer science. The usual approach to such non-well-founded structures is to represent them as coalgebraic data types.

In this paper we investigate final coalgebras in the context of dependent type theory with decidable type checking where *by coalgebras we will mean*, unless stated differently, *weakly final strictly positive coalgebras*. Decidable type checking requires that definitional equality, i.e. the equality used for type checking, is decidable. Theorem provers with decidable type checking such as Agda are very easy to use and allow one to write proofs in the same way as programs in many programming languages are written. The requirement for decidable definitional equality doesn't prevent reasoning about bisimilar coalgebras: one can define bisimilarity of coalgebras as a proposition, and prove that certain elements of coalgebras are bisimilar.

Coalgebras can be encoded using inductive types. However, in dependent type theory, it seems to be difficult or might even be impossible to get an encoding which gives the desired equalities w.r.t. decidable definitional equality. Therefore, it is of interest to add coalgebras explicitly to type theory. Coalgebras have been added in the form of codata types to both Coq [INR17] (see [Ber06, BC04] for their approach to coalgebras) and Agda [Nor07, Agd14]. The approach regarding coalgebras in Agda is described in [DA10]. Recently, the approach to define coalgebras by their elimination rules has been added as well to Agda, and used for implementing concepts from object based programming and graphical user interfaces in Agda [AAS17, AAS16].

In this article we answer the often asked question whether rules for intensional equality can be strengthened so that they permit at least one step expansion: if a stream s has head a and tail s', then it should be equal to (cons a s'). Such an equality does not necessarily imply that bisimilar streams are equal – only streams, which have the same first n elements and then are equal need to be equated. We show that indeed there are equalities which differ from bisimilarity but admit one step expansion. We give a negative answer to the initial question and show that there exists no decidable equality which allows for one step expansion. While a proof that bisimilarity on streams is undecidable is straightforward, since extensional equality on functions of type $\mathbb{N} \to \mathbb{N}$ is undecidable, this proof is much more evolved and relies on an inseparability result for sets of codes for Turing machines.

A consequence is that, if we want to stay in an intensional type theory with decidable type checking, we cannot assume that every stream is equal to (cons n s) for some n, s. Therefore, pattern matching on streams, understood literally, is not a valid principle: A definition

$$f : (s : \mathsf{Stream}) \to A[s]$$
$$f \,(\mathrm{cons}\ n\ s) = t[n, s]$$

assumes that every stream is equal to a stream of the form (cons n a).

This explains why defining coalgebras by their introduction rules led to a subject reduction problem in both Coq and Agda [Gim96, Our08, McB09, APTS13]. This problem was solved in Agda initially by disallowing the dependency of A on s. This however restricted quite severely its usefulness. Later it was solved together with the pattern matching problem by changing the type of s to be a new type (∞ Stream) of delayed computations. We discuss this approach in the conclusion. The latest approach taken in Agda is that coalgebras are defined by their elimination rules, and pattern matching is replaced by copattern matching. This approach has good properties: there are no restrictions on when to apply reductions, subject reduction holds, and we have complete duality between algebraic and coalgebraic data types.

Content of the Article. In Sect. 2 we review the notion of codata types. We discuss why decidable type checking and therefore a decidable definitional equality is useful. We review the problems of the codata approach (especially subject reduction) and review the approach of defining coalgebras by their elimination rules, which fixes this problem. We discuss as well the principle of primitive corecursion. In Sect. 3 we introduce encodings of streams which consist of a set of streams, functions head and tail, and an equality. Such encodings are universal if they admit the principle of primitive corecursion. Then we show in Theorem 9 that there is no decidable equality in such a universal encoding which fulfils the condition that ⟨head, tail⟩ is injective, i.e., that if the heads and tails of streams are equal, then the streams are equal. It follows (Corollary 11) that it is not possible to have a universal encoding of strings such that every stream is equal to a stream of the form (cons n s). We show as well (Examples 13) that there exist universal encodings for streams such that ⟨head, tail⟩ is not injective, and that injectivity of ⟨head, tail⟩ doesn't imply that the equality is bisimilarity. The proof of the main theorem makes essential use of the principle of primitive corecursion, and the question is whether the theorem still holds if we replace corecursion by coiteration. In Sect. 4 we show (Theorem 17) that this is the case. The paper ends with a conclusion, a discussion of related work, and a discussion of the use of codata types in theorem proving and programming. In particular we discuss how codata types can be reduced to coalgebras, and how notations such as the so-called "musical notation" in Agda can be understood as syntactic sugar, which allows one to keep most of the benefits of the codata approach when working with coalgebras.

2   Codata Types and Coalgebras

Codata Types. In the codata approach pioneered by Turner [Tur04, Tur95][1] one creates non-well-founded versions of algebraic data types. An example of a (well-founded) algebraic type is the type of natural numbers defined by[2]

$$\text{data } \mathbb{N} : \text{Set where}$$
$$0 \quad : \mathbb{N}$$
$$\text{suc} : \mathbb{N} \to \mathbb{N}$$

The elements of $\mathbb{N}$ are obtained by finitely many applications of the constructors. One can define a function from $\mathbb{N}$ to another type by pattern matching, i.e. by making a case distinction on whether the argument is 0 or $(\text{suc } n)$.

An example of a codata type is the set of streams of natural numbers

$$\text{codata Stream} : \text{Set where}$$
$$\text{cons} : \mathbb{N} \to \text{Stream} \to \text{Stream}$$

The keyword codata indicates that we are allowed to have infinitely many applications of cons, and therefore form infinitary terms $(\text{cons } n_0 \, (\text{cons } n_1 \, \cdots))$. As for data types one would expect pattern matching to work for codata types. We won't make this assumption in this article, and actually show that there is no decidable equality such that every element of a codata type matches a pattern, i.e. is equal to an element of the form $(\text{cons } n \, s)$. In this article by codata types we mean types which are like data types, but we allow infinite (more generally non-wellfounded) applications of the constructor.

The Need for Decidable Equality. Problems of the codata approach arise if one requires decidable type checking, as one does in most typed programming languages.

Most theorem provers use a goal-directed approach to derive proofs. One states a goal and then uses inference rules to derive that goal. If one had to write programs in normal programming languages this way one would need to derive a program by, for instance in case of Java, using a rule stating that it consists of a class with some name and some methods. Then one would have to use another rule to derive how a method is defined, etc. Using such an approach for deriving programs would be much more tedious and more difficult to learn than using the standard approach of first writing the program text and then type checking it by the compiler.

Agda is an example of a theorem prover with decidable type checking. Proving is very close to programming: instead of deriving an element of a type using rules,

---

[1] The earliest occurrence of codata types we could find is [Gim95], who called it "Coinductive". Hagino uses the notion of "codatatype" in [Hag89], but that notion refers to coalgebras defined by their elimination rules.

[2] We use in this section a notation similar to that of Agda. In particular, as common in Martin-Löf type theory, Set denotes the set of small types, and we write application in functional style, i.e. $(f \, a)$ for f applied to a.

the user types in a program text with some help from the system, which is then type checked automatically. Certain parts of the program text can be left open (called "goals"). The code is type checked automatically by the system and the user gets some help for filling in the goals. This allows the programmer to type in known parts directly and to combine in a very flexible way both forward and backward reasoning.

Decidable type checking in dependent types implies decidability of equality. This can be most easily seen when using Leibniz equality: If $a, b : A$, we have

$$(\lambda X.\lambda x.x : \Pi_{X:A\to \mathrm{Set}}(X\ a \to X\ b)) \Leftrightarrow a \text{ and } b \text{ are equal elements of } A.$$

Here, $\Pi_{X:A\to Set}(X\ a \to X\ b)$ is the polymorphic type of functions mapping any predicate on A (i.e. of type $A \to \mathrm{Set}$) to an element of $(X\ a \to X\ b)$. Therefore, in a type theory which permits the definition of the polymorphic type of Leibniz equality and which has decidable type checking we can decide using the type statement on the left hand side whether a and b are equal elements of type A. Hence decidability of type checking implies decidability of equality.

Problems of the Codata Approach. The natural equality on Streams is bisimilarity, which means that two streams (cons n_0 (cons n_1 $\cdots$)) and (cons m_0 (cons m_1 $\cdots$)) are equal if $n_i = m_i$ for all $i : \mathbb{N}$, that is, the functions $\lambda i.n_i$ and $\lambda i.m_i$ are extensionally equal. Since extensional equality on $\mathbb{N} \to \mathbb{N}$ is undecidable, bisimilarity is undecidable as well.

In order to deal with the problem of undecidability of extensional equality for function spaces, in Martin-Löf type theory ($MLTT$) one defines for type checking purposes two functions $f, g : A \to B$ as definitionally (or judgementally) equal, if f, g as λ-terms reduce to the same normal form.

Two functions are definitionally equal if the underlying programs reduce to the same normal form. In order to state that two functions are extensionally equal, one introduces a type (or proposition) expressing extensional equality, and then can prove extensional equality of functions in type theory. In the same way a decidable equality on codata types can be based on the principle that two elements of a codata type are equal if the underlying terms have the same normal form. Bisimilarity can then be introduced as a proposition which is given as a coinductive relation.

We cannot permit full expansion of codata types, since this would result in infinite and therefore non-normalising terms. The solution taken in Coq and earlier versions of Agda is to impose restrictions on when an element of a codata type can be expanded (see also the approach in [ADLO10] using lifting and boxing operators). These solutions led to a problem of subject reduction in Coq and earlier versions of Agda (see [APTS13] for a discussion on the history of this problem). As a consequence, in Agda elimination rules for codata types have been initially restricted to such extent that they are difficult to use. Later the "musical approach" was taken, which will be discussed in the conclusion. The latest approach taken in Agda uses coalgebras.

Coalgebras. A solution to this problem goes back to Hagino [Hag87, Hag89], namely to use the categorical dual of initial algebras (which correspond to algebraic data types), namely coalgebras. This approach has been further developed by Geuvers [Geu92], Howard [How96], Greiner [Gre92], Mendler [Men91]. It has been promoted for the use in MLTT by the second author in several talks and in [Set12, Set16], and by Granström [Gra08], and McBride [McB09]. See as well the work by Abbott, Altenkirch, and Ghani on containers [AAG03], and by Basold and Geuvers [BG16]. This approach has now been implemented in Agda (e.g. [AAS17, AAS16]).

Instead of defining Stream by its introduction rule, it is defined by its elimination rules

$$\text{coalg Stream : Set where}$$
$$\text{head : Stream} \to \mathbb{N}$$
$$\text{tail : Stream} \to \text{Stream}$$

The notation used in Agda is

$$\text{record Stream : Set where}$$
$$\text{coinductive}$$
$$\text{field}$$
$$\text{head : } \mathbb{N}$$
$$\text{tail : Stream}$$

Elements of Stream are terms such that head and tail applied to them return elements of $\mathbb{N}$ and Stream, respectively. A model of coalgebras as sets of natural numbers can for instance be found in [Set16].

The dual of primitive recursion is primitive corecursion (the earliest occurrence of this notion is probably Vene and Uustalu [VU98], see as well [Set12]). It corresponds to guarded recursion (see [Coq94]). Primitive corecursion means for the type Stream that if we have $A :$ Set, $h : A \to \mathbb{N}$, $t : A \to (\text{Stream} + A)$, then there exists

$$f : A \to \text{Stream}$$
$$\text{head } (f\ a) = h\ a,$$
$$\text{tail } (f\ a) = \begin{cases} s & \text{if } t\ a = \text{inl } s, \\ f\ a' & \text{if } t\ a = \text{inr } a'. \end{cases}$$

In the codata approach this principle translates as follows: Assuming h and t as before, we can define

$$f : A \to \text{Stream}$$
$$f\ a = \begin{cases} \text{cons } (h\ a)\ s & \text{if } t\ a = \text{inl } s, \\ \text{cons } (h\ a)\ (f\ a') & \text{if } t\ a = \text{inr } a'. \end{cases}$$

Essentially we can define $f\ a = \text{cons}\ n\ s$, where n and s depend on a, and s can be a stream which was defined before or $s = f\ a'$ for some $a' : A$.

Guarded recursion is widely accepted as a natural rule for coalgebras and codata types. In the POPL article [APTS13], coauthored by the second author, a simply typed recursive calculus was introduced in which the principle of primitive corecursion is represented by copattern matching, the dual of pattern matching. There it was shown that this calculus fulfils subject reduction.

Guarded Recursion for Codata Types. The principle of copattern matching and primitive corecursion for coalgebras corresponds to the principle of guarded recursion as introduced originally for codata types by Coquand [Coq94]. If we take the codata definition of Stream, guarded recursion allows one to define a function $f : A \rightarrow$ Stream by defining $f\ a = s$ for a stream s (depending on a) defined before, or by defining $f\ a = \text{cons}\ s\ (f\ t)$ for some s, t which depend on a. Nesting of constructors on the right hand side are allowed, but no other functions can be used.

An equation $f\ a = \text{cons}\ s\ (f\ t)$ corresponds in the coalgebra approach to the copattern matching equations

$$\text{head}\ (f\ a) = s$$
$$\text{tail}\ \ (f\ a) = f\ t$$

and an equation $f\ a = s$ for a stream s defined before to

$$\text{head}\ (f\ a) = \text{head}\ s$$
$$\text{tail}\ \ (f\ a) = \text{tail}\ s$$

So guarded recursion translates directly into copattern matching and primitive corecursion, and vice versa. Nested applications of constructors in a guarded recursion equation correspond to nested copattern matching: an equation $f\ a = \text{cons}\ s\ (\text{cons}\ t\ (f\ r))$ for guarded recursion corresponds to the equations

$$\text{head}\ (f\ a) \qquad\quad = s$$
$$\text{head}\ (\text{tail}\ (f\ a)) = t$$
$$\text{tail}\ \ (\text{tail}\ (f\ a)) = f\ r$$

Weakly Final Coalgebras. In final coalgebras one requires uniqueness of the function f introduced by primitive corecursion.[3] This principle is equivalent to bisimilarity as equality on coalgebras, which for streams means componentwise equality, and is therefore undecidable. We note that for final coalgebras the constructor is an isomorphism, so every element of a final coalgebra is introduced by a constructor. In order to obtain decidability of type checking, one replaces final coalgebras by weakly final coalgebras. In *weakly* final coalgebras, only the existence of functions defined by primitive corecursion is required, not their uniqueness. Elements of the coalgebra are introduced by the primitive corecursion operator $P_{\text{corec},A}$

---

[3] Actually it is only required for the principle of coiteration, where tail needs always to be of the form $f\ a'$. If one has uniqueness, one can derive the existence and uniqueness of functions defined by primitive corecursion. See [Set16] for a proof that for strictly positive coalgebras uniqueness of the functions defined by coiteration and by primitive corecursion are both equivalent to having a final coalgebra.

$$P_{\mathrm{corec},A} : (A \to \mathbb{N}) \to (A \to (\mathsf{Stream} + A)) \to A \to \mathsf{Stream}$$

$$\mathrm{head}\ (P_{\mathrm{corec},A}\ h\ t\ a) = h\ a$$

$$\mathrm{tail}\ \ (P_{\mathrm{corec},A}\ h\ t\ a) = \begin{cases} s & \text{if } t\ a = \mathrm{inl}\ s, \\ P_{\mathrm{corec},A}\ h\ t\ a' & \text{if } t\ a = \mathrm{inr}\ a'. \end{cases}$$

Elements of the coalgebra are equal if they reduce to the same normal form. MLTT style rules for coalgebras are worked out in [Set12]. Mendler [Men87] and Geuvers [Geu92] (Prop. 5.7) have shown that the polymorphic lambda calculus extended by weakly initial algebras and weakly final coalgebras for positive type schemes and higher type primitive recursion and primitive corecursion is strongly normalising. Therefore we obtain a decidable equality on coalgebras.

3   Undecidability of Weak Forms of Equality on Streams

We are going to show that, under minimal desirable conditions for streams, there is no decidable equality such that two streams with the same head and the same tail are equal. As usual when defining an undecidability result, we assume some encoding of computable streams as subsets of natural numbers. Any implementation of type theory would need some form of representing terms inside the systems, which amounts to encoding them in the computer, i.e. in binary and hence as a natural number. So we will work now in a standard setting of computability theory. As is tradition there, we will use mathematical notation for application, i.e. we write $f(x)$ instead of $(f\ x)$.

Convention 1. *(a) By a decidable relation on $A \subseteq \mathbb{N}$ we mean a subset $B \subseteq A$ such that there is a partial recursive function f such that for all $x \in A$, $f(x)$ is defined with $f(x) \in \{0,1\}$, and $x \in B$ iff $f(x) = 1$.*
(b) When writing $f : A \to B$ where $A, B \subseteq \mathbb{N}$ we man that f is a function from $\mathbb{N}$ to $\mathbb{N}$ such that $f(x) \in B$ for all $x \in A$.

Assumption 2. *(a) We assume a standard primitive recursive pairing function $\pi : \mathbb{N}^2 \to \mathbb{N}$ with projections $\pi_0, \pi_1 : \mathbb{N} \to \mathbb{N}$ satisfying $\pi_0(\pi(x,y)) = x$, $\pi_1(\pi(x,y)) = y$ for $x, y \in \mathbb{N}$.*
(b) Let $\mathrm{inl}, \mathrm{inr} : \mathbb{N} \to \mathbb{N}$, $\mathrm{inl}(n) := 2n$, $\mathrm{inr}(n) := 2n + 1$.
(c) For $A, B \subseteq \mathbb{N}$ we set
  – $A \times_{\mathbb{N}} B := \{\pi(a,b)\} \mid a \in A, b \in B\}.$
  – $A +_{\mathbb{N}} B := \{\mathrm{inl}(a) \mid a \in A\} \cup \{\mathrm{inr}(b) \mid b \in B\}.$
  (Note that $\mathbb{N} +_{\mathbb{N}} \mathbb{N} = \mathbb{N} \times_{\mathbb{N}} \mathbb{N} = \mathbb{N}$).
(d) We assume encodings of Turing machines (TM) and configurations for TMs as natural numbers. A configuration represent the finite portion of the tape currently used, the head position and the state of the TM. We assume that the working of TMs is modelled by primitive recursive functions
  – init : $\mathbb{N} \to \mathbb{N}$, which computes for a TM e its initial configuration;
  – next : $\mathbb{N}^2 \to \mathbb{N}$, which computes for a TM e and configuration c the configuration obtained after executing the next step of the TM;

- checkHalt : $\mathbb{N}^2 \to \mathbb{N}$, *which for TM* e *and configuration* c *determines whether* e *has halted (then it returns* true $:= 1$, *otherwise it returns* false $:= 0$);
- result : $\mathbb{N}^2 \to \mathbb{N}$, *such that* result$(e,c)$ *returns, if TM* e *in configuration* c *has halted, the result of this TM by reading it off the tape.*

For $e \in \mathbb{N}$, $\{e\}$ *denotes the partial recursive function (without input) corresponding to TM* e, *that is* $\{e\} \simeq$ run$(e, \text{init}(e))$ *where*

$$\text{run}(e, c) \simeq \begin{cases} \text{result}(e, c) & \text{if checkHalt}(e, c) = \text{true}, \\ \text{run}(e, \text{next}(e, c)) & \text{otherwise}. \end{cases}$$

Definition 3. *From* init, next, checkHalt, result *we derive primitive recursive functions which operate on pairs* $\pi(e, c)$ *for TMs* e *and configurations* c. *We also define a bounded variant of the function* run *that models termination after a given number* n *of computation steps:*

- init$'$: $\mathbb{N} \to \mathbb{N}$, init$'(e) = \pi(e, \text{init}(e))$.
- next$'$: $\mathbb{N} \to \mathbb{N}$, next$'(\pi(e, c)) = \pi(e, \text{next}(e, c))$.
- checkHalt$'$: $\mathbb{N} \to \mathbb{N}$, checkHalt$'(\pi(e, c)) = \text{checkHalt}(e, c)$.
- result$'$: $\mathbb{N} \to \mathbb{N}$, result$'(\pi(e, c)) = \text{result}(e, c)$.
- $\text{run}'_n(d) = \begin{cases} \text{result}'(d) + 1 & \text{if } n = 0 \text{ and checkHalt}'(d) = \text{true}, \\ \text{run}'_{n-1}(\text{next}'(d)) & \text{if } n > 0 \text{ and checkHalt}'(d) = \text{false}, \\ 0 & \text{otherwise}. \end{cases}$

$\text{run}'_n(d)$ *is a primitive recursive function of* n *and* d *such that* $\text{run}'_n(\text{init}'(e)) > 0$ *if and only if the TM encoded by* e *halts after exactly* n *steps and in that case* $\{e\} \simeq \text{run}'_n(\text{init}'(e)) - 1$.

Definition 4. *An encoding of streams* (Stream, head, tail, $==$) *is given by:*

(a) *A subset* Stream $\subseteq \mathbb{N}$.

(b) *An equivalence relation* $== \subseteq$ Stream $\times$ Stream, *called the equality of the stream encoding. We write* $s == s'$ *for* $(s, s') \in ==$, *and* $s \neq= s'$ *for* $(s, s') \notin ==$.

(c) *Functions* head : Stream $\to \mathbb{N}$, tail : Stream $\to$ Stream *that respect* $==$, *i.e.*

$$\forall s, s' : \text{Stream}\,.\, s == s' \to \text{head}(s) = \text{head}(s') \wedge \text{tail}(s) == \text{tail}(s')$$

Note that we do not impose any effectivity conditions on the set Stream *or the functions* head *and* tail.

Definition 5. *Let* (Stream, head, tail, $==$) *be an encoding of streams. For* $s, s' \in$ Stream *and a vector of natural numbers* $\boldsymbol{n}$ *we define*

$$s \xrightarrow{\boldsymbol{n}} s' \Leftrightarrow (\forall i < |\boldsymbol{n}| \text{head}_i(s) = n_i) \wedge \text{tail}^{|\boldsymbol{n}|}(s) == s'$$

where tail^k *is the k-fold iteration of* tail, *and* head$_k(s) := \text{head}(\text{tail}^k(s))$.

Definition 6. *An encoding of streams* $(\mathsf{Stream}, \mathsf{head}, \mathsf{tail}, ==)$ *is* injective *if the function*

$$\langle \mathsf{head}, \mathsf{tail}\rangle : \mathsf{Stream} \to \mathbb{N} \times \mathsf{Stream}, \quad \langle \mathsf{head}, \mathsf{tail}\rangle(s) = (\mathsf{head}(s), \mathsf{tail}(s))$$

is injective w.r.t. $==$, *that is*

$$\forall s, s' : \mathsf{Stream}\,.\,\mathsf{head}(s) = \mathsf{head}(s') \wedge \mathsf{tail}(s) == \mathsf{tail}(s') \to s == s'$$

The following easy lemma shows that every encoding of streams can be naturally turned into two injective ones that differ from the original one only in their equality.

Lemma and Definition 7. Let $(\mathsf{Stream}, \mathsf{head}, \mathsf{tail}, ==)$ be an encoding of streams. Define

$$s ==_{<\omega} s' \Leftrightarrow \exists n, t\,(s \xrightarrow{n} t \wedge s' \xrightarrow{n} t)$$
$$s \sim s' \Leftrightarrow \forall i \in \mathbb{N}\,(\mathsf{head}_i(s) = \mathsf{head}_i(s'))$$

Then $(\mathsf{Stream}, \mathsf{head}, \mathsf{tail}, ==_{<\omega})$ and $(\mathsf{Stream}, \mathsf{head}, \mathsf{tail}, \sim)$ are injective encodings of streams with $== \subseteq ==_{<\omega} \subseteq \sim$.

$==_{<\omega}$ can also be inductively defined as the least relation containing $==$ and making $\langle \mathsf{head}, \mathsf{tail}\rangle$ injective. $\sim$ is the usual bisimilarity of stream which can also be defined coinductively. If $==$ is an intensional notion of equality on streams, then the three equalities $==$, $==_{<\omega}$, $\sim$ are usually all different. We will give concrete examples where these equalities differ, after the proof of our main result, Theorem 9.

Definition 8. *An encoding of streams* $(\mathsf{Stream}, \mathsf{head}, \mathsf{tail}, ==)$ *is* universal *if for any primitive recursive functions* $h : \mathbb{N} \to \mathbb{N}$ *and* $t : \mathbb{N} \to (\mathsf{Stream} +_{\mathbb{N}} \mathbb{N})$ *there exists a primitive recursive function* $g : \mathbb{N} \to \mathsf{Stream}$ *such that*

- $\mathsf{head}(g(n)) = h(n)$
- $\mathsf{tail}(g(n)) == \begin{cases} s & \text{if } t(n) = \mathsf{inl}(s), \\ g(k) & \text{if } t(n) = \mathsf{inr}(k). \end{cases}$

We say g *is defined by* primitive corecursion *(from* h *and* t*) if* g *is primitive recursive and satisfies the equations above.*

Every constructive type theory equipped with coalgebras (or codata) and a primitive corecursion operator P gives rise to a universal encoding of streams with $g := \mathrm{P}(h, t) : \mathbb{N} \to \mathsf{Stream}$ as the function defined by primitive corecursion from h and t.

Theorem 9. Every injective universal encoding of streams has an undecidable equality.

Proof: Let $(\text{Stream}, \text{head}, \text{tail}, ==)$ be a universal encoding of streams. By universality, let $\text{const} : \mathbb{N} \to \text{Stream}$ be defined from the identity function and inr by primitive corecursion, that is, $\text{head}(\text{const}(i)) = i$ and $\text{tail}(\text{const}(i)) == \text{const}(i)$.

Claim. Assume $s \xrightarrow{0^n} \text{const}(k)$.

(a) $s == \text{const}(0)$ implies $k = 0$.
(b) If $\langle \text{head}, \text{tail} \rangle$ is injective, then $k = 0$ implies $s == \text{const}(0)$.

Proof of the Claim by induction on n: If $n = 0$, then the assumption is $s == \text{const}(k)$. For (a) assume $s == \text{const}(0)$. Then $\text{const}(k) == \text{const}(0)$ and therefore $k = \text{head}(\text{const}(k)) = \text{head}(\text{const}(0)) = 0$. Part (b) follows trivially from the assumption.

Now assume $n > 0$. The assumption now means $\text{head}(s) = 0$ and $\text{tail}(s) \xrightarrow{0^{n-1}} \text{const}(k)$. For (a) assume $s == \text{const}(0)$. Then $\text{tail}(s) == \text{tail}(\text{const}(0)) == \text{const}(0)$. Hence by induction hypothesis we get $k = 0$. For (b) assume $k = 0$. By induction hypothesis, $\text{tail}(s) == \text{const}(0)$. Hence $s == \text{const}(0)$, by injectivity and since $\text{head}(s) = 0 = \text{head}(\text{const}(0))$. This completes the proof of the Claim.

By universality, there exists a primitive recursive function $f : \mathbb{N} \to \text{Stream}$ s.t. if TM e terminates with result k after n steps, that is, $\text{run}'_n(\text{init}'(e)) = k+1$, then

$$f(e) \xrightarrow{0^{n+1}} \text{const}(k)$$

We will give a detailed argument why f exists at the end of the proof.

Now assume that $\langle \text{head}, \text{tail} \rangle$ is injective. Then we have by the Claim, applied to $s = f(e)$ where e is a TM that halts with result k, that $f(e) == \text{const}(0)$ iff $k = 0$. Therefore, if $==$ were decidable, then the function $\lambda e \, . \, f(e) == \text{const}(0)$ would be recursive and it would separate the TMs which terminate with result 0 from the TMs terminating with result > 0. But there is no recursive function separating these two sets, by the following well-known result in computability theory (part of the proof of Theorem II.2.5 on p. 148 in Odifreddi [Odi92]; references to originators are due to Odifreddi; the result can be found as well in Gasarch 1998 [Gas98], p. 1047, Note 2.8.):

Theorem 10. (Rosser [Ros36]**, Kleene** [Kle50]**, Novikov, Trakhtenbrot** [Tra53]**).** Let $A := \{e \mid \{e\} \simeq 0\}$ and $B := \{e \mid \{e\} \simeq 1\}$. Then A and B are recursively inseparable, that is, there is no (total) recursive function $f : \mathbb{N} \to \{0, 1\}$ such that $f(0) = 0$ for all $e \in A$, and $f(e) = 1$ for all $e \in B$.

We complete the proof of Theorem 9 by showing that a function f with the property specified above exists. Define primitive recursive functions $h : \mathbb{N} \to \mathbb{N}$ and $t : \mathbb{N} \to (\text{Stream} +_{\mathbb{N}} \mathbb{N})$ by

$$h(d) = 0$$
$$t(d) = \begin{cases} \text{inl}(\text{const}(\text{result}'(d))) & \text{if checkHalt}'(d) = \text{true,} \\ \text{inr}(\text{next}'(d)) & \text{otherwise.} \end{cases}$$

Let g be defined by primitive corecursion from h and t. We have

$$\text{head}(g(d)) \;=\; 0$$
$$\text{tail}(g(d)) \;==\; \begin{cases} \text{const}(\text{result}'(d)) & \text{if checkHalt}'(d) = \text{true}, \\ g(\text{next}'(d)) & \text{otherwise.} \end{cases}$$

Let $f : \mathbb{N} \to \mathsf{Stream}$, $f(e) = g(\text{init}'(e))$. We show that f is as required, that is, if $\text{run}'_n(\text{init}'(e)) = k + 1$, then $f(e) \overset{0^{n+1}}{\to} \text{const}(k)$. We show more generally if $\text{run}'_n(d) = k + 1$, then $g(d) \overset{0^{n+1}}{\to} \text{const}(k)$, by induction on n.

If $n = 0$, then $\text{checkHalt}'(d) = \text{true}$ (since $\text{run}'_n(d) > 0$). Therefore $\text{run}'_n(d) = \text{result}'(d) + 1$ and $k = \text{result}'(d)$. It follows $\text{head}(g(d)) = 0$, $\text{tail}(g(d)) == \text{const}(\text{result}'(d)) = \text{const}(k)$, and therefore $g(d) \overset{0^{n+1}}{\to} \text{const}(k)$.

If $n > 0$, then $\text{checkHalt}'(d) = \text{false}$ (since $\text{run}'_n(d) > 0$). Therefore $\text{run}'_n(d) = \text{run}'_{n-1}(\text{next}'(d)) = k+1$. By induction hypothesis $g(\text{next}'(d)) \overset{0^{n}}{\to} \text{const}(k)$. Since $\text{head}(g(d)) = 0$ and $\text{tail}(g(d)) == g(\text{next}'(d))$ it follows $g(d) \overset{0^{n+1}}{\to} \text{const}(k)$.

Corollary 11. Assume a universal encoding of streams $(\mathsf{Stream}, \text{head}, \text{tail}, ==)$. Assume a function $\text{cons} : \mathbb{N} \times \mathsf{Stream} \to \mathsf{Stream}$ that respects $==$, that is,

$$\forall n, s, s'\,.\, s == s' \to \text{cons}(n, s) == \text{cons}(n, s')$$

(a) Assume

$$\forall s : \mathsf{Stream}\,.\, s == \text{cons}(\text{head}(s), \text{tail}(s))$$

that is, cons is a left-inverse of $\langle \text{head}, \text{tail}\rangle$ w.r.t. $==$. Then $==$ is undecidable.

(b) Assume

$$\forall s : \mathsf{Stream}\,.\, \text{head}(\text{cons}(n, s)) = n \wedge \text{tail}(\text{cons}(n, s)) == s$$

that is, cons is a right-inverse of $\langle \text{head}, \text{tail}\rangle$ w.r.t. $==$. Assume further

$$\forall s : \mathsf{Stream}\,.\, \exists n.\exists s' : \mathsf{Stream}\,.\, s == \text{cons}(n, s')$$

that is, cons is surjective w.r.t. $==$. Then $==$ is undecidable.

Proof of Corollary 11: (a) If $\langle \text{head}, \text{tail}\rangle$ has a left-inverse, it is injective, hence Theorem 9 applies. (b) A surjective right-inverse is also a left-inverse.

Corollary 12. For every universal encoding of streams the equalities $==_{<\omega}$ and $\sim$ defined in Lemma 7 are undecidable.

Examples 13. Let $(\mathsf{Stream}, \text{head}, \text{tail}, ==)$ be a universal encoding of streams that is derived from some intensional constructive type theory with primitive corecursion (like for example the theory underlying Agda) such that $==$ corresponds to definitional equality.

First, we argue that $==_{<\omega}$ is not the same as bisimilarity by constructing bisimilar streams that are not related by $==_{<\omega}$: Let $f : \mathbb{N} \to$ Stream be defined by primitive corecursion such that $\mathrm{head}(f(x)) = 0$ and $\mathrm{tail}(f(x)) == f(x)$ for all $x \in \mathbb{N}$. Since $f(0)$ and $f(1)$ come from different terms in normal form we have $f(0) \neq= f(1)$. Since for all $n \in \mathbb{N}$

$$\mathrm{tail}^n(f(0)) == f(0) \neq= f(1) == \mathrm{tail}^n(f(1))$$

it follows that $f(0) \neq=_{<\omega} f(1)$. However, clearly $f(0)$ and $f(1)$ are bisimilar.

Next, we construct streams witnessing the fact that $==$ and $==_{<\omega}$ are different. From Theorem 9 we know that these two relation cannot coincide since $==$ is decidable but $==_{<\omega}$ isn't, however, it is interesting to see the difference by an example. We simply modify the above example slightly. Let $f' : \mathbb{N} \to$ Stream be defined by primitive corecursion such that $\mathrm{head}(f'(x)) = 0$ and $\mathrm{tail}(f'(x)) == f(0)$ for all $x \in \mathbb{N}$. With the same argument as before, $f'(0) \neq= f'(1)$. However $f'(0) \overset{0}{\to} f(0)$ and $f'(1) \overset{0}{\to} f(0)$, therefore $f'(0) ==_{<\omega} f'(1)$.

Remark 14. In the definitions and proofs above one may replace the class of primitive recursive functions by any other class of recursive functions satisfying some minimal closure conditions, for example all recursive functions, elementary functions, or polynomial time computable functions. Then Theorem 9 is still valid with the same proof.

4   Extension of Theorem 9 to Coiteration

For coalgebras we have the principles of primitive corecursion and coiteration which are the dual of primitive recursion and iteration for algebraic data types. A detailed discussion of these concepts and why they are dual can for instance be found in [Set16]. When we define a function $f : A \to$ Stream by primitive corecursion, we have the choice of defining $\mathrm{tail}(f(a)) = f(a')$ or $\mathrm{tail}(f(a)) = s$ for some given stream s. Coiteration restricts this choice by demanding that $\mathrm{tail}(f(a))$ always needs to be equal to $f(a')$ for some a'. An encoding of streams is coiteratively universal if it is closed under the coiteration operator:

Definition 15. *An encoding of streams* (Stream, head, tail, $==$) *is coiteratively universal if for any primitive recursive functions* $h : \mathbb{N} \to \mathbb{N}$ *and* $t : \mathbb{N} \to \mathbb{N}$ *there exists a primitive recursive function* $g : \mathbb{N} \to$ Stream *such that*

- $\mathrm{head}(g(n)) = h(n)$
- $\mathrm{tail}(g(n)) == g(t(n))$.

We say g *is defined by coiteration (from* h *and* t*), if* g *is primitive recursive and satisfies the equations above.*

Note that the functions f and f' in Example 13 are in fact defined by coiteration. However, our main Theorem 9 above relied essentially on the fact that we have primitive corecursion. This allowed us to escape once the TM has terminated into the streams $\mathrm{const}(i)$, and it is important that it was the same stream and not only a stream bisimilar to $\mathrm{const}(i)$. We will show that the main theorem applies as well to coiteratively universal encodings of streams, and that we can overcome the problem of not being able to escape into $\mathrm{const}(i)$ directly. But let us first repeat the standard argument that coiteration can simulate primitive corecursion up to bisimilarity:

Lemma 16. Let $(\mathsf{Stream}, \mathrm{head}, \mathrm{tail}, ==)$ be a coiteratively universal encoding of streams. Assume head and tail are primitive recursive (which are therefore defined on $\mathbb{N}$). Let h, t as in the definition of "universal encoding of streams", that is, $h : \mathbb{N} \to \mathbb{N}$ and $t : \mathbb{N} \to (\mathsf{Stream} +_{\mathbb{N}} \mathbb{N})$. Then there exist a primitive recursive function $g : \mathbb{N} \to \mathsf{Stream}$ such that g behaves up to $\sim$ like a function defined by primitive corecursion from h, t, more precisely,

(a) $\mathrm{head}(g(n)) = h(n)$,
(b) if $t(n) = \mathrm{inl}(s)$, then $\mathrm{tail}(g(n)) \sim s$,
(c) if $t(n) = \mathrm{inr}(m)$, then $\mathrm{tail}(g(n)) == g(m)$.

Proof: Define

$$h' : \mathbb{N} \to \mathbb{N} \quad (\text{recall that } \mathbb{N} = \mathbb{N} +_{\mathbb{N}} \mathbb{N})$$
$$h'(\mathrm{inl}(n)) = \mathrm{head}(n)$$
$$h'(\mathrm{inr}(n)) = h(n)$$

$$t' : \mathbb{N} \to \mathbb{N}$$
$$t'(\mathrm{inl}(n)) = \mathrm{inl}(\mathrm{tail}(n))$$
$$t'(\mathrm{inr}(n)) = t(n)$$

Let g' be defined by coiteration from h' and t', that is, for all $n \in \mathbb{N}$

$$\mathrm{head}(g'(n)) = h'(n),$$
$$\mathrm{tail}(g'(n)) == g'(t'(n)).$$

Let $g(n) := g'(\mathrm{inr}(n))$. Then g is primitive recursive and satisfies the conditions (a), (b), (c) as we show now. Conditions (a) and (c) are easy:

$$\mathrm{head}(g(n)) = \mathrm{head}(g'(\mathrm{inr}(n))) = h'(\mathrm{inr}(n)) = h(n),$$

and if $t(n) = \mathrm{inr}(m)$, then

$$\mathrm{tail}(g(n)) == \mathrm{tail}(g'(\mathrm{inr}(n))) == g'(t'(\mathrm{inr}(n'))) == g'(t(n)) = g(m).$$

For condition (b) we show first that $g'(\mathrm{inl}(s)) \sim s$ for all $s \in \mathsf{Stream}$. In fact, for all $n \in \mathbb{N}$ $\mathrm{tail}^n(g'(\mathrm{inl}(s))) == g'(t'^n(\mathrm{inl}(s))) = g'(\mathrm{inl}(\mathrm{tail}^n(s)))$ and therefore

$$\mathrm{head}(\mathrm{tail}^n(g'(\mathrm{inl}(s)))) = \mathrm{head}(g'(\mathrm{inl}(\mathrm{tail}^n(s))))$$
$$= h'(\mathrm{inl}(\mathrm{tail}^n(s)))$$
$$= \mathrm{head}(\mathrm{tail}^n(s)).$$

Now, if $t(n) = \mathrm{inl}(s)$, then

$$\mathrm{tail}(g(n)) == \mathrm{tail}(g'(\mathrm{inr}(n))) == g'(t'(\mathrm{inr}(n))) == g'(t(n)) = g'(\mathrm{inl}(s)) \sim s.$$

Theorem 17. Every injective coiteratively universal encoding of streams has an undecidable equality.

Proof: First note that although Lemma 16 reduces primitive corecursion to coiteration it cannot be used to reduce Theorem 17 to Theorem 9 since the reduction (b) in Lemma 16 is only with respect to bisimilarity. Therefore, we need a new proof, which however can be obtained by suitably modifying the proof of Theorem 9.

We replace the function g used in Theorem 9 by a function g' which on arguments $\mathrm{inl}(n)$ behaves like the function g before, and on arguments $\mathrm{inr}(n)$ behaves like the constant stream with elements in n. Now we can replace escaping into $\mathrm{const}(k)$ by a recursive call to $g'(\mathrm{inr}(n))$: More precisely, we define by coiteration

$$
\begin{aligned}
&g'' : \mathbb{N} \to \mathsf{Stream} \\
&\mathsf{head}(g''(\mathrm{inl}(d))) = 0 \\
&\mathsf{tail}(g''(\mathrm{inl}(d))) \quad = \begin{cases} g''(\mathrm{inr}(\mathrm{result}'(d))) & \text{if } \mathsf{checkHalt}'(d) = \mathrm{true}, \\ g''(\mathrm{inl}(\mathrm{next}'(d))) & \text{otherwise.} \end{cases} \\
&\mathsf{head}(g''(\mathrm{inr}(k))) = k \\
&\mathsf{tail}(g''(\mathrm{inr}(k))) \quad = g''(\mathrm{inr}(k))
\end{aligned}
$$

We define now

$$
\begin{aligned}
&g' : \mathbb{N} \to \mathsf{Stream} \\
&g'(k) \qquad = g''(\mathrm{inl}(k))
\end{aligned}
$$

$$
\begin{aligned}
&\mathrm{const}' : \mathbb{N} \to \mathsf{Stream} \\
&\mathrm{const}'(k) = g''(\mathrm{inr}(k))
\end{aligned}
$$

We obtain

$$
\begin{aligned}
&\mathsf{head}(g'(d)) \qquad = 0 \\
&\mathsf{tail}(g'(d)) \qquad = \begin{cases} \mathrm{const}'(\mathrm{result}'(d)) & \text{if } \mathsf{checkHalt}'(d) = \mathrm{true}, \\ g'(\mathrm{next}'(d)) & \text{otherwise.} \end{cases} \\
&\mathsf{head}(\mathrm{const}'(k)) = k \\
&\mathsf{tail}(\mathrm{const}'(k)) \quad = \mathrm{const}'(k)
\end{aligned}
$$

Now by replacing const by const′ and g by g' in the proof of Theorem 9, and using the equations above, we obtain a proof of Theorem 17.

Corollary 18. Corollaries 11 and 12 hold for iteratively universal encodings of streams as well.

5   Conclusion and Related Work

Codata Types and Coalgebras in Programming and Theorem Proving.
This paper shows that codata types are problematic in dependent type theory if one requires decidability of type checking. Codata types can still be used in a

simply typed settings in functional programming since type checking there does not require checking of equalities. They can be used as well in systems such as Nuprl where type statements are derived by the user and therefore decidability of type checking is not required. Otherwise, the best approach known at the moment is to define coalgebraic types as defined by their elimination rules.

Programming with coalgebras is very natural in a situation where a corresponding codata type would only have one constructor. The main example is the type of streams defined by having observations head and tail as defined earlier. As an example demonstrating that copattern matching is very natural consider the function enum : $\mathbb{N} \to$ Stream enumerating the natural numbers from n onwards which can be defined by the copattern equations head (enum n) = n and tail (enum n) = enum $(n + 1)$.

When we define a coalgebra where the corresponding codata type has more than one constructor we face the problem that several constructors in a codata type correspond to a disjoint union whereas several observations in a coalgebra correspond to a product. For instance, the observations head and tail of Stream can be replaced by one observation elim : Stream $\to \mathbb{N} \times$ Stream. Several observations in a coalgebra therefore do not allow to simulate several constructors of a codata type directly. Consider the example of colists (i.e. potentially infinite lists) which are defined as codata as

$$\begin{aligned} &\text{codata coList : Set where} \\ &\quad \text{nil} \quad : \text{coList} \\ &\quad \text{cons} : \mathbb{N} \to \text{coList} \to \text{coList} \end{aligned}$$

The eliminator for a corresponding coalgebra needs to determine for a colist whether it is nil or (cons n s). It can be done by defining

$$\begin{aligned} &\text{coalg coList : Set where} \\ &\quad \text{elim : coList} \to \top + \mathbb{N} \times \text{coList} \end{aligned}$$

Here $\top$ is the one element type with element tt, $+$ the disjoint union. elim l = inl tt means that l is of the form nil, and elim l = inr $(n$, $l')$ means that l is of the form (cons n l').

For programming it is more convenient to replace $\top + \mathbb{N} \times$ Stream by an extra type. A good notation is to replace the name coList by ∞coList and use coList for the extra type. We obtain the simultaneous definition of two types coList and ∞coList (using notations inspired by the "musical approach" in Agda see below):

$$\begin{aligned} &\text{coalg } \infty\text{coList : Set where} \\ &\quad \flat : \infty\text{coList} \to \text{coList} \end{aligned}$$

$$\begin{aligned} &\text{data coList : Set where} \\ &\quad \text{cons} : \mathbb{N} \to \infty\text{coList} \to \text{coList} \\ &\quad \text{nil} \quad : \text{coList} \end{aligned}$$

Every element of coList is of the form (cons n s) or nil, and one can make case distinction on elements of coList. But one cannot pattern match on ∞coList and therefore not pattern match on the second argument of cons – in order to unfold it further one needs to apply $\flat$ to it.

Decidability of equalities for coalgebras mentioned at the end of Sect. 2 holds in this situation as well. Our proof regarding undecidability of equality (Theorem 9) wouldn't go through in this situation, since it required that if we unfold l, l' : coList finitely many times and get the same heads and tail, then $l == l'$. For example, the case of unfolding the elements l and l' twice doesn't mean that they are both equal to (cons n (cons n' s)). It only means that $l = $ cons n l_0, where $\flat$ $l_0 = $ cons n' s, and $l' = $ cons n l'_0, where $\flat$ $l'_0 = $ cons n' s. But these equations do not imply $l_0 == l'_0$ and therefore neither $l == l'$.

Using the "Musical Approach" in Agda to Simulate Codata Types by Coalgebras (and Related Work). In Agda there exists, apart from the coalgebra approach, an approach which can be considered as introducing syntactic sugar for the above way of simulating codata types by coalgebras [Agd11, Dan09]. In that approach Agda generates for every name A for a type automatically a builtin type (∞ A), which is a type defined simultaneously with A.[4] Note that we *should not* have ∞ : Set $\to$ Set.[5] The type (∞ A) can be considered as a coalgebra defined simultaneously with A by

$$\text{coalg } (\infty \, A) : \text{Set where}$$
$$\flat : \infty \, A \to A$$

Agda provides as well a builtin function $\sharp$ which is defined by copattern matching as

$$\sharp : A \to \infty \, A$$
$$\flat \, (\sharp \, a) = a$$

With this approach we can replace ∞coList by (∞ coList), omit its definition (since it is builtin) and get a definition which is close to that of a codata type:

$$\text{data coList : Set where}$$
$$\text{cons} : \mathbb{N} \to \infty \text{ coList} \to \text{coList}$$
$$\text{nil} \quad : \text{coList}$$

We can now define enum : $\mathbb{N} \to \infty$ coList by copattern matching in a way which is very close to the definition for codata types:[6]

$$\flat \, (\text{enum } n) = \text{cons } n \, (\text{enum } (n + 1))$$

---

[4] There are various options of how to deal with types depending on parameters – this is left as future work.

[5] Actually, a constant of this type exists in Agda – the reason is that the musical approach is introduced via a library rather than a direct syntactic extension of Agda.

[6] That's how we believe Agda should behave. In fact, in Agda one defines instead enum : $\mathbb{N} \to$ coList by enum $n = $ cons n ($\sharp$ (enum $(n + 1)$)), an equation which, considered verbally, is not normalising and brings back the problems avoided by the coalgebra approach.

In [Agd11, Dan09], the type (∞ A) is considered as the type of delayed computations, and $\sharp : A \to \infty$ A forms a delayed computation from an element of A. Conversely, $\flat$ triggers a delayed computation.

This approach works well in situations where one needs to simulate pure codata types, which occur quite often. However, especially the work of the second author with Bashar Igried on CSP-Agda [IS18, IS17, IS16] has shown that it can be useful to have coalgebras with several observations. Even in a situation where one has a type which has a main observation similar to $\flat$ above, one often needs additional observations (in CSP-Agda there was the need to add an additional string component to the type of processes).

The musical approach in Agda is also a way of interpreting the approach by Altenkirch et al. [ADLO10] who introduce the language $\Pi\Sigma$, which has the type $[A]$ of delayed computations, which require ! (similar to $\flat$) in order to unfold them further. Note that $\Pi\Sigma$ is, as stated in [ADLO10], designed as a partial language which permits general recursion.

Further Related Work. McBride states at the end of Sect. 3 of [McB09] that there is no equality on colists such that every colist is introduced by a constructor. He gives some argument, but that argument relies on the undecidabilty of the Turing halting problem and doesn't work if one takes into account that there exist, as constructed by us, an (undecidable) equality on colists (we constructed it for streams) such that every element is introduced by a constructor, but which is not equal to bisimilarity. We needed to use a deeper theorem from computability theory in order to give a full mathematical proof for our theorem.

Conclusion. We have reviewed the two approaches for introducing non-well-founded data types, namely codata types given by introduction rules, and coalgebras given by elimination rules. We have shown that under weak assumptions, which are very natural for both approaches, there exists no decidable equality on Stream such that every element of Stream is introduced by a constructor. This causes at least conceptual problems for the codata approach. The theory of coalgebras seems to be simpler, avoids this problem and appears to be a conceptually superior approach to codata types. Reduction rules are easier in coalgebras since there are no special restrictions on when to apply reductions. Elements of coalgebras are finite objects which unfold to infinite objects only when applying destructors to them iteratively.

Overall, our results suggest that the future of codata types in dependent type theory with decidable type checking lies in its role as a useful derived concept based on coalgebras defined by observations. The musical notation in Agda can be seen as a realisation of this idea which makes it easy to work with the very commonly occurring situation of coalgebras which originate from codata types.

Acknowledgements. We would like to thank the three anonymous referees for many useful suggestions which led us to a more thorough discussion of the codata vs. coalgebra question regarding its origin, current literature and future development.

This research was supported by the projects CORCON (Correctness by Construction, FP7 Marie Curie International Research Project, PIRSES-GA-2013-612638),

COMPUTAL (Computable Analysis, FP7 Marie Curie International Research Project, PIRSES-GA-2011-294962), CID (Computing with Infinite Data, Marie Curie RISE project, H2020-MSCA-RISE-2016-731143), and by CA COST Action CA15123 European research network on types for programming and verification (EUTYPES).

References

[AAG03] Abbott, M., Altenkirch, T., Ghani, N.: Categories of containers. In: Gordon, A.D. (ed.) FoSSaCS 2003. LNCS, vol. 2620, pp. 23–38. Springer, Heidelberg (2003). https://doi.org/10.1007/3-540-36576-1_2

[AAS16] Abel, A., Adelsberger, S., Setzer, A.: ooAgda. Agda Library (2016). https://github.com/agda/ooAgda

[AAS17] Abel, A., Adelsberger, S., Setzer, A.: Interactive programming in Agda - Objects and graphical user interfaces. J. Funct. Program. **27**, January 2017. https://doi.org/10.1017/S0956796816000319

[ADLO10] Altenkirch, T., Danielsson, N.A., Löh, A., Oury, N.: $\Pi\Sigma$: dependent types without the sugar. In: Blume, M., Kobayashi, N., Vidal, G. (eds.) FLOPS 2010. LNCS, vol. 6009, pp. 40–55. Springer, Heidelberg (2010). https://doi.org/10.1007/978-3-642-12251-4_5

[Agd11] Agda Wiki: Coinductive data types, 1 January 2011. http://wiki.portal.chalmers.se/agda/pmwiki.php?n=ReferenceManual.Codatatypes

[Agd14] Agda team: The Agda Wiki (2014). http://wiki.portal.chalmers.se/agda/pmwiki.php

[Alt04] Altenkirch, T.: Codata. Talk given at the TYPES Workshop in Jouy-en-Josas, December 2004. http://www.cs.nott.ac.uk/~txa/talks/types04.pdf

[APTS13] Abel, A., Pientka, B., Thibodeau, D., Setzer, A.: Copatterns: programming infinite structures by observations. In: Giacobazzi, R., Cousot, R., (eds.) Proceedings of POPL 2013, pp. 27–38. ACM, New York (2013). https://doi.org/10.1145/2429069.2429075

[BC04] Bertot, Y., Castéran, P.: Interactive theorem proving and program development. Coq'Art: The Calculus of Inductive Constructions. Texts in Theoretical Computer Science. An EATCS Series. Springer (2004). ISBN 3-540-20854-2

[Ber06] Bertot, Y.: CoInduction in Coq, March 2006. arxiv:0603119

[BG16] Basold, H., Geuvers, H.: Type theory based on dependent inductive and coinductive types. In: Proceedings of the 31st Annual ACM/IEEE Symposium on Logic in Computer Science, LICS 2016, pp 327–336. ACM, New York (2016). https://doi.org/10.1145/2933575.2934514

[CF92] Cockett, R., Fukushima, T.: About charity. Technical report, Department of Computer Science, The University of Calgary, June 1992. Yellow Series Report No. 92/480/18. ftp://ftp.cpsc.ucalgary.ca/pub/projects/charity/literature/papers_and_reports/about_charity.ps.gz

[Coq94] Coquand, T.: Infinite objects in type theory. In: Barendregt, H., Nipkow, T. (eds.) TYPES 1993. LNCS, vol. 806, pp. 62–78. Springer, Heidelberg (1994). https://doi.org/10.1007/3-540-58085-9_72

[DA10] Danielsson, N.A., Altenkirch, T.: Subtyping, declaratively. In: Bolduc, C., Desharnais, J., Ktari, B. (eds.) MPC 2010. LNCS, vol. 6120, pp. 100–118. Springer, Heidelberg (2010). https://doi.org/10.1007/978-3-642-13321-3_8

[Dan09] Danielsson, N.A.: Changes to coinduction, 17 March 2009. Message posted on gmane.comp.lang.agda. http://article.gmane.org/gmane.comp. lang.agda/763/

[Gas98] Gasarch, W.: Chapter 16: A survey of recursive combinatorics. In: Ershov, Yu.L., Goncharov, S.S., Nerode, A., Remmel, J.B., Marek, V.W. (eds.) Handbook of Recursive Mathematics - Volume 2: Recursive Algebra, Analysis and Combinatorics, vol. 139, pp. 1041–1176. Elsevier (1998). https:// doi.org/10.1016/S0049-237X(98)80049-9

[Geu92] Geuvers, H.: Inductive and coinductive types with iteration and recursion. In: Nordström, B., Petersson, K., Plotkin, G. (eds.) Informal Proceedings of the 1992 Workshop on Types for Proofs and Programs, Bastad 1992, Sweden, pp. 183–207 (1992). http://www.lfcs.inf.ed.ac.uk/research/types-bra/proc/proc92.ps.gz

[Gim95] Giménez, E.: Codifying guarded definitions with recursive schemes. In: Dybjer, P., Nordström, B., Smith, J. (eds.) TYPES 1994. LNCS, vol. 996, pp. 39–59. Springer, Heidelberg (1995). https://doi.org/10.1007/3-540-60579-7_3

[Gim96] Giménez, C.E.: Un calcul de constructions infinies et son application à la vérification de systèmes communicants. (English: A calculus of infinite constructions and its application to the verification of communicating systems). Ph.D. thesis, Ecole normale supérieure de Lyon, Lyon, France (1996). http://citeseerx.ist.psu.edu/viewdoc/summary?doi=10.1.1. 16.9849&rank=1

[Gra08] Granström, J.G.: Reference and Computation in Intuitionistic Type Theory. Ph.D. thesis, Department of Mathematics, Uppsala University, Sweden (2008). http://intuitionistic.files.wordpress.com/2010/07/theses_published_uppsala.pdf

[Gre92] Greiner, J.: Programming with inductive and co-inductive types. Technical report CMU-CS-92-109, ADA249562, Dept. of Computer Science, Carnegie-Mellon University Pittburgh, Pittburgh, PA, USA, January 27 1992. 37 p. http://www.dtic.mil/docs/citations/ADA249562

[Hag87] Hagino, T.: A Categorical Programming Language. Ph.D. thesis, Laboratory for Foundations of Computer Science, University of Edinburgh (1987). http://www.tom.sfc.keio.ac.jp/~hagino/thesis.pdf

[Hag89] Hagino, T.: Codatatypes in ML. J. Symb. Comput. **8**(6), 629–650 (1989). https://doi.org/10.1016/S0747-7171(89)80065-3

[How96] Howard, B.T.: Inductive, coinductive, and pointed types. In: Proceedings of the First ACM SIGPLAN International Conference on Functional Programming, ICFP 1996, pp. 102–109. ACM, New York (1996). https://doi. org/10.1145/232627.232640

[HS00]   Hancock, P., Setzer, A.: Interactive programs in dependent type theory. In: Clote, P.G., Schwichtenberg, H. (eds.) CSL 2000. LNCS, vol. 1862, pp. 317–331. Springer, Heidelberg (2000). https://doi.org/10.1007/3-540-44622-2_21

[HS04]   Hancock, P., Setzer, A.: Interactive programs and weakly final coalgebras (extended version). In: Altenkirch, T., Hofmann, M., Hughes, J. (eds.) Dependently typed programming, number 04381 in Dagstuhl Seminar Proceedings. Internationales Begegnungs- und Forschungszentrum (IBFI), Schloss Dagstuhl, Germany (2004). http://drops.dagstuhl.de/opus/

[HS05] Hancock, P., Setzer, A.: Interactive programs and weakly final coalgebras in dependent type theory. In: Crosilla, L., Schuster, P. (eds.) From Sets and Types to Topology and Analysis. Towards Practicable Foundations for Constructive Mathematics, pp. 115–134. Oxford. Clarendon Press (2005). https://doi.org/10.1093/acprof:oso/9780198566519.003.0007

[INR17] INRIA. The Coq Proof Assistant Reference Manual. INRIA, version 8.7.1 edition (2017). https://coq.inria.fr/refman/

[IS16] Igried, B., Setzer, A.: Programming with monadic CSP-style processes in dependent type theory. In: Proceedings of the 1st International Workshop on Type-Driven Development, TyDe 2016, pp. 28–38. ACM, New York (2016). https://doi.org/10.1145/2976022.2976032

[IS17] Igried, B., Setzer, A.: Trace and stable failures semantics for CSP-Agda. In: Komendantskaya, E., Power, J. (eds.) Proceedings of the First Workshop on Coalgebra, Horn Clause Logic Programming and Types. Electronic Proceedings in Theoretical Computer Science, vol. 258, Edinburgh, UK, 28–29 November 2016, pp. 36–51. Open Publishing Association (2017). https://doi.org/10.4204/EPTCS.258.3

[IS18] Igried, B., Setzer, A.: Defining trace semantics for CSP-Agda, 30 Jan 2018. Accepted for publication in Postproceedings TYPES 2016, 23 pp. http://www.cs.swan.ac.uk/~csetzer/articles/types2016PostProceedings/igriedSetzerTypes2016Postproceedings.pdf

[JR97] Jacobs, B., Rutten, J.: A tutorial on (co)algebras and (co)induction. EATCS Bull. **62**, 62–222 (1997). http://www.cs.ru.nl/B.Jacobs/PAPERS/JR.pdf

[Kle50] Kleene, S.C.: A symmetric form of Gödel's theorem. Ind. Math. **12**, 244–246 (1950). https://doi.org/10.2307/2266709

[McB09] McBride, C.: Let's see how things unfold: reconciling the infinite with the intensional (extended abstract). In: Kurz, A., Lenisa, M., Tarlecki, A. (eds.) CALCO 2009. LNCS, vol. 5728, pp. 113–126. Springer, Heidelberg (2009). https://doi.org/10.1007/978-3-642-03741-2_9

[Men87] Mendler, N.P.: Recursive types and type constraints in second-order lambda calculus. In: Gries, D. (eds.) Proceedings of the Second Annual IEEE Symposium on Logic in Computer Science, LICS 1987, pp. 30–36. IEEE Computer Society Press, June 1987

[Men91] Mendler, N.P.: Inductive types and type constraints in the second-order lambda calculus. Ann. Pure Appl. Logic **51**(1–2), 159–172 (1991). https://doi.org/10.1016/0168-0072(91)90069-X

[Nor07] Norell, U.: Towards a practical programming language based on dependent type theory. Ph.D. thesis, Department of Computer Science and Engineering, Chalmers University of Technology, Göteborg, Sweden, September 2007. http://www.cse.chalmers.se/~ulfn/papers/thesis.pdf

[Odi92] Odifreddi, P.: Classical Recursion Theory, vol. 125. Elsevier (1992). https://doi.org/10.1016/S0049-237X(08)70011-9

[Our08] Oury, N.: Coinductive types and type preservation. Email posted 6 June 2008 at science.mathematics.logic.coq.club. (2008).https://sympa-roc.inria.fr/wws/arc/coq-club/2008-06/msg00022.html?checked_cas=2

[Ros36] Rosser, B.: Extensions of some theorems of Gödel and church. J. Symb. Logic **1**(3), 87–91 (1936). http://www.jstor.org/stable/2269028

[Rut00] Rutten, J.: Universal coalgebra: a theory of systems. Theor. Comput. Sci. **249**(1), 3–80 (2000). https://doi.org/10.1016/S0304-3975(00)00056-6

[Set12] Setzer, A.: Coalgebras as types determined by their elimination rules. In: Dybjer, P., Lindström, S., Palmgren, E., Sundholm, G. (eds.) Epistemology versus Ontology. Logic, Epistemology, and the Unity of Science, vol. 27, pp. 351–369. Springer (2012). https://doi.org/10.1007/978-94-007-4435-6_16

[Set16] Setzer, A.: How to reason coinductively informally. In: Kahle, R., Strahm, T., Studer, T. (eds.) Advances in Proof Theory. PCSAL, vol. 28, pp. 377–408. Springer, Cham (2016). https://doi.org/10.1007/978-3-319-29198-7_12

[Tra53] Trakhtenbrot, B.A.: On recursive separability. Dokl. Acad. Nauk **88**, 953–956 (1953)

[Tur95] Turner, D.A.: Elementary strong functional programming. In: Hartel, P.H., Plasmeijer, R. (eds.) FPLE 1995. LNCS, vol. 1022, pp. 1–13. Springer, Heidelberg (1995). https://doi.org/10.1007/3-540-60675-0_35

[Tur04] Turner, D.A.: Total functional programming. J. Univ. Comput. Sci. **10**(7), 751–768 (2004). http://www.jucs.org/jucs_10_7/total_functional_programming

[VU98] Vene, V., Uustalu, T.: Functional programming with apomorphisms (corecursion). Proc. Estonian Acad. Sci. Phys. Math. **47**(3), 147–161 (1998). http://cs.ioc.ee/~tarmo/papers/nwpt97-peas.pdf

Predicate Liftings and Functor Presentations in Coalgebraic Expression Languages

Ulrich Dorsch$^{(\boxtimes)}$ (ID), Stefan Milius (ID), Lutz Schröder (ID),
and Thorsten Wißmann (ID)

Friedrich-Alexander-Universität Erlangen-Nürnberg, Erlangen, Germany
{ulrich.dorsch,stefan.milius,lutz.schroeder,thorsten.wissmann}@fau.de

Abstract. We introduce a generic expression language describing behaviours of finite coalgebras over sets; besides relational systems, this covers, e.g., weighted, probabilistic, and neighbourhood-based system types. We prove a generic Kleene-type theorem establishing a correspondence between our expressions and finite systems. Our expression language is similar to one introduced in previous work by Myers but has a semantics defined in terms of a particular form of predicate liftings as used in coalgebraic modal logic; in fact, our expressions can be regarded as a particular type of modal fixed point formulas. The predicate liftings in question are required to satisfy a natural preservation property; we show that this property holds in particular for the Moss liftings introduced by Marti and Venema in work on lax extensions.

1 Introduction

Expression languages that support the syntactic description of system behaviour are one of the classical topics in computer science. The prototypic example are regular expressions; further examples include Kleene algebra with tests [17] and expression languages for labelled transition systems [1].

There has been recent interest in phrasing such expression languages generically, obtaining their syntax and semantics as well as meta-theoretic results including Kleene theorems by instantiation of a parametrized framework. This is achieved by abstracting the type of systems as coalgebras for a given type functor. This line of work originates with expression languages for a specific class of functors that essentially covers relational systems, so-called *Kripke polynomial functors* [34], and was subsequently extended to cover also weighted systems [32]. A generic expression language for arbitrary finitary functors can be based on algebraic functor presentations [25]. Here, we introduce a similar and, as it will turn out, in fact largely equivalent generic expression language

---

Work forms part of the DFG project COAX (MI 717/5-1/SCHR 1118/11-1).

C. Cîrstea (Ed.): CMCS 2018, LNCS 11202, pp. 56–77, 2018.
https://doi.org/10.1007/978-3-030-00389-0_5

for finitary functors, which we base on coalgebraic modalities in predicate lifting style, following the paradigm of coalgebraic logic [9]; on predicate liftings, we impose strong conditions, notably including preservation of singletons. Marti and Venema [20] have shown that for functors admitting a *lax extension* (in particular for functors that admit a separating set of monotone predicate liftings), one can convert operations from the functor presentation into predicate liftings, the so-called *Moss liftings*. We show that the Moss liftings preserve singletons; the converse does not hold in general, i.e. not all singleton-preserving predicate liftings are Moss liftings under a given lax extension.

We thus arrive at a generic expression language that covers, e.g., various flavours of relational, weighted, and probabilistic systems, as well as monotone neighbourhood systems as in the semantics of game logic [26] and concurrent dynamic logic [29]. We prove a Kleene theorem stating that every expression denotes the behavioural equivalence class of some state in a finite system, and that conversely every such behavioural equivalence class is denoted by some expression.

We make no claim to novelty for the design of a generic expression language as such, and in fact the expression language developed by Myers in his PhD dissertation [25] appears to be even more general. In particular, unlike Myers' language our expression language is currently restricted to describing behavioural equivalence classes in set-based coalgebras, and does not yet support algebraic operations (e.g. a join semilattice structure as in Silva et al.'s language for Kripke-polynomial functors [34] or in fact in standard regular expressions). The main point we are making is, in fact, a different one: we show that

coalgebraic expression languages embed into coalgebraic logic,

specifically into (the conjunctive fragment of) the coalgebraic μ-calculus [8], extending the classical result that every bisimilarity class of states in finite labelled transition systems is expressible by a *characteristic formula* in the μ-calculus [2,10,14,35]. This result provides a direct link between descriptions of processes and their property-oriented specification; as indicated above, the key to lifting it to a coalgebraic level of generality are singleton-preserving predicate liftings.

Related Work. As mentioned above, we owe much to work by Marti and Venema on Moss liftings [20], and moreover we use a notion of Λ-*bisimulation* [12] that turns out to be an instance of their definition of bisimulation via lax extensions. Besides the mentioned work on generic expression languages for Kripke polynomial [34], weighted [32], and finitary [25] functors, there is work on expression languages for *reactive T-automata* [11], which introduce an orthogonal dimension of genericity: The coalgebra functor as such remains fixed but the computational capacities of the automaton model at hand are encapsulated as a computational monad [23]. Venema [38] proves that for weak-pullback preserving functors, every bisimilarity class of finite coalgebras is expressible in coalgebraic fixpoint logic over Moss' ∇ modality.

2   Preliminaries

In the standard paradigm of universal coalgebra, types of state-based systems are encapsulated as endofunctors. We recall details on presentations of set functors and on their property-oriented description via predicate-lifting based coalgebraic modalities.

Functor Presentations describe set functors by signatures of operations and a certain restricted form of equations, so-called *flat* equations, alternatively by a suitable natural surjection. A *signature* is a sequence $\Sigma = (\Sigma_n)_{n \in \omega}$ of sets. Elements of Σ_n are regarded as n-ary operation symbols (we write $\tau/n \in \Sigma$ for $\tau \in \Sigma_n$). Every signature Σ determines the corresponding polynomial endofunctor T_Σ on **Set**, which maps a set X to the set

$$T_\Sigma X = \coprod_{n \in \omega} \Sigma_n \times X^n$$

and similarly on maps.

Definition 2.1. A *presentation* of a functor $T : \mathbf{Set} \to \mathbf{Set}$ is a pair (Σ, α) consisting of a signature Σ and a natural transformation $\alpha : T_\Sigma \twoheadrightarrow T$ with surjective components α_X. In the following, we abuse notation and denote, for every $\tau/n \in \Sigma$, the corresponding coproduct component of $\alpha : T_\Sigma \twoheadrightarrow T$ again by $\tau : (-)^n \to T$, and refer to it as an *operation* of T.

Most of our results concern finitary set functors. Recall that a functor is *finitary* if it preserves filtered colimits. Over **Set**, we have the following equivalent characterizations:

Theorem 2.2 (Adámek and Trnkova [3]). *Let $T : \mathbf{Set} \to \mathbf{Set}$ be a functor. Then the following are equivalent:*

1. *T is finitary;*
2. *T is bounded, i.e. for every element $x \in TX$ there exists a finite subset $m : Y \hookrightarrow X$ and an element $y \in TY$ such that $x = Tm(y)$;*
3. *T has a presentation.*

Indeed, for the equivalence of (1) and (3) note that every polynomial functor T_Σ is finitary, and finitary functors are closed under taking quotient functors. Conversely, given a finitary functor T, let $\Sigma_n = Tn$ and define $\alpha_X : T_\Sigma X \to TX$ by $\alpha_X(\tau, t) = Tt(\tau)$, where $t \in X^n$ is considered as a function $n \to X$. It is easy to show that this yields a natural transformation with surjective components.

Remark 2.3. As indicated above, the natural surjection α in a functor presentation (Σ, α) can be replaced with a set of flat equations over Σ, where an equation is called *flat* if both sides consist of an operation symbol applied to variables [3]. Incidentally, this (standard) term should not be confused with the same term introduced in the context of our expression language in Sect. 5.

Example 2.4. (1) Let A be an input alphabet. The functor $TX = 2 \times X^A$, whose coalgebras are deterministic automata, is polynomial, and finitary if A is finite. Thus, T has a presentation (Σ, α) by a signature Σ with two $|A|$-ary operations and no equations, i.e. α is the natural isomorphism $T_\Sigma \cong 2 \times (-)^A$.

(2) For a commutative monoid $(M, +, 0_M)$ the monoid-valued functor $M^{(-)}$: **Set** $\to$ **Set** is defined by

$$M^{(X)} = \{\mu : X \to M \mid \mu(x) = 0_M \text{ for all but finitely many } x \in X\}$$

and by $M^{(h)}(\mu) = y \mapsto \sum_{h(x)=y} \mu(x)$ on maps $h : X \to Y$. We view elements of $M^{(X)}$ as finitely supported additive measures on X, and in particular write $\mu(A) = \sum_{x \in A} \mu(x)$ for $A \subseteq X$; in this view, maps $M^{(h)}$ just take image measures. For a set $G \subseteq M$ of generators (i.e. there exists a surjective monoid morphism $G^* \twoheadrightarrow M$), $M^{(-)}$ is represented by

$$\alpha_X : \coprod_{n \in \omega} G^n \times X^n \twoheadrightarrow M^{(X)}, \qquad \alpha_X(\tau, t) = M^{(t)}(\tau),$$

where $\tau \in G^n$ is considered as an element of $M^{(n)}$.

(3) The finite powerset functor $\mathcal{P}_\omega$ (with $\mathcal{P}_\omega(X)$ being the set of finite subsets of X) is the monoid-valued functor for the monoid $(\{0,1\}, \vee, 0)$. Since this is generated by $G = \{1\}$, we have one n-ary operation symbol for each $n \in \omega$:

$$\alpha_X : \coprod_{n \in \omega} X^n \twoheadrightarrow \mathcal{P}_\omega X, \qquad \alpha_X(x_1, \ldots, x_n) = \{x_1, \ldots x_n\};$$

e.g. α identifies the tuples (x_1, x_1, x_2) and (x_1, x_2).

(4) For the monoid $\mathbb{N}$ of natural numbers with addition, one obtains the bag functor $\mathcal{B}$ as $\mathbb{N}^{(-)}$. Concretely, $\mathcal{B}$ maps a set X to the set $\mathcal{B}X$ of bags (i.e. finite multisets) on X. Since $(\mathbb{N}, +, 0)$ is generated by $G = \{1\}$, we have the same signature as for $\mathcal{P}_\omega$, namely one n-ary operation symbol per $n \in \omega$; of course, the presentation α now identifies fewer tuples, e.g. distinguishes (x_1, x_2, x_1) and (x_1, x_2).

(5) The *finite distribution functor* $\mathcal{D}$ is a subfunctor of the monoid-valued functor $\mathbb{R}_{\geq 0}^{(-)}$ for the additive monoid of the non-negative reals, given by $\mathcal{D}X = \{\mu \in \mathbb{R}_{\geq 0}^{(-)} \mid \sum_{x \in X} \mu(x) = 1\}$. Note that elements of $\mathcal{D}X$ can be represented as formal convex combinations $\sum_{i=1}^n p_i x_i$, $p_i \in \mathbb{R}_{\geq 0}, x_i \in X$ for $i = 1, \ldots, n$, with $p_1 + \cdots + p_n = 1$. Taking $\mathbb{R}_{\geq 0}$ itself as the set of generators and restricting to $\mathcal{D}$, we obtain a presentation (Σ, α) with an n-ary operation symbol for each n-tuple $(p_1, \ldots, p_n) \in \mathbb{R}_{\geq 0}^n$ such that $p_1 + \cdots + p_n = 1$, and α_X maps $((p_1, \ldots, p_n), (x_1, \ldots, x_n))$ to the formal convex combination $\sum_{i=1}^n p_i x_i$.

(6) The *finitary monotone neighbourhood functor* $\mathcal{M}_\omega$, i.e. the finitary part of the standard monotone neighbourhood functor $\mathcal{M}$, can be described as follows. To begin, $\mathcal{M}$ is the subfunctor of the double contravariant powerset functor $\mathcal{Q}\mathcal{Q}^{\mathrm{op}}$ given on objects by

$$\mathcal{M}X = \{\mathfrak{A} \subseteq \mathcal{Q}(X) \mid \mathfrak{A} \text{ upwards closed under } \subseteq\}.$$

We can then describe $\mathcal{M}_\omega X$ as consisting of all $\mathfrak{A} \in \mathcal{M}X$ having finitely many minimal elements, all of them finite, such that every element of $\mathfrak{A}$ is above a minimal one. We have the following presentation of $\mathcal{M}_\omega$: For every choice of numbers $n \geq 0$, $k_1, \ldots, k_n \geq 0$, we have a $\sum_{i=1}^n k_i$-ary operation mapping $(x_{ij})_{i=1,\ldots,n;j=1,\ldots,k_i}$ to the upwards closure of the set system

$$\{\{x_{i1}, \ldots, x_{ik_i}\} \mid i = 1, \ldots, n\}.$$

Coalgebraic Logic. Since coalgebras serve as generic models of reactive systems, it is natural to specify properties of coalgebras in terms of suitable modalities. The semantics of coalgebraic modalities can be defined using *predicate liftings* [27,30], which specify how a predicate on a base set X induces a predicate on the set TX where T is the coalgebraic type functor:

Definition 2.5. For $n \in \omega$ an n-ary *predicate lifting* for a functor $T : \mathbf{Set} \to \mathbf{Set}$ is a natural transformation

$$\lambda : \mathcal{Q}^n \to \mathcal{Q}T^{\mathrm{op}}$$

where $\mathcal{Q} : \mathbf{Set}^{\mathrm{op}} \to \mathbf{Set}$ is the contravariant powerset functor, with $\mathcal{Q}f$ taking preimages, i.e.

$$\mathcal{Q}f(A) = f^{-1}[A].$$

We write λ/n to indicate that λ has arity n. A predicate lifting λ is *monotone* if it preserves set inclusion in every argument. A set Λ of predicate liftings is *separating* [28,30] if every $t \in TX$ is uniquely determined by the set

$$T_\Lambda(t) = \{(\lambda, A_1, \ldots, A_n) \mid \lambda/n \in \Lambda, A_i \in \mathcal{Q}X \text{ and } t \in \lambda_X(A_1, \ldots, A_n)\}.$$

Example 2.6. The basic example is the interpretation of the standard box modality $\square$ over the covariant powerset functor $\mathcal{P}$ (with $\mathcal{P}f$ taking direct images), given by the monotone unary predicate lifting λ defined by

$$\lambda_X(A) = \{B \in \mathcal{P}(X) \mid B \subseteq A\}.$$

For a further monotone example, we interpret the box modality over the monotone neighbourhood functor $\mathcal{M}$ (Example 2.4) by the monotone unary predicate lifting

$$\lambda_X(A) = \{\mathfrak{A} \in \mathcal{M}X \mid A \in \mathfrak{A}\}.$$

It is easy to see that in both these examples, the predicate lifting for $\square$ alone is separating.

Predicate-lifting-based modalities can be embedded into *coalgebraic logics* of varying degrees of expressiveness. Our expression language introduced in Sect. 5 will live inside the *coalgebraic μ-calculus* [8], more precisely its *conjunctive fragment* [13]. We defer details to Sect. 5.

3   Singleton-Preserving Predicate Liftings

Our generic expression language will depend on a specific type of predicate liftings, as well as on a strengthening of separation:

Definition 3.1. An n-ary predicate lifting λ *preserves singletons* if

$$|\lambda_X(\{x_1\}, \ldots, \{x_n\})| = 1$$

for all $x_1, \ldots, x_n \in X$. Moreover, a set Λ of predicate liftings is *strongly expressive* if for every $t \in TX$ there exist $\lambda/n \in \Lambda$ and $x_1, \ldots, x_n \in X$ such that

$$\{t\} = \lambda_X(\{x_1\}, \ldots, \{x_n\}).$$

Singleton preservation will serve to ensure that expressions of our language denote unique behaviours, while strong expressivity will guarantee that all (finite) behaviours are expressible. The following is immediate:

Lemma 3.2. *Every strongly expressive set of predicate liftings is separating.*

Example 3.3. The predicate liftings in Example 2.6 both fail to preserve singletons. Our main source of singleton-preserving predicate liftings are Moss liftings as introduced in general terms in the next section. For the finite powerset functor $\mathcal{P}_\omega$ consider the predicate liftings λ^n/n given by

$$\lambda_X^n(A_1, \ldots, A_n) = \{B \in \mathcal{P}_\omega X \mid B \subseteq \bigcup_{i=1}^n A_i \text{ and} \\ B \cap A_i \neq \emptyset \text{ for } i = 1, \ldots, n\} \tag{3.1}$$

(which can be seen as arising from the above lifting for $\Box$ by Boolean combination). Then $\lambda_X^n(\{x_1\}, \ldots, \{x_n\}) = \{\{x_1, \ldots, x_n\}\}$ for $x_1, \ldots, x_n \in X$, which shows that the λ^n preserve singletons and that the set $\{\lambda^n \mid n \in \omega\}$ is strongly expressive.

Remark 3.4. Singleton-preserving predicate liftings should not be confused with Kurz and Leal's *singleton liftings* [18,19]. The definition of the latter is based on the one-to-one correspondence between subsets of $T(2^n)$ and n-ary predicate liftings for T [30], which maps an n-ary predicate lifting λ to $\lambda_{2^n}(\pi_1^{-1}(\{\top\}), \ldots, \pi_n^{-1}(\{\top\})) \subseteq T(2^n)$, and $C \subseteq T(2^n)$ to the lifting λ defined by $\lambda_X(A_1, \ldots, A_n) = \{t \in TX \mid T\langle \chi_{A_1}, \ldots, \chi_{A_n}\rangle(t) \in C\}$, where $\pi_i : 2^n \to 2$ is the i-th projection and $\chi_A : X \to 2$ denotes the characteristic function of $A \subseteq X$. An n-ary predicate lifting is a *singleton lifting* if it corresponds to a singleton subset of $T(2^n)$.

It is then indeed immediate that every *unary* singleton-preserving predicate lifting λ is a singleton lifting, since the above correspondence maps λ to the singleton $\lambda_2(\{\top\})$. The following examples show that this implication breaks down at higher arities, and that the converse also fails in general.

Example 3.5. (1) The unary singleton lifting for $\mathcal{P}$ corresponding to $\{\{\bot\}\} \subseteq$ $\mathcal{P}2$ fails to preserve singletons. Of course, this lifting fails to be monotone.

(2) Binary monotone singleton liftings need not preserve singletons. E.g. for the distribution functor $\mathcal{D}$, the monotone singleton lifting λ corresponding to $\{1 \cdot (\top, \top)\} \subseteq \mathcal{D}(2^2)$ is given by $\lambda(A, B) = \{\mu \mid \mu(A) = \mu(B) = 1\}$, so $\lambda(\{x\}, \{y\}) = \varnothing$ for $x \neq y$. We leave it as an open question whether unary monotone singleton liftings preserve singletons.

(3) The binary singleton-preserving predicate lifting

$$\lambda(A, B) = \{\mu \mid \mu(A) \geq 1/2, \mu(B) \geq 1/2, \mu(A \cup B) = 1\}$$

for the distribution functor $\mathcal{D}$ (see Example 4.7 for details) is not a singleton lifting, as it corresponds to the following infinite subset of $\mathcal{D}(2^2)$:

$$\{\mu \mid \mu(2 \times \{\top\}) \geq 1/2, \mu(\{\top\} \times 2) \geq 1/2, \mu(2 \times \{\top\} \cup \{\top\} \times 2) = 1\}.$$

It is not hard to see that we can recover operations for a functor from *monotone* singleton preserving predicate liftings; in detail:

Lemma 3.6. *Let $T : \mathbf{Set} \to \mathbf{Set}$. Then the following hold.*

1. *For each monotone singleton-preserving predicate lifting λ/n,*

$$\{\tau_{\lambda,X}(x_1, \ldots, x_n)\} := \lambda_X(\{x_1\}, \ldots, \{x_n\}) \tag{3.2}$$

   defines a natural transformation $\tau_\lambda : (-)^n \to T$.
2. *If Λ is a strongly expressive set of monotone singleton-preserving predicate liftings, then taking operation symbols τ_λ for each $\lambda \in \Lambda$, with associated interpretation as per (3.2), yields a functor presentation of T.*

Example 3.7. The singleton-preserving predicate liftings λ^n from Example 2.6 induce, according to the above construction, the operations $X^n \to \mathcal{P}_\omega(X)$, $(x_1, \ldots, x_n) \mapsto \{x_1, \ldots, x_n\}$.

The other direction, generating predicate liftings from functor presentations, is more involved, and treated next.

4   Moss Liftings

Marti and Venema [20] introduce *Moss liftings*, predicate liftings that are constructed from functor presentations with the help of a generalized form of the nabla operator, extending an earlier construction for weak-pullback preserving functors by Kurz and Leal [18]. Recall that for a weak-pullback-preserving functor T, Moss' [24] classical nabla operator $\nabla : T\mathcal{Q} \Rightarrow \mathcal{Q}T^{\mathrm{op}}$ is the natural transformation defined by

$$\nabla(\Phi) = \{t \in TX \mid (t, \Phi) \in \overline{T}(\in_X)\}.$$

Here, $\in_X \subseteq X \times QX$ is the element-of relation for X, and $\overline{T}$ is the *Barr extension* of T, viz. the functor $\overline{T}$ on the category of sets and relations defined on a relation $R \subseteq X \times Y$ by $\overline{T}R = \{(T\pi_1(r), T\pi_2(r)) \mid r \in TR\}$, where $\pi_1 : R \to X$ and $\pi_2 : R \to Y$ are the projection maps (cf. [24]). Barr [5] (see also Trnková [37]) proved that $\overline{T}$ is a functor if and only if T preserves weak pullbacks.

Further recall that the *converse* of a relation $R \subseteq X \times Y$ is the relation $R^\circ = \{(y,x) \mid x\,R\,y\} \subseteq Y \times X$. We denote the composite of two relations $R \subseteq X \times Y$ and $S \subseteq Y \times Z$ diagrammatically by $R; S \subseteq X \times Z$. Also, for $A \subseteq X$ we denote by $R[A] \subseteq Y$ the relational image $R[A] = \{y \mid \exists x \in A.\, xRy\}$. The construction $T \mapsto \overline{T}$ is generalized and abstracted in the notions of *relation lifting* and, more specifically, *lax extension* of a functor, as recalled next.

Definition 4.1 (Relation lifting, lax extension [20]). A *relation lifting* L for a functor T is an assignment mapping every relation $R \subseteq X \times Y$ to a relation $LR \subseteq TX \times TY$ such that converses are preserved: $L(S^\circ) = (LS)^\circ$. A relation lifting L is a *lax extension* if for all relations $R, R' \subseteq X \times Z$, $S \subseteq Z \times Y$ and functions $f : X \to Z$ (identified with their graph relation) the following hold:

$$R' \subseteq R \Rightarrow LR' \subseteq LR,$$
$$LR; LS \subseteq L(R; S),$$
$$Tf \subseteq Lf.$$

A lax extension L *preserves diagonals* if for all sets X

$$L\Delta_X \subseteq \Delta_{TX}.$$

Proposition 4.2 (Properties of Lax Extensions [20]). *Let L be a lax extension for a functor T. Then for all functions $f : X \to Z$, $g : Y \to Z$ and relations $R \subseteq X \times Z$, $S \subseteq Z \times Y$,*

(i) $\Delta_{TX} \subseteq L\Delta_X$,
(ii) $Tf; LS = L(f; S)$ *and* $LR; (Tg)^\circ = L(R; g^\circ)$,

and if L preserves diagonals, then

(iii) $\Delta_{TX} = L\Delta_X$ *and* $Tf = Lf$,
(iv) $Tf; (Tg)^\circ = L(f; g^\circ)$.

One use of relation liftings is to determine coalgebraic notions of bisimulation:

Definition 4.3 (L-Bisimulation [20]). Let L be a relation lifting for a functor $T : \mathbf{Set} \to \mathbf{Set}$, and let (X, ξ), (Y, ζ) be T-coalgebras. A relation $S \subseteq X \times Y$ is an *L-simulation* if for all $x \in X$ and $y \in Y$,

$$x\,S\,y \quad \text{implies} \quad \xi(x)\,LS\,\zeta(y).$$

An *L-bisimulation* is a relation S such that S and S° are L-simulations. Two states are *L-bisimilar* if there exists an L-bisimulation relating them.

Marti and Venema [20, Theorem 11] show that if L is a lax extension that preserves diagonals, then L-bisimilarity coincides with behavioural equivalence.

Assumption 4.4. From now on we fix a finitary endofunctor $T : \mathbf{Set} \to \mathbf{Set}$ having a diagonal-preserving lax extension L and a presentation (Σ, α) of T.

Another key feature of lax extensions is that they induce canonical modalities, generalizing Moss' coalgebraic logic [24]:

Definition 4.5 (Lax Nabla [20]). The *lax nabla* of L is the family of functions

$$\nabla_X^L : TQX \to QT^{\mathrm{op}}X$$
$$\Phi \quad \mapsto \{t \in TX \mid (t, \Phi) \in L(\in_X)\},$$

where $\in_X \subseteq X \times QX$ is the element-of relation for X.

As shown by Marti and Venema [20], the lax nabla is in fact a natural transformation $\nabla^L : TQ \Rightarrow QT^{\mathrm{op}}$, and coincides with Moss' classical ∇ for L being the Barr extension of T (and T preserving weak pullbacks). In combination with a functor presentation, the lax nabla gives rise to a family of predicate liftings:

Definition 4.6 (Moss Liftings [20]). Every operation symbol $\tau/n \in \Sigma$ yields a predicate lifting λ defined by

$$\lambda = (Q^n \overset{\tau Q}{\Rightarrow} TQ \overset{\nabla^L}{\Rightarrow} QT^{\mathrm{op}}),$$

that is,

$$\lambda_X(X_1, \ldots X_n) = \{t \in TX \mid (t, \tau_{QX}(X_1, \ldots, X_n)) \in L(\in_X)\}.$$

These predicate liftings are called the *Moss liftings* of T.

Example 4.7. Some standard functor presentations are converted into Moss liftings as follows.

(1) For the deterministic automata functor $TX = 2 \times X^A$ consider the Barr extension $L = \overline{T}$. Then elements of TQX are pairs $(b, (Y_a)_{a \in A})$, where each Y_a is a subset of X, and

$$\nabla_X(b, (Y_a)_{a \in A}) = \{(b, (x_a)_{a \in A}) \mid \forall a \in A : x_a \in Y_a\} \qquad \text{for } b = 0, 1.$$

The two Moss liftings $\lambda^0, \lambda^1 : Q^A \to Q(2 \times (-)^A)$ corresponding to the two $|A|$-ary operation symbols from the presentation in Example 2.4(1) are thus defined (slightly abusing notation) by

$$\lambda^i((Y_a)_{a \in A}) = \{(i, (x_a)_{a \in A}) \mid \forall a \in A : x_a \in Y_a\} \qquad \text{for } i = 0, 1.$$

(2) As indicated in Example 2.4, the finite powerset functor $\mathcal{P}_\omega$ has operations τ^n/n given by $\tau^n(x_1, \ldots, x_n) = \{x_1, \ldots, x_n\}$. The Moss lifting λ^n associated to τ^n when using the Barr extension is exactly the one given by (3.1) above.

(3) Recall from Example 2.4 that the operations of the finite distribution functor $\mathcal{D}$ take formal convex combinations. Via the Barr extension, such an operation, determined by coefficients $p_1, \ldots, p_n$ such that $\sum p_i = 1$, induces the predicate lifting λ given by $\lambda_X(A_1, \ldots, A_n)$ consisting of all $\mu \in \mathcal{D}X$ such that there exists a distribution on $\in_X$ (a subset of $X \times \mathcal{Q}(X)$) whose marginal distributions are μ (on X) and the distribution ν on $\mathcal{Q}(X)$ given by $\nu(\{A_i\}) = p_i$, respectively. In fact, however, this description can be substantially simplified; e.g. one readily checks that in the case $n = 2$, we actually have

$$\lambda(A_1, A_2) = \{\mu \in \mathcal{D}(X) \mid \mu(A_1) \geq p_1, \mu(A_2) \geq p_2, \mu(A_1 \cup A_2) = 1\}.$$

(The generalization to higher arities is via what is nowadays known as the *splitting lemma* [36, Theorem 11].)

(4) For the finitary monotone neighbourhood functor $\mathcal{M}_\omega$ (Example 2.4), we obtain Moss liftings as follows. Marti and Venema [20] define a diagonal-preserving lax extension L for $\mathcal{M}$ (which, then, restricts to $\mathcal{M}_\omega$) by means of nested Egli-Milner liftings. An explicit description of L is

$$LR = \{(\mathfrak{A}, \mathfrak{B}) \in \mathcal{M}X \times \mathcal{M}Y \mid \forall A \in \mathfrak{A}.\, R[A] \in \mathfrak{B}, \forall B \in \mathfrak{B}.\, R^\circ[B] \in \mathfrak{A}\}$$

for $R \subseteq X \times Y$. In particular, for $\mathfrak{A} \in \mathcal{M}X$ and $\Phi \in \mathcal{M}\mathcal{Q}X \subseteq \mathcal{Q}\mathcal{Q}\mathcal{Q}X$, we have

$$\mathfrak{A} \in \nabla^L_X(\Phi) \quad \text{iff} \quad \mathfrak{A}\, L(\in)\, \Phi \quad \text{iff} \quad \forall \beta \in \Phi.\, \bigcup \beta \in \mathfrak{A} \text{ and}$$
$$\forall A \in \mathfrak{A}.\, \{B \in \mathcal{Q}X \mid B \cap A \neq \emptyset\} \in \Phi.$$

Combining ∇^L with the presentation of $\mathcal{M}_\omega$ (Example 2.4) produces, for each choice of numbers $n \geq 0$ and $k_1, \ldots, k_n \geq 0$, a $\sum_{i=1}^n k_i$-ary Moss lifting λ given by

$$\lambda((A_{ij})_{i=1,\ldots,n;j=1,\ldots,k_i}) = \{\mathfrak{A} \in \mathcal{M}_\omega X \mid \forall i.\, \bigcup_j A_{ij} \in \mathfrak{A} \text{ and}$$
$$\forall B \in \mathfrak{A}. \exists i. \forall j.\, B \cap A_{ij} \neq \emptyset\}.$$

Since $\mathcal{M}_\omega$ preserves finite sets and the box modality $\square$ as described in Example 2.6 is separating, it is clear that the Moss liftings are expressible using $\square$ and Boolean operators. Concretely, this works as follows. For readability, we denote the predicate lifting interpreting $\square$ by $\square$ as well, similarly for the dual modality $\lozenge$, so that $\lozenge_X(A) := \mathcal{M}X \setminus \square_X(X \setminus A) = \{\mathfrak{A} \in \mathcal{M}X \mid \forall B \in \mathfrak{A}.\, B \cap A \neq \emptyset\}$. Then the Moss lifting λ as described above can be written as

$$\lambda((A_{ij})) = \bigcap_i \square_X\left(\bigcup_j A_{ij}\right) \cap \bigcap_\pi \lozenge_X\left(\bigcup_i A_{i\pi(i)}\right)$$

where π ranges over all selection functions assigning to each $i \in \{1, \ldots, n\}$ an index $\pi(i) \in \{1, \ldots, k_i\}$.

Moss liftings are always monotone [20, Proposition 24]. We show that they also preserve singletons:

Proposition 4.8. *Moss liftings preserve singletons. More specifically, let λ be the Moss lifting induced by $\tau/n \in \Sigma$. Then for all $x_1, \ldots, x_n \in X$,*

$$\lambda_X(\{x_1\}, \ldots, \{x_n\}) = \{\tau_X(x_1, \ldots, x_n)\}.$$

Marti and Venema already establish that the Moss liftings are separating [20, Proposition 25]; we show that they are even strongly expressive:

Proposition 4.9. *The set Λ of all Moss liftings of T is strongly expressive.*

Remark 4.10. Incidentally, this also means that for finitary functors the existence of a separating set of monotone predicate liftings is equivalent to the existence of a strongly expressive set of monotone singleton-preserving predicate liftings. The right-to-left implication is trivial; the converse follows from Propositions 4.8 and 4.9, and the fact that for finitary functors the existence of a separating set of monotone predicate liftings is equivalent to the existence of a lax extension [20].

We have thus seen that given a fixed diagonal-preserving lax extension, from every natural transformation $\tau : (-)^n \to T$ we obtain the corresponding Moss lifting λ^τ/n, which is a monotone singleton-preserving predicate lifting. Conversely, every monotone singleton-preserving predicate lifting λ yields a natural transformation $\tau^\lambda : (-)^n \to T$ (Lemma 3.6(1)). From Proposition 4.8, it is immediate that for $\tau : (-)^n \to T$ we have

$$\tau = \tau^{(\lambda^\tau)}.$$

In particular, taking Moss liftings is an injection from functor operations to monotone singleton-preserving predicate liftings. Conversely, however, $\lambda = \lambda^{(\tau^\lambda)}$ need not hold in general – recall that the construction of Moss liftings depends on the choice of a diagonal-preserving lax extension, and a functor may have more than one such extension. We report an example due to Paul Levy:

Example 4.11. Let M be the monoid of non-negative reals. This monoid in fact forms a division semiring in the expected sense (e.g. [39]), i.e. it is a semiring, and its non-zero elements form a multiplicative group. We note that every division semiring is *refinable* in the sense of Gumm and Schröder [15], i.e. n specified row sums $b_1, \ldots, b_n$ and k specified column sums $c_1, \ldots, c_k$ that induce the same total sum $d = \sum b_i = \sum c_j$ can always be realized by some $n \times k$-matrix (a_{ij}) – in fact, one can just put $a_{ij} = b_i c_j/d$. Now let $b \in (0,1)$ be a transcendental number, and let $N \subseteq M$ be generated by b in M as a division semiring. Concretely, elements of N have the form $f(b)/g(b)$ where $f(X)$ and $g(X) \neq 0$ are polynomials with non-negative rational coefficients. In particular, $1 - b \notin N$: If we could write $1 - b$ in the prescribed form $f(b)/g(b)$, then by transcendentality of b, $f(X)/g(X) = 1 - X$, in contradiction to the leading coefficients of f and g being positive.

Both M and N are *positive* ($x + y = 0$ implies $x = y = 0$) and refinable, so that the monoid-valued functors $F = M^{(-)}$ and $G = N^{(-)}$ both preserve weak pullbacks [15]. As recalled above, it follows that in both cases, the Barr extension is functorial, in particular is a diagonal-preserving lax extension. Now diagonal-preserving lax extensions are easily seen to be inherited by subfunctors, so that the Barr extension $\overline{F}$ induces a diagonal-preserving lax extension L

of G. This extension differs from the Barr extension $\overline{G}$; we immediately cast the counterexample in the form that interests us here:

Let $X = \{u, v\}$. Representing elements of GX as formal linear combinations, we have a binary functor operation $\tau(x, y) = x + by$ for G. We write λ^1 and λ^2 for the Moss liftings induced from τ via $\overline{G}$ and via L, respectively (by the above, both λ^1 and λ^2 induce τ). Then $u + bv \in \lambda^1(\{u, v\}, \{u\})$ but $u + bv \notin \lambda^2(\{u, v\}, \{u\})$: For the former, we have a unique witnessing element of $F \in_X$, namely $(1 - b)(u, \{u, v\}) + b(v, \{u, v\}) + b(u, \{u\})$; but in $G \in_X$, there is no witnessing element since $1 - b \notin N$.

Summing up, even for weak-pullback preserving functors, singleton-preserving monotone predicate liftings are not in general uniquely determined by the functor operation they induce. In the above example, both singleton predicate liftings inducing the given functor operation arise as Moss liftings, via different diagonal-preserving lax extensions; we currently do not know whether every singeleton-preserving monotone predicate lifting is a Moss lifting for some diagonal-preserving lax extension.

Remark 4.12. It is fairly easy to see that for monotone singleton-preserving *unary* predicate liftings λ, we do have $\lambda = \lambda^{(\tau^\lambda)}$.

5 Generic Expressions

We proceed to define, given a set of monotone and singleton-preserving predicate liftings for a functor T, syntactic expressions describing the behaviour of states of T-coalgebras. Our main result is a Kleene-type theorem stating that for every state of a T-coalgebra there exists an equivalent expression, and conversely, every expression describes the behaviour of some state of a finite T-coalgebra. As indicated above, our expression language is a small fragment of the coalgebraic μ-calculus, essentially restricted to modalities and greatest fixed points $\nu z. \phi$.

Definition 5.1 (Expressions). We fix a set V of *fixed point variables* and a set $\mathcal{L}$ of *modalities* equipped with an arity function $\mathrm{ar} : \mathcal{L} \to \omega$; we write $L/n \in \mathcal{L}$ if $L \in \mathcal{L}$ and $\mathrm{ar}(L) = n$. The set $\mathcal{E}$ of *expressions* $\phi, \dots$ is then defined by the grammar

$$\phi ::= z \mid \nu z. \phi \mid L(\phi_1, \dots \phi_n) \qquad (z \in \mathsf{V}, L/n \in \mathcal{L}).$$

An expression is *closed* if all its fixed point variables are bound by a fixed point operator. An expression is *guarded* if all its fixed point variables are separated from their binding fixed point operator by at least one modality. We write $\mathcal{E}_0$ for the set of closed and guarded expressions. We have the usual notion of α-*equivalence* of expressions modulo renaming of bound variables. An occurrence of a fixed point operator in an expression is *top-level* if it is not in scope of a modality.

We next define the semantics of expressions, which agrees with their interpretation as formulas in coalgebraic logic. We fix the requisite data:

Assumption 5.2. For the rest of the paper, we fix a set $\mathcal{L}$ of modalities and an assignment of a singleton-preserving monotone n-ary predicate lifting $[\![L]\!]$ for T to each $L/n \in \mathcal{L}$ such that the set $\Lambda := \{[\![L]\!] \mid L \in \mathcal{L}\}$ is strongly expressive.

By the results of the previous section, these assumptions imply that T has a presentation and is thus finitary (Theorem 2.2).

Definition 5.3 (Semantics). Given a T-coalgebra $C = (X, \xi)$ and a valuation $\kappa : V \to QX$, the semantics $[\![\phi]\!]_C^\kappa \subseteq X$ of expressions $\phi \in \mathcal{E}$ is given by

$$[\![z]\!]_C^\kappa = \kappa(z)$$
$$[\![L(\phi_1, \ldots \phi_n)]\!]_C^\kappa = \xi^{-1}[[\![L]\!]_X([\![\phi_1]\!]_C^\kappa, \ldots [\![\phi_n]\!]_C^\kappa)]$$
$$[\![\nu z.\phi]\!]_C^\kappa = \nu Y.[\![\phi]\!]_C^{\kappa[z \mapsto Y]}$$

where as usual, we use ν to denote greatest fixed points of monotone maps. When ϕ is closed, we simply write $[\![\phi]\!]_C$ in lieu of $[\![\phi]\!]_C^\kappa$, and we drop the subscript C whenever C is clear from the context.

Note that since the predicate liftings $[\![L]\!]$ are monotone and ξ^{-1} is a monotone map, the requisite greatest fixed points exist by the Knaster-Tarski fixed point theorem. Moreover, the assumption that the predicate liftings are singleton-preserving will ensure that every expression describes exactly one behavioural equivalence class (see Theorem 5.15).

By dint of the fact that our expression language is contained in the coalgebraic μ-calculus, the following is an immediate consequence of the fact that the latter is invariant under behavioural equivalence (e.g. [31]):

Lemma 5.4 (Invariance under behavioural equivalence). *For every closed expression ϕ and coalgebras $C = (X, \xi)$, $D = (Y, \zeta)$, if states $x \in X$ and $y \in Y$ are behaviourally equivalent, then $x \in [\![\phi]\!]_C$ iff $y \in [\![\phi]\!]_D$.*

Lemma 5.5. *For all expressions $\phi \in \mathcal{E}$, $[\![\nu z.\phi]\!] = [\![\phi[\nu z.\phi/z]]\!]$.*

Example 5.6. (1) For the deterministic automaton functor $TX = 2 \times X^A$ with $A = \{a, b\}$, we let $\mathcal{L}$ be the set of two binary modalities $\langle 0, a.(-), b.(-)\rangle$ and $\langle 1, a.(-), b.(-)\rangle$ (corresponding to the two Moss liftings of Example 4.7(1)). We interpret expressions in the final T-coalgebra νT carried by all formal languages over A. Here are a few closed and guarded expressions and their semantics in νT (as usual $|w|_b$ denotes the number of b's in w):

$$[\![\nu v.\langle 0, a.v, b.v\rangle]\!] = \{\emptyset\}$$
$$[\![\nu z.\langle 1, a.z, b.z\rangle]\!] = \{A^*\}$$
$$[\![\nu x.\langle 1, a.x, b.\nu y.\langle 0, a.y, b.x\rangle\rangle]\!] = \{\{w \in A^* \mid |w|_b \text{ even}\}\}$$

Note that the semantics of each of these expressions is a singleton (up to behavioural equivalence); in fact, for an arbitrary T-coalgebra X, the semantics of the above expressions is the set of states accepting the language in the singleton on the right. In Lemma 5.12 further below we prove that this holds in general.

(2) Consider $T = \mathcal{P}_\omega(A \times -)$ where A is a finite set of labels. A presentation of T is given by the signature containing for each n-tuple $\vec{a} = (a_1, \ldots, a_n) \in A^n$ one n-ary operation symbol, and the corresponding natural transformation $\tau^{\vec{a}} : (-)^n \to T$ is defined by

$$\tau_X^{\vec{a}} : (x_1, \ldots, x_n) \mapsto \{(a_1, x_1), \ldots, (a_n, x_n)\}.$$

The corresponding Moss lifting is $\lambda^{\vec{a}}/n$ given by

$$\lambda_X^{\vec{a}}(Y_1, \ldots, Y_n) = \{Z \in \mathcal{P}_\omega(A \times X) \mid Z \subseteq \bigcup_{i=1}^n(\{a_i\} \times Y_i) \\ \text{and } Z \cap \{a_i\} \times Y_i \neq \emptyset \text{ for } i = 1, \ldots, n\}$$

(cf. (3.1)). Now put $\mathcal{L} = \{[\vec{a}]/n \mid \vec{a} \in A^n, n \in \omega\}$ and interpret each $[\vec{a}]$ by $\lambda^{\vec{a}}$. For example, for $A = \{a, b\}$ the expression $\nu x.[a]([a, b, a](x, [()], [()]))$, where $[()]$ is the unique nullary modality in $\mathcal{L}$, describes the left-hand state in the following labelled transition system

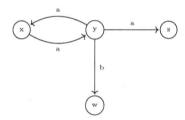

(3) For $T = \mathcal{D}$ we have the presentation with an n-ary operation $\tau^{\vec{p}}$ for every $\vec{p} = (p_1, \ldots, p_n)$ with $\sum_{i=1}^n p_i = 1$ and corresponding Moss liftings as described in Example 4.7(3). For each such $\vec{p}$, we introduce a modality $[\vec{p}]/n \in \mathcal{L}$, and interpret it as $\lambda^{\vec{p}}$. Now consider the Markov chain (i.e. $\mathcal{D}$-coalgebra)

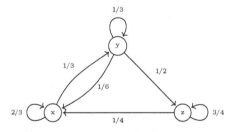

The behaviour of the left-hand state is described by the expression

$$\nu x.[2/3, 1/3](x, \nu y.[1/6, 1/3, 1/2](x, y, \nu z.[1/4, 3/4](x, z))).$$

Remark 5.7. The syntax of our expressions is determined purely by the finitary coalgebraic type functor, more precisely, by a given strongly expressive set Λ of monotone singleton-preserving predicate lifting. In contrast, existing expression calculi such as standard regular expressions for deterministic automata or the coalgebraic expression calculi in [32,34] use extra operations (e.g. expressing union or concatenation of languages). These operations are not dictated by the setting, viz. an endofunctor on **Set**. Rob Myers' PhD thesis [25] explains nicely how such extra operations are obtained naturally in an expression calculus when one works over an algebraic category (such as the one of join-semilattices or vector spaces over the reals, i.e. algebras for the monad $\mathbb{R}^{(-)}$). We leave the extension of our expression language to this more general setting for future work.

Our Kleene theorem requires a number of technical lemmas:

Lemma 5.8. *Let λ/n and λ'/n' be monotone singleton-preserving predicate liftings for T. Let S be an equivalence relation on a set X, let $A_1, \ldots, A_n$, be S-equivalence classes or empty, and let $B_1, \ldots, B_{n'}$ be S-closed subsets of X. Then the following holds.*

(1) $\lambda_X(A_1, \ldots A_n) \subseteq \lambda'_X(B_1, \ldots B_{n'})$ or $\lambda_X(A_1, \ldots A_n) \cap \lambda'_X(B_1, \ldots B_{n'}) = \varnothing$.
(2) If the $B_1, \ldots, B_{n'}$ are even S-equivalence classes or empty, then

$$\lambda_X(A_1, \ldots A_n) = \lambda'_X(B_1, \ldots B_{n'}) \text{ or } \lambda_X(A_1, \ldots A_n) \cap \lambda'_X(B_1, \ldots B_{n'}) = \varnothing.$$

Proof (Sketch). Apply naturality to the quotient map $q : X \twoheadrightarrow X/S$.   □

In the proof of Lemma 5.12 further below, we will make use of a Λ-bisimulation. We briefly recall the essentials of this notion [12]:

Definition 5.9 (Λ-Simulation). Given a pair of T-coalgebras (X, ξ) and (Y, ζ), a Λ-*simulation* is a relation $S \subseteq X \times Y$ such that for all predicate liftings $\lambda \in \Lambda$ and $X_i \subseteq X$, $x \, S \, y$ implies

$$\xi(x) \in \lambda_X(X_1, \ldots, X_n) \Rightarrow \zeta(y) \in \lambda_Y(S[X_1], \ldots, S[X_n]).$$

A Λ-*bisimulation* is a Λ-simulation S such that S° is also a Λ-simulation. Elements $(x, y) \in X \times Y$ are Λ-*bisimilar* if there is a Λ-bisimulation relating x and y.

Theorem 5.10. *Λ-bisimilarity conincides with behavioural equivalence.*

Remark 5.11. In fact, for Theorem 5.10 it is sufficient that Λ is separating and the predicate liftings in Λ are monotone. It turns out that Theorem 5.10 is actually a special case of [20, Theorem 11], applied to the case where the lax extension is induced by a separating set of monotone predicate liftings.

Lemma 5.12. *Let (X, ξ) be a T-coalgebra, let $\lambda_i/k \in \Lambda$, $i = 1, \ldots, k$, and let $(A_1, \ldots, A_k)$ be the greatest fixed point of the map $h : (QX)^k \to (QX)^k$ defined by*

$$
\begin{pmatrix} X_1 \\ \vdots \\ X_k \end{pmatrix} \mapsto \begin{pmatrix} \xi^{-1}[\lambda_{1,X}(X_1, \ldots, X_k)] \\ \vdots \\ \xi^{-1}[\lambda_{k,X}(X_1, \ldots, X_k)] \end{pmatrix}
\tag{5.1}
$$

Then for each i, all elements of A_i are behaviourally equivalent, and for all i, j, either $A_i \cap A_j = \varnothing$ or $A_i = A_j$.

(In the above lemma, we restrict to all λ_i having full arity k and using their arguments in the given order only in the interest of readability; this is w.l.o.g. since we can just reorder arguments and add dummy arguments.)

Proof (Sketch). Let $S \subseteq X \times X$ be the relation

$$
S = \{(x_1, x_2) \mid \exists A_i \,.\, x_1 \in A_i \wedge x_2 \in A_i\} \cup \Delta_X.
$$

Using Lemma 5.8 one shows first that S is an equivalence relation, which already takes care of the second part of the claim, and then that S is a Λ-bisimulation. The first claim of the lemma then follows by Theorem 5.10. $\qquad\square$

The final ingredient of our Kleene-type correspondence is the following adaptation of Bekič's bisection lemma [6]:

Lemma 5.13. *For complete lattices $(X, \leq)$, $(Y, \leq)$ and for every pair of monotone maps $f : X \times Y \to X$ and $g : X \times Y \to Y$, we have*

$$
\nu(x, y).(f(x, y), g(x, y)) = (x_0, y_0) \quad with \quad \begin{matrix} x_0 = \nu x.f(x, \nu y.g(x, y)) \\ y_0 = \nu y.g(x_0, y). \end{matrix}
$$

Although in [6] this lemma only covers least fixed points in a slightly different setting, the proof is the same.

Using Lemma 5.13 we can transform every expression $\phi \in \mathcal{E}_0$ into a system of flat equations $(z_1 = \phi_1, \ldots, z_k = \phi_k)$ for some $k \in \omega$, i.e. equations without nested modalities or fixed point operators: This is done by first ensuring that every fixed point operator uses a different fixed point variable and then binding every modality that is not nested directly under a fixed point operator with a new fixed point operator using a fresh variable. Thus we can rewrite every expression $\phi \in \mathcal{E}_0$ in the form

$$
\phi \equiv \nu z_1.\, L_1(z_1, \nu z_2.L_2(\ldots), \ldots, \nu z_k.L_k(\ldots))
$$

for some modalities $L_i \in \mathcal{L}$, $i = 1, \ldots, k$. If we now inductively apply Lemma 5.13 and, for readability, additionally normalize every modality to have as many arguments as there are different fixed point variables in such an expression, introducing dummy arguments where necessary, then we can write ϕ as a system

$$
\begin{aligned}
z_1 &= L_1(z_1, z_2, \ldots, z_k) \\
z_2 &= L_2(z_1, z_2, \ldots, z_k) \\
&\;\;\vdots \\
z_k &= L_n(z_1, z_2, \ldots, z_k)
\end{aligned}
\tag{5.2}
$$

of flat equations. Given any coalgebra $C = (X, \xi)$, the above system induces an obvious map of the form (5.1) (replacing z_i by X_i and L_i by $\lambda_i = [\![L_i]\!]$), and the first components of its greatest fixed point is the semantics $[\![\phi]\!]_C$. The following example shows a concrete case.

Example 5.14 (Applying Bekič's bisection lemma).   Consider the expression

$$\phi = \nu x. L_1(x, L_2(x), \nu y. L_3(y, \nu z. L_2(z)))$$

In order to transform it as per the procedure indicated, we first need to add a fixed point operator with a fresh variable to the first occurrence of L_2:

$$\phi = \nu x. L_1(x, \nu w. L_2(x), \nu y. L_3(y, \nu z. L_2(z)))$$

Then we can form the equation system for the variables x, w, y, z

$$x = \overline{L}_1(x, w, y, z) = L_1(x, w, y)$$
$$w = \overline{L}_2(x, w, y, z) = L_2(x)$$
$$y = \overline{L}_3(x, w, y, z) = L_3(y, z)$$
$$z = \overline{L}_4(x, w, y, z) = L_2(z)$$

where we extend $\mathcal{L}$ with additional operators $\overline{L}_i$ having dummy arguments, defined as indicated. The semantics of this equation system in a coalgebra $C = (X, \xi)$ is defined as the greatest fixpoint (A_0, A_1, A_2, A_3) of the map $h : \mathcal{Q}^n X \to \mathcal{Q}^n X$ defined by

$$h : \begin{pmatrix} X_1 \\ X_2 \\ X_3 \\ X_4 \end{pmatrix} \mapsto \begin{pmatrix} \xi^{-1}[\![L_1]\!]_X(X_1, X_2, X_3)] \\ \xi^{-1}[\![L_2]\!]_X(X_1)] \\ \xi^{-1}[\![L_1]\!]_X(X_2, X_4)] \\ \xi^{-1}[\![L_2]\!]_X(X_4)] \end{pmatrix}.$$

The semantics of ϕ in C is then $[\![\phi]\!]_C = A_0$.

The following two results together establish a Kleene-type correspondence for the generic expressions of Definition 5.1.

Theorem 5.15. *Every expression $\phi \in \mathcal{E}_0$ describes exactly one behavioural equivalence class, which is moreover realized in a finite coalgebra. Explicitly: there exists a state x in a finite coalgebra such that for every coalgebra C, $[\![\phi]\!]_C$ contains precisely the states of C that are behaviourally equivalent to x.*

Proof. (Sketch). By Lemma 5.4, it suffices to show that any two states (w.l.o.g. in the same coalgebra, using coproducts) satisfying ϕ are bisimilar. Since ϕ can transformed into a system (5.2) of flat equations, this follows by Lemma 5.12. Realization in a finite coalgebra follows from the finite model property of the coalgebraic μ-calculus [8], and alternatively is shown by constructing a model from the variables in a flat equation system. □

Theorem 5.16. *Let $C = (X, \xi)$ be a finite T-coalgebra. For every $x \in X$, there exists an expression $\phi \in \mathcal{E}_0$ such that $x \in [\![\phi]\!]_C$.*

Proof. Let $X = \{x_1, \ldots, x_k\}$ and w.l.o.g. $x = x_1$. Since Λ is strongly expressive, for every $x_i \in X$ there is a modality L_i, w.l.o.g. with arity k and prescribed argument ordering, such that

$$\{\xi(x_i)\} = [\![L_i]\!]_X(\{x_1\}, \ldots, \{x_k\}).$$

That is, the $\{x_i\}$ solve the system $(x_i = L_i(x_1, \ldots, x_k))_{i=1,\ldots,k}$ of flat fixed point equations, so for the greatest fixed point $(A_1, \ldots, A_k)$ of the system, we have $x_i \in A_i$ for every i, in particular $x = x_1 \in A_1$. It now just remains to convert the equation system into an equivalent single expression in the standard manner [7] (incurring exponential blow-up); then $x \in [\![\phi]\!]_C$ as desired.   □

Corollary 5.17. *Every expression denotes a behavioural equivalence class of a state in a finite coalgebra, and conversely every such class is denoted by some expression.*

Example 5.18. (1) For the functor $TX = 2 \times X^A$ for $A = \{a, b\}$ consider the coalgebra with carrier $X = \{x_1, x_2\}$ and with coalgebra structure $\xi : X \to 2 \times X^A$ with $\xi(x_0) = (1, (a \mapsto x_0, b \mapsto x_1))$ and $\xi(x_1) = (0, (a \mapsto x_1, b \mapsto x_0))$. Then we clearly have $\{\xi(x_1)\} = \lambda^1(\{x_1\}, \{x_2\})$ and $\{\xi(x_2)\} = \lambda^1(\{x_2\}, \{x_1\})$. Using the syntax of Example 5.6(1) and following the proof of Theorem 5.16, we obtain the following expression for the behavioural equivalence class (i.e. formal language) for x_1:

$$\nu x_1.\langle 1, a.x_1, b.\nu x_2.\langle 0, a.x_2, b.x_1 \rangle \rangle.$$

Note that this is the same expression (modulo α-equivalence) as the third expression from Example 5.6(1).

   (2) For the functor $\mathcal{P}_\omega(A \times -)$ and $A = \{a, b\}$ the coalgebra $C = (\{x, y, z, w\}, \xi)$ depicted in Example 5.6(2) satisfies the following equations:

$$\{\xi(x)\} = \lambda_C^{(a)}(\{y\}), \quad \{\xi(y)\} = \lambda_C^{(a,b,c)}(\{x, w, z\}), \quad \{\xi(w)\} = \lambda_C^{()}(), \quad \{\xi(w)\} = \lambda_C^{()}()$$

By Theorem 5.16 $\{\{x\}, \{y\}, \{z\}, \{w\}\}$ solves the following system, reusing the same variable names,

$$x = [a](y), \quad y = [a, b, a](x, w, z), \quad w = [()], \quad z = [()]$$

which can be transformed as demonstrated in Example 5.14 to the expression given in Example 5.6(2), describing the behaviour of the state x.

   (3) For the functor $T = \mathcal{D}$ consider the expression from Example 5.6(3): $\nu x.[2/3, 1/3](x, \nu y.[1/6, 1/3, 1/2](x, y, \nu z.[1/4, 3/4](x, z)))$, which transforms to the system

$$x = [2/3, 1/3](x, y), \quad y = [1/6, 1/3, 1/2](x, y, z), \quad z = [1/4, 3/4](x, z).$$

By Theorem 5.15 we can construct a coalgebra $C = (\{x, y, z\}, \xi)$ defined by:

$$\{\xi(x)\} = \lambda_X^{(2/3, 1/3)}(\{x\}, \{y\})$$
$$\{\xi(y)\} = \lambda_X^{(1/6, 1/3, 1/2)}(\{x\}, \{y\}, \{z\})$$
$$\{\xi(z)\} = \lambda_X^{(1/4, 3/4)}(\{x\}, \{z\})$$

which is exactly the coalgebra depicted in Example 5.6(3) where x is in the behavioural equivalence class of the above expression.

An alternative approach to defining the semantics of expressions is to construct a T-coalgebra structure on the set $\mathcal{E}_0$ of closed and guarded expressions, similarly as in the work of Silva et al. [34] and also Myers [25]. In Theorem 5.21 below we show that this new semantics coincides with the previous one.

Definition 5.19. We define a T-coalgebra $\varepsilon : \mathcal{E}_0 \to T\mathcal{E}_0$ inductively by

$$\varepsilon(L(\phi_1, \ldots, \phi_n)) \in [\![L]\!](\{\phi_1\}, \ldots, \{\phi_n\}) \tag{5.3}$$
$$\varepsilon(\nu x.\phi) = \varepsilon(\phi[\nu x.\phi/x]). \tag{5.4}$$

This is actually a definition of ε because (a) in (5.3), $[\![L]\!]$ preserves singletons and thus there is only one element in $[\![L]\!](\{\phi_1\}, \ldots, \{\phi_n\})$, and (b) for the inductive part (5.4), one can use the number of top-level fixed point operators as a termination measure, which decreases in each step because the fixed points are guarded.

Now recall that a coalgebra $\xi : X \to TX$ is *locally finite* if every $x \in X$ is contained in a finite subcoalgebra of ξ. Locally finite coalgebras are precisely the (directed) unions of finite coalgebras (see [21]). Thus, it follows from Theorem 5.16 that for any $x \in X$ in a locally finite coalgebra $\xi : X \to TX$, there exists a $\phi \in \mathcal{E}_0$ with $x \in [\![\phi]\!]_X$.

Moreover, $\mathcal{E}_0$ is obviously not finite; however, arguing via finiteness of the Fischer-Ladner closure [16] we obtain

Proposition 5.20. *The T-coalgebra $(\mathcal{E}_0, \varepsilon)$ is locally finite.*

The following theorem says that $(\mathcal{E}_0, \varepsilon)$ serves as a canonical model of the expression language:

Theorem 5.21. *For every closed and guarded expression $\phi \in \mathcal{E}_0$ and every state x in a T-coalgebra C, $x \in [\![\phi]\!]_C$ iff x is behaviourally equivalent to ϕ as a state in $(\mathcal{E}_0, \varepsilon)$.*

In particular, the above implies that

$$\phi \in [\![\phi]\!]_{\mathcal{E}_0} \qquad \text{for all } \phi \in \mathcal{E}_0, \tag{5.5}$$

essentially a truth lemma for $\mathcal{E}_0$. For the proof of Theorem 5.21, we note:

Lemma 5.22. *α-Equivalent expressions are behaviourally equivalent as states in $(\mathcal{E}_0, \varepsilon)$.*

*Proof (*Theorem 5.21, *sketch).* It suffices to prove (5.5): The 'if' direction of the claim then follows from invariance of ϕ under behavioural equivalence (Lemma 5.4), and 'only if' is by Theorem 5.15. We generalize (5.5) to expressions ϕ with free variables: Whenever σ is a substitution of the free variables of ϕ and κ a valuation such that $\sigma(v) \in \kappa(v)$ for every free variable v of ϕ, then

$$\phi\sigma \in [\![\phi]\!]_{\mathcal{E}_0}^\kappa.$$

We proceed by induction on ϕ, using Lemma 5.22 in the fixpoint case. $\square$

Remark 5.23. To give a concrete example use of the connection between expression languages and modal fixed point logics afforded by the above results, we note that we now obtain an alternative handle on equivalence of expressions that complements the standard approach via partition refinement: Expressions ϕ, ψ are equivalent iff some state described by ϕ (obtained, e.g., via the one of the model constructions in Theorems 5.15 and 5.21) satisfies ψ. Note that the latter is fairly easy to check as long as the modalities are computationally tractable, since ψ otherwise involves only greatest fixed points. This approach is similar to reasoning algorithms in the lightweight description logic $\mathcal{EL}$ [4], where checking validity of $\phi \to \psi$ is reduced to model checking ψ in a minimal model of ϕ; we leave a more detailed analysis to future work.

6 Conclusion and Further Work

We have defined a generic expression language for behaviours of finite set coalgebras based on predicate liftings, specifically on a strongly expressive set of singleton-preserving predicate liftings. There are mutual conversions between such sets of predicate liftings and functor presentations, one direction being via the Moss liftings introduced by Marti and Venema [20]; we have however demonstrated that these fail to be mutually inverse in one direction, i.e. in general not all singleton-preserving predicate liftings are Moss liftings. Our language is presumably equivalent to the set-based instance of Myer's expression language [25]; our alternative presentation is aimed primarily at showing that expression languages embed naturally into the coalgebraic μ-calculus, generalizing well-known results on the relational μ-calculus [2,10,14,35]. The benefit of this insight is to tighten the connection between expression languages and specification logics, e.g. it allows for combining model checking, equivalence checking, and reasoning within a single formalism. On a more technical note, we show, e.g., that one can provide an alternative semantics of expressions by defining a coalgebra structure on expressions, an approach pioneered by Silva et al. [34] and used also by Myers [25]; in the light of the expressions/logic correspondence, this construction is now seen as a canonical model construction for a fragment of the coalgebraic μ-calculus, and the core part of the proof that the two semantics agree becomes just a truth lemma.

An important point for further work is to extend the current setup from the base category **Set** to algebraic categories (such as join semi-lattices or positive convex algebras) in order to generalize our results to expression calculi involving convenient additional operations (reflecting the ambient algebraic theory) such as addition. A closely related point is the connection with coalgebraic determinization [33]; it should be interesting to see whether our ideas can lead to expression calculi for coarser system equivalences than bisimilarity, such as trace equivalence for transition systems or distribution bisimilarity for Segala systems. Such a generalization might be based on our recent approach to coalgebraic trace semantics via graded monads [22].

References

1. Aceto, L., Hennessy, M.: Termination, deadlock, and divergence. J. ACM **39**, 147–187 (1992)
2. Aceto, L., Ingólfsdóttir, A., Larsen, K., Srba, J.: Reactive Systems: Modelling, Specification and Verification. Cambridge University Press, New York (2007)
3. Adámek, J., Trnková, V.: Automata and Algebras in Categories, Mathematics and Its Applications, vol. 37. Kluwer, Dordrecht (1990)
4. Baader, F., Brandt, S., Lutz, C.: Pushing the $\mathcal{EL}$ envelope. In: International Joint Conference on Artificial Intelligence, IJCAI 2005. Morgan-Kaufmann (2005)
5. Barr, M.: Relational algebras. In: Reports of the Midwest Category Seminar. LNM, vol. 137. Springer (1970)
6. Bekič, H.: Definable operations in general algebras, and the theory of automata and flowcharts. In: Jones, C.B. (ed.) Programming Languages and Their Definition. LNCS, vol. 177, pp. 30–55. Springer, Heidelberg (1984). https://doi.org/10.1007/BFb0048939
7. Bradfield, J., Stirling, C.: Modal logics and mu-calculi. In: Handbook of Process Algebra, pp. 293–332. Elsevier (2001)
8. Cîrstea, C., Kupke, C., Pattinson, D.: EXPTIME tableaux for the coalgebraic mu-calculus. Log. Methods Comput. Sci. **7**(3:3), 33 (2011)
9. Cîrstea, C., Kurz, A., Pattinson, D., Schröder, L., Venema, Y.: Modal logics are coalgebraic. Comput. J. **54**, 31–41 (2011)
10. Godskesen, J., Ingólfsdóttir, A., Zeeberg, M.: Fra Hennessy-Milner logik til CCS-processer. Master's thesis, Aalborg University (1987)
11. Goncharov, Sergey, Milius, Stefan, Silva, Alexandra: Towards a coalgebraic Chomsky hierarchy. In: Diaz, Josep, Lanese, Ivan, Sangiorgi, Davide (eds.) TCS 2014. LNCS, vol. 8705, pp. 265–280. Springer, Heidelberg (2014). https://doi.org/10.1007/978-3-662-44602-7_21
12. Gorín, Daniel, Schröder, Lutz: Simulations and bisimulations for coalgebraic modal logics. In: Heckel, Reiko, Milius, Stefan (eds.) CALCO 2013. LNCS, vol. 8089, pp. 253–266. Springer, Heidelberg (2013). https://doi.org/10.1007/978-3-642-40206-7_19
13. Gorín, D., Schröder, L.: Subsumption checking in conjunctive coalgebraic fixpoint logics. In: Advances in Modal Logic, AiML 2014. pp. 254–273. College Publications (2014)
14. Graf, S., Sifakis, J.: A modal characterization of observational congruence on finite terms of CCS. Inf. Control **68**, 125–145 (1986)
15. Gumm, H.P., Schröder, T.: Monoid-labeled transition systems. In: Coalgebraic Methods in Computer Science, CMCS 2001. ENTCS, vol. 44, pp. 185–204. Elsevier (2001)

16. Kozen, D.: Results on the propositional μ-calculus. Theor. Comput. Sci. **27**, 333–354 (1983)
17. Kozen, D.: Kleene algebra with tests. ACM Trans. Program. Lang. Syst. **19**, 427–443 (1997)
18. Kurz, A., Leal, R.: Equational coalgebraic logic. In: Mathematical Foundations of Programming Semantics, MFPS 2009. ENTCS, vol. 249, pp. 333–356. Elsevier (2009)
19. Leal, R.: Predicate liftings versus nabla modalities. In: Coalgebraic Methods in Computer Science, CMCS 2008. ENTCS, vol. 203, pp. 195–220. Elsevier (2008)
20. Marti, J., Venema, Y.: Lax extensions of coalgebra functors and their logic. J. Comput. Syst. Sci. **81**(5), 880–900 (2015)
21. Milius, S.: A sound and complete calculus for finite stream circuits. In: Proceedings of the 25th Annual Symposium on Logic in Computer Science (LICS 2010). pp. 449–458. IEEE Computer Society (2010)
22. Milius, S., Pattinson, D., Schröder, L.: Generic trace semantics and graded monads. In: Coalgebraic and Algebraic Methods in Computer Science, CALCO 2015. LIPIcs, vol. 35, pp. 253–269 (2015)
23. Moggi, E.: Notions of computation and monads. Inf. Comput. **93**(1), 55–92 (1991)
24. Moss, L.: Coalgebraic logic. Ann. Pure Appl. Log. **96**, 277–317 (1999)
25. Myers, R.: Rational coalgebraic machines in varieties: languages, completeness and automatic proofs. Ph.D. thesis, Imperial College London (2013)
26. Parikh, R.: Propositional game logic. In: Foundations of Computer Science, FOCS 1983. IEEE (1983)
27. Pattinson, D.: Coalgebraic modal logic: soundness, completeness and decidability of local consequence. Theor. Comput. Sci. **309**, 177–193 (2003)
28. Pattinson, D.: Expressive logics for coalgebras via terminal sequence induction. Notre Dame J. Formal Log. **45**, 19–33 (2004)
29. Peleg, D.: Concurrent dynamic logic. J. ACM **34**, 450–479 (1987)
30. Schröder, L.: Expressivity of coalgebraic modal logic: the limits and beyond. Theor. Comput. Sci. **390**, 230–247 (2008)
31. Schröder, L., Venema, Y.: Completeness of flat coalgebraic fixpoint logics. ACM Trans. Comput. Log. **19**, 4:1–4:34 (2018)
32. Silva, A., Bonchi, F., Bonsangue, M., Rutten, J.: Quantitative Kleene coalgebras. Inf. Comput. **209**, 822–849 (2011)
33. Silva, A., Bonchi, F., Bonsangue, M., Rutten, J.: Generalizing determinization from automata to coalgebras. Log. Methods Comput. Sci. **9**(1:9), 27 (2013)
34. Silva, A., Bonsangue, M., Rutten, J.: Non-deterministic Kleene coalgebras. Log. Methods Comput. Sci. **6**(3:23), 39 (2010)
35. Steffen, B., Ingólfsdóttir, A.: Characteristic formulae for processes with divergence. Inf. Comput. **110**, 149–163 (1994)
36. Strassen, V.: The existence of probability measures with given marginals. Ann. Math. Stat. **36**, 423–439 (1965)
37. Trnková, V.: General theory of relational automata. Fund. Inform. **3**, 189–234 (1980)
38. Venema, Y.: Automata and fixed point logic: a coalgebraic perspective. Inf. Comput. **204**, 637–678 (2006)
39. Weinert, H.J.: On 0-simple semirings, semigroup semirings, and two kinds of division semirings. Semigroup Forum **28**, 313–333 (1984)

Long-Term Values in Markov Decision Processes, (Co)Algebraically

Frank M. V. Feys[1], Helle Hvid Hansen[1(✉)], and Lawrence S. Moss[2]

[1] Department of Engineering Systems and Services, TPM,
Delft University of Technology, Delft, The Netherlands
{f.m.v.feys,h.h.hansen}@tudelft.nl
[2] Department of Mathematics, Indiana University, Bloomington, IN 47405, USA
lsm@cs.indiana.edu

Abstract. This paper studies Markov decision processes (MDPs) from the categorical perspective of coalgebra and algebra. Probabilistic systems, similar to MDPs but without rewards, have been extensively studied, also coalgebraically, from the perspective of program semantics. In this paper, we focus on the role of MDPs as models in optimal planning, where the reward structure is central. The main contributions of this paper are (i) to give a coinductive explanation of policy improvement using a new proof principle, based on Banach's Fixpoint Theorem, that we call contraction coinduction, and (ii) to show that the long-term value function of a policy with respect to discounted sums can be obtained via a generalized notion of corecursive algebra, which is designed to take boundedness into account. We also explore boundedness features of the Kantorovich lifting of the distribution monad to metric spaces.

Keywords: Markov decision process · Long-term value
Discounted sum · Coalgebra · Algebra · Corecursive algebra
Fixpoint · Metric space

1 Introduction

Markov Decision Processes (MDPs) [23] are a family of probabilistic, state-based models used in planning under uncertainty and reinforcement learning. Informally stated, an MDP models a situation in which an agent (the decision maker) has to make choices at each state of a process, and each choice leads to some reward and a probabilistic transition to a next state. The aim of the agent is to find an optimal policy, i.e., a way of choosing actions that maximizes future expected rewards. In this paper, we consider a simple version of MDPs known as time-homogeneous, infinite-horizon MDPs in which the set of states and actions are finite, and future rewards are computed according to the discounted summation criterion.

Probabilistic systems of similar type have been studied extensively, also coalgebraically, in the area of program semantics (see for instance [8,9,27,28]). Our

© IFIP International Federation for Information Processing 2018
Published by Springer Nature Switzerland AG 2018. All Rights Reserved
C. Cîrstea (Ed.): CMCS 2018, LNCS 11202, pp. 78–99, 2018.
https://doi.org/10.1007/978-3-030-00389-0_6

focus is not so much on the observable behavior of MDPs viewed as computations, but on their role in solving optimal planning problems.

The classic theory of MDPs with discounting is well-developed (see [23, Chapter 6]), and indeed we do not prove any new results about MDPs as such. Our work is inspired by Bellman's principle of optimality, which states the following: "An optimal policy has the property that whatever the initial state and initial decision are, the remaining decisions must constitute an optimal policy with regard to the state resulting from the first decision" [4, Chapter III.3]. This principle has clear coinductive overtones, and our aim is to situate it in a body of mathematics that is also concerned with infinite behavior and coinductive proof principles, i.e., in coalgebra.

The main contributions of this paper are the following. First, we present a coinductive proof of the correctness of a classic iterative procedure known as policy iteration [12]. This leads us to formulate a coinductive proof principle that we have named *contraction (co)induction*, and which is closely related to Kozen's metric coinduction [17]. We believe contraction (co)induction should have applications far beyond the topic of MDPs. Second, we show that long-term values of policies can be obtained from the universal property of a generalized notion of corecursive algebra. The technical challenge here is to encode boundedness information in order to ensure the unique existence of certain fixpoints. This leads us to introduce the notions of b-categories and b-corecursive algebras (bcas). Combining these with well-known techniques from coinductive specification and trace semantics [3,14], we obtain the desired universal maps.

Contents of This Paper. In Sect. 2 we give a brief introduction to MDPs and the classical results that we aim to categorify. In Sect. 3, we present contraction coinduction and apply it to prove the correctness of policy iteration and related results. In Sect. 4, we describe the (set-based) coalgebraic and algebraic structures that we use to model MDPs and discounted sums. In Sect. 5, we move to a category of metric spaces, we introduce b-categories and b-corecursive algebras (bcas), and we show that the long-term value of a policy as well as the optimal value arise as universal arrows. We briefly discuss extensions of our work in Sect. 6. Finally, we conclude and discuss related and future work in Sect. 7.

2   Markov Decision Processes

We refer to [23] for a comprehensive overview of MDPs, including numerous applications to planning problems such as inventory management and highway maintenance. Here, we confine ourselves to a brief introduction.

An MDP models a situation in which an agent in each state $s \in S$ chooses to execute an action $a \in Act$, and this choice results in a probabilistic transition to a new state $s' \in S$. That is, for every state s and every action a, there is a probability distribution $t(s)(a)$ over states. Furthermore, in each state s, the agent collects a reward (or utility) specified by a real number $u(s)$. The aim of the agent is to find a policy that will maximize his expected long-term rewards.

Letting ΔS denote the set of probability distributions on a finite set S, MDPs and policies are formally defined as follows.

Definition 1. *Let Act be a finite set of* actions. *A* Markov decision process (MDP) $m = \langle S, u, t \rangle$ *consists of a finite set S of* states, *a* reward function $u: S \to \mathbb{R}$, *and a probabilistic transition structure $t: S \to (\Delta S)^{Act}$. We often omit S and simply write $m = \langle u, t \rangle$. A* policy *is a function $\sigma: S \to Act$.*

More generally, MDPs are considered with respect to a time evolution which may be discrete or continuous, and the transition structure and reward function may depend on the time step. If the time evolution is assumed to end after finitely many steps, the MDP is called *finite-horizon*. In our definition of MDPs, time evolution is implicitly assumed to be discrete, but t and u do not depend on the time step, making them *time-homogeneous*, and the time evolution is not assumed to end, making them *infinite-horizon*.

Similarly, there are more general notions of policy in which the policy may depend on the time step. A policy that does not depend on the time step is called *stationary*. The choices prescribed by a non-stationary policy may depend on the entire history of the system up until the present time step, but if each choice depends only on the current state and not the history, then the policy is called *Markovian* or *memoryless*. Finally, a policy may also be randomized, i.e., of type $S \to \Delta Act$, as opposed to *deterministic*. That means, in this paper we consider *stationary* (and therefore *memoryless*), *deterministic policies*.

Example 1. Consider the MDP m shown in Fig. 1, taken from [20]. A startup company can be in one of four states that we abbreviate by PU, PF, RU, and RF. In each state, the company receives an *immediate reward* $u(s)$, and chooses to either *advertise* (**A**) or *save* (**S**). The effect of an action in a state is in general probabilistic, as indicated by the arrows. We take a *discount factor* $\gamma = 0.9$.

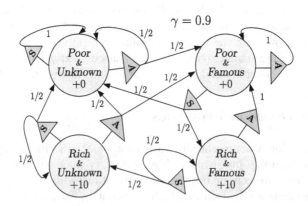

Fig. 1. Example of an MDP modeling of a startup (taken from [20]).

There are several criteria for evaluating the long-term rewards expected by following a given policy. A classic one found in the literature takes the long-term rewards to be the *discounted infinite sum of expected rewards*. The idea is that rewards collected tomorrow are worth less than rewards collected today. Before we state the definition, we need some notation. Given a probabilistic transition structure $t \colon S \to (\Delta S)^{Act}$ and a policy $\sigma \in Act^S$, we get a map $t_\sigma \colon S \to \Delta S$ by letting $t_\sigma(s) = t(s)(\sigma(s))$. The pair $\langle u, t_\sigma \rangle$ is sometimes called a *Markov reward process*. The map t_σ corresponds to a column-stochastic $|S| \times |S|$-matrix P_σ. Viewing $u \in \mathbb{R}^S$ as a row $|S|$-vector and a start state s as a column $|S|$-vector v_s with 1 in position s and 0 everywhere else, the probability that the agent is in a state s' at time step n is found in position s' of the column-stochastic vector $P_\sigma^n v_s$, and the expected reward $r_n^\sigma(s)$ at time step n is the scalar $uP_\sigma^n v_s$.

Definition 2. *Let γ be a fixed real number with $0 \le \gamma < 1$. Such a γ is called a* discount factor. *Let an MDP $m = \langle u, t \rangle$ be given. The* long-term value *of a policy σ (for m) according to the* discounted sum criterion *is the function* $\mathrm{LTV}_\sigma \colon S \to \mathbb{R}$ *defined as follows:*

$$\mathrm{LTV}_\sigma(s) = r_0^\sigma(s) + \gamma \cdot r_1^\sigma(s) + \cdots + \gamma^n \cdot r_n^\sigma(s) + \cdots \tag{1}$$

where $r_n^\sigma(s)$ is the expected reward at time step n. A policy σ is optimal *if for all $s \in S$, $\mathrm{LTV}_\sigma(s) = \max_{\tau \in Act^S} \mathrm{LTV}_\tau(s)$.*

Note that $r_0^\sigma(s) = u(s)$ for all $s \in S$, and since S is finite, $\max_s r_0^\sigma(s) < \infty$. This boundedness property entails that the infinite sum in (1) is convergent.

It will be convenient to work with the map ℓ_σ that takes the expected value of LTV_σ relative to some distribution. Formally, $\ell_\sigma \colon \Delta S \to \mathbb{R}$ is defined for all $\varphi \in \Delta S$ by

$$\ell_\sigma(\varphi) = \sum_{s \in S} \varphi(s) \cdot \mathrm{LTV}_\sigma(s). \tag{2}$$

Observe that for each state s, $\mathrm{LTV}_\sigma(s)$ is equal to the immediate rewards plus the discounted future expected rewards. Seen this way, (1) may be re-written to the corecursive equation

$$\mathrm{LTV}_\sigma(s) = u(s) + \gamma \cdot \left(\sum_{s' \in S} t_\sigma(s)(s') \cdot \mathrm{LTV}_\sigma(s') \right) = u(s) + \gamma \cdot \ell_\sigma(t_\sigma(s)). \tag{3}$$

Viewing LTV_σ as a column vector in $\mathbb{R}^S$, the equation in (3) represents a linear system, $\mathrm{LTV}_\sigma = u + \gamma P_\sigma \mathrm{LTV}_\sigma$. We find LTV_σ by solving it: $\mathrm{LTV}_\sigma = (I - \gamma P_\sigma)^{-1} u$, where I is the identity matrix.

Equivalently, LTV_σ is defined as the unique fixpoint of the (linear), contractive, monotone (for the pointwise order) operator

$$\Psi_\sigma \colon \mathbb{R}^S \to \mathbb{R}^S \qquad \Psi_\sigma(v) = u + \gamma P_\sigma v. \tag{4}$$

Note that Ψ_σ is contractive since P_σ is column-stochastic and we multiply with γ, and $\mathbb{R}^S$ is a complete metric space. Hence by the Banach Fixpoint Theorem, the unique fixpoint exists. Moreover, Ψ_σ is monotone, because P_σ has all non-negative entries.

Example 2. We continue with Example 1. An example of a policy is the "miserly" σ given by $\sigma(s) = \mathbf{S}$ for all states s, i.e., the startup chooses to save in each state. The equations that describe the probabilistic system m_σ resulting from following σ are given in the equations on the left below. To compute LTV_σ, the expression from (1) for this policy may be rewritten to the equations on the right in (5) below. (Recall that the discount factor is $\gamma = 0.9$.)

$$
\begin{aligned}
m_\sigma(\text{PU}) &= (0, 1 \cdot \text{PU}) \\
m_\sigma(\text{PF}) &= (0, \tfrac{1}{2} \cdot \text{PU} + \tfrac{1}{2} \cdot \text{RF}) \\
m_\sigma(\text{RU}) &= (10, \tfrac{1}{2} \cdot \text{PU} + \tfrac{1}{2} \cdot \text{RU}) \\
m_\sigma(\text{RF}) &= (10, \tfrac{1}{2} \cdot \text{RU} + \tfrac{1}{2} \cdot \text{RF})
\end{aligned}
\quad
\begin{aligned}
\text{LTV}_\sigma(\text{PU}) &= 0 + \gamma \cdot \text{LTV}_\sigma(\text{PU}) \\
\text{LTV}_\sigma(\text{PF}) &= 0 + \gamma \cdot (\tfrac{1}{2} \cdot \text{LTV}_\sigma(\text{PU}) + \tfrac{1}{2} \cdot \text{LTV}_\sigma(\text{RF})) \\
\text{LTV}_\sigma(\text{RU}) &= 10 + \gamma \cdot (\tfrac{1}{2} \cdot \text{LTV}_\sigma(\text{PU}) + \tfrac{1}{2} \cdot \text{LTV}_\sigma(\text{RU})) \\
\text{LTV}_\sigma(\text{RF}) &= 10 + \gamma \cdot (\tfrac{1}{2} \cdot \text{LTV}_\sigma(\text{RU}) + \tfrac{1}{2} \cdot \text{LTV}_\sigma(\text{RF}))
\end{aligned}
\quad (5)
$$

Solving this linear system, we get $\text{LTV}_\sigma(\text{PU}) = 0$, $\text{LTV}_\sigma(\text{PF}) = 14.876$, $\text{LTV}_\sigma(\text{RU}) = 18.182$, and $\text{LTV}_\sigma(\text{RF}) = 33.058$.

The long-term value induces an ordering on policies: $\sigma \leq \tau$ if $\text{LTV}_\sigma \leq \text{LTV}_\tau$ in the pointwise order on $\mathbb{R}^S$. It is a classic result [23, Theorem 6.2.7] that for our simple model of MDPs with discounting, the best stationary, memoryless, deterministic policy is as good as any policy. In other words, one cannot do better by allowing time-dependence, memory, or randomization in policies. This result is also the theoretical basic for finding optimal policies via policy iteration, as we describe further below.

Before we move on to policy iteration, we recall the notion of optimal value function. Given an MDP m, the *optimal value of m* is the map $V^*: S \to \mathbb{R}$ that for each state gives the best long-term value that can be obtained for any policy [23]:

$$
V^*(s) = \max_{\sigma \in Act^S} \{\text{LTV}_\sigma(s)\}.
$$

We note that a transition structure $t: S \to (\Delta S)^{Act}$ corresponds to an Act-indexed set of maps $t_a: S \to \Delta S$, $a \in Act$, each of which in turn corresponds to a column-stochastic $|S| \times |S|$-matrix. It is an important classic result that V^* is the unique (bounded) map $v: S \to \mathbb{R}$ that satisfies *Bellman's optimality equation* [4,23]:

$$
v(s) = u(s) + \gamma \cdot \max_{a \in Act} \left\{ \sum_{s' \in S} t_a(s)(s') \cdot v(s') \right\}.
$$

In other words, V^* is a fixpoint of the (non-linear) contractive, monotone *Bellman operator*, given by

$$
\Psi^*: \mathbb{R}^S \to \mathbb{R}^S \qquad \Psi^*(v) = u + \gamma \cdot \max_{a \in Act} \{t_a v\},
$$

where the maximum is taken in the pointwise order on $\mathbb{R}^S$.

3 Policy Improvement via Contraction Coinduction

3.1 Policy Iteration

The optimality equation together with the above mentioned result that an optimal policy may be found among the stationary, deterministic policies is the basis for an effective algorithm for finding optimal policies, known as *policy iteration* [12]. The algorithm starts from any policy $\sigma \in Act^S$, and iteratively improves σ to some τ such that $\sigma \leq \tau$. This leads to an increasing sequence of policies in the preorder of all policies $(S^{Act}, \leq)$. Since this preorder is finite, this process will at some point stabilize. The policy improvement step of the algorithm is obtained via the following definition.

Definition 3. (Improved Policy). *A policy τ is called an* improvement *of a policy σ if for all $s \in S$ it holds that*

$$\tau(s) = argmax_{a \in Act} \{\ell_\sigma(t_a(s))\}. \tag{6}$$

Informally, $\tau(s)$ is an action a that maximizes the expected future rewards obtained by doing a now, and then continuing with σ. However, it is not prima facie clear that τ is an improvement, since following τ means to also "continue with τ" (not with σ). Proving that $\sigma \leq \tau$ is the content of the Policy Improvement Theorem (Theorem 2) which we prove below.

Note that improved policies need not be unique, because there could be different actions a that maximize $\ell_\sigma(t_a(s))$, but (6) describes a procedure for improving a policy σ assuming that LTV_σ has been computed (e.g., by solving the associated linear system).

Example 3. We return to Example 1 and to the "miserly" policy σ in Example 2. To determine $\tau(PF)$ where τ is an improved policy, we compare

$$\ell_\sigma(t(PF)(\mathbf{S})) = E \circ \Delta LTV_\sigma \left(\frac{1}{2} PU + \frac{1}{2} RF \right) = \left(\frac{1}{2} \cdot 0 \right) + \left(\frac{1}{2} \cdot 33.058 \right)$$

and

$$\ell_\sigma(t(PF)(\mathbf{A})) = E \circ \Delta LTV_\sigma(1 \cdot PF) = 1 \cdot 14.876.$$

Since the latter is smaller, we have $\tau(PF) = \mathbf{S}$.

Classically, policy improvement is proved [12,23] using that $(I - \gamma P_\sigma)^{-1}$ is a monotone operator. This in turn follows from the matrix $(I - \gamma P_\sigma)^{-1}$ having only non-negative entries, a property which we show in Example 5 below using contraction coinduction.

3.2 The Contraction Coinduction Principle

We now introduce the contraction coinduction principle. We only assume basic knowledge of metric spaces, as can be found in, e.g., [21]. Here we just recall a few basic definitions and fix notation. A *metric space* (X, d_X) is a set X equipped with a metric $d_X \colon X \to \mathbb{R}$. Sometimes the metric is left implicit and we simply refer to the metric space X. We always assume the standard Euclidean metric on the set of real numbers $\mathbb{R}$. Any set X can be equipped with the discrete metric, given by $d_X(x, y) = 1$ if $x \neq y$, and $d_X(x, y) = 0$ if $x = y$, for all $x, y \in X$.

A function $f\colon X \to Y$ between metric spaces is *bounded* if there is a real number C such that for all $x, y \in X$, it holds that $d_Y(f(x), f(y)) \le C$. We write $B(X, Y)$ for the set of all bounded $f\colon X \to Y$. The set $B(X, Y)$ can be equipped with the *supremum metric*: for all $f, g \in B(X, Y)$,

$$d(f, g) = \sup\{d_Y(f(x), g(x)) \mid x \in X\}. \tag{7}$$

When Y is a complete space (for example, $\mathbb{R}$), so is $B(X, Y)$. We recall that a function $f\colon X \to X$ on a metric space X is *contractive* if there is a $C < 1$ such that for all $x_1, x_2 \in X$, we have $d_X(f(x_1), f(x_2)) \le C \cdot d_X(x_1, x_2)$. A *fixpoint* of f is an element x^* such that $f(x^*) = x^*$.

The contraction coinduction principle is a variation of the classic Banach Fixpoint Theorem, asserting that any contractive mapping has a unique fixpoint. We need a version of this theorem which, in addition to a complete metric, also has an order. For this reason we introduce the following definition.

Definition 4. *An* ordered metric space *is a structure* $(M, d, \le)$ *such that d is a metric on M and $\le$ is a partial order on M, satisfying the extra property that for all $y \in M$, $\{z \mid z \le y\}$ and $\{z \mid y \le z\}$ are closed sets in the metric topology. This space is said to be* complete *if it is complete as a metric space.*[1]

Example 4. For any set X, $B(X, \mathbb{R})$ with the pointwise order (and supremum metric, as in (7) above) is a complete ordered metric space.

We can now state contraction (co)induction. It will lead to elegant proofs of order statements concerning fixpoints, as we shall see below.

Theorem 1. (Contraction (Co)Induction). *Let M be a non-empty, complete ordered metric space. If $f\colon M \to M$ is both contractive and order-preserving, then the fixpoint x^* of f is a least pre-fixpoint (if $f(x) \le x$, then $x^* \le x$), and also a greatest post-fixpoint (if $x \le f(x)$, then $x \le x^*$).*

Proof. We only verify the first assertion; the second is similar. Suppose that $f(x) \le x$. By induction on n and monotonicity of f, we have for all $n \ge 0$, $f^n(x) \le x$. Since f is contractive, the proof of the Banach Fixpoint Theorem shows that $\{f^n(x)\}_n$ is a convergent sequence. But $\{z \mid z \le x\}$ is closed and contains this sequence, so $\lim_n f^n(x) \le x$. The proof of the Banach Fixpoint Theorem also shows that $\lim_n f^n(x)$ equals the fixpoint x^*. Thus, $x^* \le x$. □

Remark 1. Theorem 1 follows from the Metric Coinduction Principle [17,24]. E.g., to derive contraction induction, let x be such that $f(x) \le x$. The set $A = \{y \in X \mid y \le x\}$ is non-empty, since $x \in A$, and closed by our assumption that X is an ordered metric space. Moreover, $f[A] \subseteq A$ by monotonicity. Hence by metric coinduction, $x^* \in A$.

---

[1] We could weaken the partial order in the definition of an ordered metric space to a transitive relation. However, our aim is not the highest level of generality. Rather, we see contraction (co)induction as an instance of metric coinduction (see Remark 1) that suffices to prove interesting results about MDPs.

Example 5. We recover a fact that comes up frequently in the area (e.g., [12] uses it to justify policy improvement): if P is a column-stochastic $n \times n$ matrix, then $(I - \gamma P)^{-1}$ has all non-negative entries. We do not really need this fact, and we mainly mention it to point out that contraction coinduction might streamline the proofs of known results. To see this, let $M = \mathbb{R}^{n \times n}$. We order M pointwise, and as metric we use $d(X, Y) = \|X - Y\|$, where $\|X\| = \max_{i,j} |x_{i,j}|$ (so that $\|PX\| \le \|X\|$). This gives a complete ordered metric space. Let $f \colon M \to M$ be $f(X) = I + (\gamma P)X$. Easily, f is a monotone contraction, and its fixpoint is $(I - \gamma P)^{-1}$. Note that $f(0) \ge 0$, where 0 is the zero matrix. By contraction coinduction, we conclude that $(I - \gamma P)^{-1} \ge 0$.

We now give our proof of policy improvement using contraction coinduction.

Theorem 2 (Policy Improvement). *Let an MDP be given by $t \colon S \to (\Delta S)^{Act}$ and $u \colon S \to \mathbb{R}$. Let σ and τ be policies. If $\ell_\sigma \circ t_\tau \ge \ell_\sigma \circ t_\sigma$, then* $\mathrm{LTV}_\tau \ge \mathrm{LTV}_\sigma$. *Similarly, if $\ell_\sigma \circ t_\tau \le \ell_\sigma \circ t_\sigma$, then* $\mathrm{LTV}_\tau \le \mathrm{LTV}_\sigma$.

Proof. Assuming that $\ell_\sigma \circ t_\tau \ge \ell_\sigma \circ t_\sigma$, we have for all $s \in S$,

$$u(s) + \gamma \sum_{s' \in S} t_\sigma(s)(s') \cdot \mathrm{LTV}_\sigma(s') \le u(s) + \gamma \sum_{s' \in S} t_\tau(s)(s') \cdot \mathrm{LTV}_\sigma(s').$$

Since LTV_σ and LTV_τ are the fixpoints of the contractive, monotone operators Ψ_σ and Ψ_τ, respectively (cf. (4)), the above inequality may be recast to say that $\Psi_\tau(\mathrm{LTV}_\sigma) \ge \Psi_\sigma(\mathrm{LTV}_\sigma) = \mathrm{LTV}_\sigma$. By contraction coinduction (Theorem 1), $\mathrm{LTV}_\tau \ge \mathrm{LTV}_\sigma$. This completes the proof of the first assertion. The second one is proved similarly. □

Next, we use contraction coinduction to show the classic result that V^* is an upper bound for the long-term value of all policies, and that a so-called *greedy policy* is optimal [4,23]. Lemma 1 below is standard, and essentially the same proof as ours appears as Lemma 5.2 in Kozen and Ruozzi [24].

Lemma 1. *For all policies σ,* $\mathrm{LTV}_\sigma \le V^*$.

Proof. A straightforward calculation and monotonicity argument shows that for all $f \in B(S, \mathbb{R})$, $\Psi_\sigma(f) \le \Psi^*(f)$. In particular, $\mathrm{LTV}_\sigma = \Psi_\sigma(\mathrm{LTV}_\sigma) \le \Psi^*(\mathrm{LTV}_\sigma)$. By contraction coinduction we conclude that $\mathrm{LTV}_\sigma \le V^*$. □

Define the *greedy policy* $\sigma^* \colon S \to Act$ by

$$\sigma^*(s) = \mathrm{argmax}_{a \in Act} \left\{ \sum_{s' \in S} t_a(s)(s') \cdot V^*(s') \right\}. \tag{8}$$

Due to ties, this policy is strictly speaking not unique. But following standard usage, we speak of "the" greedy policy when we mean "a" greedy one.

Proposition 1. *The greedy policy is optimal. That is,* $\mathrm{LTV}_{\sigma^*} = V^*$.

Proof. Observe that $\Psi_{\sigma^*}(V^*) \ge V^*$ (in fact, equality holds). By contraction coinduction, $V^* \le \mathrm{LTV}_{\sigma^*}$. The other direction follows from Lemma 1. □

It is a direct consequence of Proposition 1 that the optimal value is attained by some policy.

4   Coalgebras and Algebras for MDPs

In this section, we present the coalgebraic and algebraic structures that we use to model MDPs and their long-term values. We assume the reader is familiar with basic notions in coalgebra [25] and category theory [18], but we briefly recall some definitions and results (see, e.g., [3,13,16]) related to monads and distributive laws that are needed for this paper.

4.1   Algebras, Monads, and Distributive Laws

Given a functor $T \colon \mathsf{C} \to \mathsf{C}$ on a category C, a *T-algebra* is a pair (A, α) where A is a C-object and $\alpha \colon TA \to A$ is a C-arrow. A *T-algebra homomorphism* from (A, α) to (B, β) is a C-arrow $f \colon A \to B$ such that $f \circ \alpha = \beta \circ Tf$.

A *monad* (on C) is a triple (T, η, μ) where T is a C-endofunctor, and $\eta \colon \mathrm{Id} \Rightarrow T$ and $\mu \colon TT \Rightarrow T$ are natural transformations such that $\mu \circ T\eta = \mathrm{id} = \mu \circ \eta_T$ and $\mu \circ \mu_T = \mu \circ T\mu$. Given a monad (T, η, μ), an *Eilenberg-Moore T-algebra* is a T-algebra (A, ω) such that $\omega \circ \eta_A = \mathrm{id}$ and $\omega \circ \mu_A = \omega \circ T\omega$. We denote the category of Eilenberg-Moore T-algebras and T-algebra homomorphisms by $\mathcal{EM}(T)$. Note that (TX, μ_X) is an Eilenberg-Moore T-algebra.

Example 6. The well-known *distribution monad* is the discrete variant of the *Giry monad* [11]. The functor part $\Delta \colon \mathsf{Set} \to \mathsf{Set}$ maps a set X to the finitely supported probability distributions on X:

$$\Delta X = \{\varphi \colon X \to [0,1] \mid \mathrm{supp}(\varphi) \text{ is finite and } \textstyle\sum_x \varphi(x) = 1\},$$
$$(\Delta f)(\varphi)(y) = \textstyle\sum_{x \in f^{-1}(y)} \varphi(x) \qquad \text{for all } f \colon X \to Y.$$

It is sometimes convenient to write an element φ of ΔX as a formal linear combination $\varphi = r_1 x_1 + \cdots + r_n x_n$, where $\mathrm{supp}(\varphi) = \{x_1, \ldots, x_n\}$ and $\varphi(x_i) = r_i$, or also $\varphi = \sum_{x \in X} \varphi(x)\, x$. In this notation, $(\Delta f)(\varphi) = r_1 f(x_1) + \cdots + r_n f(x_n)$ for $f \colon X \to Y$, where coefficients of identical $f(x_i)$-values are summed implicitly. Equivalently stated, we have

$$(\Delta f)(\varphi) = \sum_{x \in X} f(x)\varphi(x). \tag{9}$$

The unit $\delta \colon \mathrm{Id} \Rightarrow \Delta$ is $\delta_X(x) = 1x$ (the Dirac distribution at x), and the multiplication $\mu \colon \Delta\Delta \Rightarrow \Delta$ is given as follows. For $\psi = r_1 \varphi_1 + \cdots + r_n \varphi_n \in \Delta\Delta X$, we have $\mu_X(\psi)(x) = \sum_{i=1}^{n} r_i \varphi_i(x)$, i.e., $\mu_X(\psi) = \sum_{x \in X} \left(\sum_{\varphi \in \Delta X} \psi(\varphi) \cdot \varphi(x) \right) x$. The category $\mathcal{EM}(\Delta)$ is also known as the category of convex sets and affine (or linear) maps, since an Eilenberg-Moore Δ-algebra can be seen as a set X in which convex combinations $r_1 x_1 + \cdots + r_n x_n$ can be evaluated.

Let (T, η, μ) be a monad and F an endofunctor, both on C. A *distributive law* of (T, η, μ) over F is a natural transformation $\lambda \colon TF \Rightarrow FT$ that is compatible with the monad structure, meaning that $\lambda \circ \eta_F = F\eta$ and $\lambda \circ \mu_F = F\mu \circ \lambda_T \circ T\lambda$. We recall (see, e.g., [15,16]) that such a distributive law corresponds to a lifting

F_λ of F to the category $\mathcal{EM}(T)$, and equivalently to a lifting T_λ of T to the category $\mathrm{Coalg}_C(F)$ of F-coalgebras. The functors F_λ and T_λ are defined as follows:

$$F_\lambda(A, \omega\colon TA \to A) = (FA, F\omega \circ \lambda_A) \qquad F_\lambda(f) = Ff,$$
$$T_\lambda(B, \beta\colon B \to FB) = (TB, \lambda_B \circ T\beta) \qquad T_\lambda(f) = Tf.$$

We also recall (cf. [3,14]) that such a distributive law induces an operation $(-)^\sharp\colon \mathrm{Coalg}_C(FT) \to \mathrm{Coalg}_{\mathcal{EM}(T)}(F_\lambda)$, which is often referred to as an abstract form of *determinization* (cf. [14,26]). For every FT-coalgebra $c\colon X \to FTX$, $c^\sharp$ is defined as

$$c^\sharp = F\mu_X \circ \lambda_{TX} \circ Tc\colon (TX, \mu_X) \to F_\lambda(TX, \mu_X), \text{ and we have } c^\sharp \circ \eta_X = c. \quad (10)$$

Determinization $(-)^\sharp$ is a functor, but we shall not use this fact. Note that the underlying F-coalgebra of $c^\sharp$ is of type $TX \to FTX$.

We write $\mathrm{E}\colon \Delta\mathbb{R} \to \mathbb{R}$ for the map that computes expected value. That is, viewing an element $\varphi \in \Delta\mathbb{R}$ as a formal linear combination, E evaluates φ by interpreting the formal expression in $\mathbb{R}$, i.e., $\mathrm{E}(\varphi) = \sum_{x \in \mathbb{R}} \varphi(x) \cdot x$.

Note that for $f\colon X \to \mathbb{R}$, by (9) we have, for all $\varphi \in \Delta X$, that

$$\mathrm{E}((\Delta f)(\varphi)) = \sum_{x \in X} f(x) \cdot \varphi(x). \quad (11)$$

Lemma 2. *The expected value* $\mathrm{E}\colon \Delta\mathbb{R} \to \mathbb{R}$ *is an Eilenberg-Moore* Δ*-algebra:* $\mathrm{E} \circ \delta_\mathbb{R} = \mathrm{id}_\mathbb{R}$ *and* $\mathrm{E} \circ \Delta\mathrm{E} = \mathrm{E} \circ \mu_\mathbb{R}$.

4.2 Coalgebraic Modeling of MDPs

As we saw in Definition 2, long-term values arise by summing infinite sequences (or streams) of real numbers. It is well-known [25] that such streams form a final coalgebra for the Set-endofunctor $H = \mathbb{R} \times \mathrm{Id}$. The final H-coalgebra structure is given by mapping a stream $x = (x_0, x_1, x_2, \ldots)$ to $(\mathsf{head}(x), \mathsf{tail}(x))$, where $\mathsf{head}(x) = x_0$ and $\mathsf{tail}(x) = (x_1, x_2, \ldots)$.

Given an MDP $m = \langle u, t \rangle$ and a policy $\sigma\colon S \to Act$, the resulting Markov reward process $m_\sigma = \langle u, t_\sigma \rangle$ is easily seen to be an $H\Delta$-coalgebra

$$m_\sigma = \langle u, t_\sigma \rangle\colon S \to \mathbb{R} \times \Delta S,$$

where, as we recall from from Sect. 2, $t_\sigma(s) = t(s)(\sigma(s))$.

Similarly, it is not hard to see that an MDP $m = \langle u, t \rangle$ is a $K\Delta$-coalgebra $\langle u, t \rangle\colon S \to \mathbb{R} \times (\Delta S)^{Act}$, where $K = H \circ (-)^{Act}$ and $(-)^{Act}$ is the covariant hom-functor.

Since $\mathrm{E}\colon \Delta\mathbb{R} \to \mathbb{R}$ is an Eilenberg-Moore Δ-algebra, there is a distributive law χ of (Δ, δ, μ) over H (cf. [13]) specified by

$$\chi_X\colon \Delta(\mathbb{R} \times X) \xrightarrow{\langle \Delta\pi_1, \Delta\pi_2 \rangle} \Delta\mathbb{R} \times \Delta X \xrightarrow{\mathrm{E} \times \mathrm{id}} \mathbb{R} \times \Delta X,$$

i.e.,

$$\chi_X = \langle E \circ \varDelta\pi_1, \varDelta\pi_2 \rangle. \tag{12}$$

The lifted functor $H_\chi \colon \mathcal{EM}(\varDelta) \to \mathcal{EM}(\varDelta)$ is concretely given as

$$
\begin{aligned}
H_\chi(A, \omega) &= (\mathbb{R} \times A, (\mathbb{R} \times \omega) \circ \langle E \circ \varDelta\pi_1, \varDelta\pi_2 \rangle) \\
&= (\mathbb{R} \times A, \langle E \circ \varDelta\pi_1, \omega \circ \varDelta\pi_2 \rangle).
\end{aligned}
$$

Using the distributive law χ from (12), the determinization $m_\sigma^\sharp \colon \varDelta S \to \mathbb{R} \times \varDelta S$ is given for each $\varphi \in \varDelta S$ by

$$
\begin{aligned}
m_\sigma^\sharp(\varphi) &= ((E \circ \varDelta u)(\varphi), (\mu_S \circ \varDelta t_\sigma)(\varphi)) \\
&= (\textstyle\sum_{s \in S} u(s) \cdot \varphi(s),\, s \mapsto \sum_{s' \in S} t_\sigma(s)(s') \cdot \varphi(s')).
\end{aligned}
$$

Considering φ as a probabilistic state, the first component of the pair $m_\sigma^\sharp(\varphi)$ is the expected reward given φ, and the second component is the expected next probabilistic state. The morphism $\mu_S \circ \varDelta t_\sigma \colon \varDelta S \to \varDelta S$ is the Kleisli extension of $t_\sigma \colon S \to \varDelta S$, which can be seen as a column-stochastic $|S| \times |S|$-matrix. Viewing $u \in \mathbb{R}^S$ as a row $|S|$-vector and a distribution $\varphi \in \varDelta S$ as a column-stochastic $|S|$-vector, we have that $m_\sigma^\sharp(\varphi) = \langle u\varphi, t_\sigma\varphi \rangle$, where juxtaposition denotes matrix-vector multiplication. The unique H-coalgebra morphism from $m_\sigma^\sharp$ to the final H-coalgebra of streams maps a distribution φ to the stream of expected rewards $(u\varphi, ut_\sigma\varphi, ut_\sigma^2\varphi, \dots)$.

The distributive law given by χ is an $\mathcal{EM}$-law in the terminology of [14], where determinization was studied for the purpose of obtaining trace semantics. The trace semantics of $m_\sigma \colon S \to \mathbb{R} \times \varDelta S$ is the function that maps a state s to the stream of expected rewards $(r_0^\sigma(s), r_1^\sigma(s), r_2^\sigma(s), \dots)$ from (1).

4.3   Algebraic Modeling of Discounted Sums

The long-term value of a policy σ in state s is the discounted infinite sum of the stream $\rho(s) = (r_0^\sigma(s), r_1^\sigma(s), r_2^\sigma(s), \dots)$ of expected rewards. Due to S being finite, the values in this stream are bounded, which ensures that the discounted sum converges. A leading observation of this paper is that we can re-express the recursive equation (3) for LTV_σ by saying that $\mathrm{LTV}_\sigma \colon S \to \mathbb{R}$ makes the following diagram commute:

$$
\begin{array}{ccc}
S & \xrightarrow{\ m_\sigma\ } & \mathbb{R} \times \varDelta S \\
{\scriptstyle \mathrm{LTV}_\sigma}\big\downarrow & & \big\downarrow{\scriptstyle \mathbb{R} \times \varDelta(\mathrm{LTV}_\sigma)} \\
\mathbb{R} \xleftarrow[\ \alpha_\gamma\ ]{} \mathbb{R} \times \mathbb{R} & \xleftarrow[\ \mathbb{R} \times E\ ]{} & \mathbb{R} \times \varDelta\mathbb{R}
\end{array} \tag{13}
$$

Here, $\alpha_\gamma \colon \mathbb{R} \times \mathbb{R} \to \mathbb{R}$ is the H-algebra

$$\alpha_\gamma \colon H\mathbb{R} \to \mathbb{R} \qquad \alpha_\gamma(x, y) = x + \gamma \cdot y. \tag{14}$$

Notice that LTV$_\sigma$ is an $H\Delta$-coalgebra-to-algebra map. We naturally wonder whether the $H\Delta$-algebra at the bottom of the diagram is a *corecursive algebra* [6]: for every coalgebra $f\colon X \to H\Delta X$ (where X is possibly infinite), is there a unique map $f^\dagger\colon S \to \mathbb{R}$ making the diagram commute? As suggested by the previous discussion, problems can arise if the reward values in f are unbounded. But the question can be framed in an even more basic way. Namely, by [5, Theorem 19], $\alpha_\gamma \circ (\mathbb{R} \times E)$ is a corecursive algebra for $H\Delta$ if and only if α_γ is a corecursive algebra for H. But the latter is not the case. Consider an infinite system of equations

$$x_n = a_n + \gamma \cdot x_{n+1}, \qquad n = 0, 1, \ldots, \tag{15}$$

where a_n are fixed real numbers. Then (15) corresponds uniquely to a H-coalgebra $g\colon X \to \mathbb{R} \times X$. Solutions to (15) in turn correspond to maps $g^\dagger$ such that $g^\dagger = \alpha_\gamma \circ (\mathbb{R} \times g^\dagger) \circ g$, i.e., to coalgebra-to-algebra maps from (X, g) to $(\mathbb{R}, \alpha_\gamma)$. The reason why α_γ is not a corecursive algebra is that (15) always has continuum many solutions. Namely, the solution value for x_0 may be chosen arbitrarily, and the rest are determined from it. Note however, if $(a_n)_n$ is unbounded then all solutions are unbounded. [To see this, let $(b_n)_n$ be a solution. We have: $|a_n| > 2K \Rightarrow |b_{n+1}| = |a_n + \gamma \cdot b_n| \geq |a_n| - \gamma|b_n| \Rightarrow |b_{n+1}| + \gamma|b_n| > 2K \Rightarrow |b_n| > K$ or $|b_{n+1}| > K$. Hence, for each K there is some i such that $|b_i| > K$.]

If, on the other hand, $(a_n)_n$ is bounded, then there is a *unique bounded solution* to (15), namely $x_n = \sum_{i=0}^{\infty} \gamma^i \cdot a_{n+i}$ for all n. Boundedness is used in asserting that the sum converges, and the detailed verification that this solution works and is unique follows from Proposition 6 below. In summary, uniqueness is only obtained by restricting to bounded solutions.

We end this section by noting that α_γ is an algebra for the lifted functor H_χ, essentially because α_γ is affine. We will need this result in Sect. 5.3.

Lemma 3. *$((\mathbb{R}, E), \alpha_\gamma)$ is an H_χ-algebra in $\mathcal{EM}(\Delta)$, that is, we have the equality $E \circ \Delta\alpha_\gamma = \alpha_\gamma \circ \langle E \circ \Delta\pi_1, E \circ \Delta\pi_2 \rangle$.*

5   Long-Term Values via b-Corecursive Algebras

In this section, we will develop some categorical notions in order to capture boundedness properties, and eventually show that long-term values can be characterized via a universal property of a notion of corecursive algebra for bounded maps.

5.1   MDPs in Metric Spaces

The first step is to identify the appropriate category of metric spaces. There are several types of functions on metric spaces that are of interest. In this paper, we shall consider the following. Let (X, d_X) and (Y, d_Y) be metric spaces and $f\colon X \to Y$ a function (not necessarily continuous). Then f is said to be *Lipschitz* if $d_Y(f(x_1), f(x_2)) \leq C \cdot d_X(x_1, x_2)$ for all $x_1, x_2 \in X$, for some fixed real number

C. A Lipschitz function that satisfies the above inequality for $C = 1$ is called *non-expansive* (or *short*). It is clear that the composition of two Lipschitz functions is again Lipschitz, and the composition of non-expansive functions again non-expansive.

Bounded functions need not be Lipschitz, and vice versa. Although bounded maps are of particular interest to us, we point out the fact that metric spaces with bounded maps do not form a category, since the identity on a space of infinite diameter is not bounded. Our main interest in Lipschitz functions is that if g is bounded and f is Lipschitz, then $f \circ g$ is bounded; also, they are used in the Kantorovich metric just below.

We write Met for the category that has metric spaces as objects and all functions as arrows. Usually, the morphisms of metric spaces are taken to be the non-expanding functions or continuous functions. The reason we take *all* set functions is that we are going to use the metric structure only in connection with *boundedness*, and so our (non-standard) choice will become more sensible. (In Sect. 6.2, we hint that with additional results we can indeed work with a "real" metric-type category, the Polish metric spaces.)

We lift our Set-endofunctors H and Δ to Met using the maximum and Kantorovich(-Wasserstein) metrics (cf. [2, 29]). This last metric is usually defined in the measure-theoretic setting, so discrete probability measures are a special case.

Definition 5 (Product and Kantorovich Metrics).   *Let* (X, d_X) *and* (Y, d_Y) *be metric spaces.*

- *The product* $(X, d_X) \times (Y, d_Y) = (X \times Y, d_X \times d_Y)$ *has the maximum metric*

$$(d_X \times d_Y)((x_1, y_1), (x_2, y_2)) = \max\{d_X(x_1, x_2), d_Y(y_1, y_2)\}.$$

- *The Kantorovich lifting of* d_X *is the metric* $d_{\Delta X}$ *on* ΔX, *defined by*

$$d_{\Delta X}(\varphi, \psi) = \sup\{d_{\mathbb{R}}((\mathrm{E} \circ \Delta f)(\varphi), (\mathrm{E} \circ \Delta f)(\psi)) \mid f \colon X \to \mathbb{R} \text{ is non-expansive}\}.$$

Remark 2. See [10] for ten choices for the metric d on ΔX. Incidentally, very little is known concerning the question of whether each d leads to a functor on the category of *all* metric spaces and continuous functions. However, it follows from Theorem 1 of [11] that for the related category of Polish spaces, the Kantorovich lifting does lead to a functor.

We can view a Markow reward process $\langle u, t_\sigma \rangle \colon S \to \mathbb{R} \times \Delta S$ as a coalgebra in Met for the lifted functor $H\Delta$, by equipping the state space S with a metric. (The discrete metric is the canonical choice, but any metric will do.)

The next lemma will be frequently used in Sect. 5.2 to prove boundedness preservation properties.

Lemma 4. *If* $f \colon X \to Y$ *is Lipschitz, so is* Hf. *If* $f \colon X \to Y$ *is Lipschitz with constant* C, *then so is* Δf.

5.2 Categorical Structure for Bounded Maps

This section aims at a sparse categorification of boundedness that will permit us to re-work the notion of a corecursive algebra to a *bounded corecursive algebra* in Sect. 5.3 below. To this aim, we introduce the notion of *b*-category and related concepts.

Definition 6. *Let* C *be a category and* $\mathcal{B}$ *a class of morphisms in* C. *We call* $\mathcal{B}$ *a b-structure[2] on* C *if for all* $f \in \mathcal{B}$ *and all arrows* g *in* C, *if* $f \circ g$ *is defined, then* $f \circ g \in \mathcal{B}$. *A b-category is a pair* $(C, \mathcal{B})$, *where* C *is a category and* $\mathcal{B}$ *is a b-structure on* C. *We frequently call a morphism* $f \in \mathcal{B}$ *a* $\mathcal{B}$*-morphism. We denote the collection of all* C*-morphisms* $X \to Y$ *that are also in* $\mathcal{B}$ *by* $\mathcal{B}(X, Y)$.

The key feature of Lipschitz and bounded functions for our purposes is captured in the following definition.

Definition 7. *We say that a* C*-arrow* f *preserves* $\mathcal{B}$ *if whenever* $g \in \mathcal{B}$ *and* $f \circ g$ *is defined, then* $f \circ g \in \mathcal{B}$.

It is easy to see that for every category C, $(C, \mathcal{M})$ is a *b*-category, where $\mathcal{M}$ is the collection of morphisms of C. If $(C, \mathcal{B})$ is a *b*-category, then every morphism in $\mathcal{B}$ preserves $\mathcal{B}$.

Example 7. Our primary example of a *b*-category is (Met, B), where Met is the category of metric spaces and all functions, and B is the collection of bounded maps of metric spaces. While the metric structure is not used in the Met-morphisms, it figures in the *b*-structure.

Every Lipschitz function preserves B. For any metric spaces X_1 and X_2, the projections $\pi_i \colon X_1 \times X_2 \to X_i$ preserve B. The algebras E: $\Delta\mathbb{R} \to \mathbb{R}$ and α_γ, from (14), both preserve B.

Next, we formulate definitions of functors and natural transformations which incorporate *b*-structures. The main motivation for the definitions below are the examples which follow and also the properties that we shall see at the end of this section, in Proposition 5 and Example 8.

Definition 8. *Let* $(C, \mathcal{B})$ *and* $(C', \mathcal{B}')$ *be b-categories. A functor* $F \colon C \to C'$ *is a b-functor, written* $F \colon (C, \mathcal{B}) \to (C', \mathcal{B}')$, *if whenever* $f \in \mathcal{B}$, *then* Ff *preserves* $\mathcal{B}'$; *and* F *is a* strong *b-functor if whenever* $f \in \mathcal{B}$, *then* $Ff \in \mathcal{B}'$.

If $F, G \colon C \to C'$ *are functors (not necessarily b-functors), then a b-natural transformation* $\sigma \colon F \Rightarrow G$ *is a natural transformation in the usual sense such that every component* σ_X *preserves* $\mathcal{B}'$.

Proposition 2. *(1) Constant functors are b-functors. (2) The identity on a b-category is a b-endofunctor. (3) If* F *is a strong b-functor, then* F *is a b-functor.*

We now investigate how the functor H, monad (Δ, δ, μ), and distributive law χ interact with the *b*-structure B of bounded maps on Met.

---

[2] During CMCS 2018, we learned from Henning Urbat that a *b*-structure is also known as a *sieve*. We currently do not know how to put this fact to use.

Proposition 3. $H:$ Met $\rightarrow$ Met *is a b-endofunctor, but not a strong b-endofunctor, on* (Met, B).

Proposition 4. $\Delta:$ Met $\rightarrow$ Met *is a strong b-endofunctor on* (Met, B).

Lemma 5. *If* $f: X \rightarrow Y$ *preserves* B, *then so does* $Hf: \mathbb{R} \times X \rightarrow \mathbb{R} \times Y$.

Lemma 6. *For all metric spaces* (X, d_X), *the following hold.*

1. δ_X *is an isometric embedding.*
2. μ_X *is non-expanding.*
3. χ_X *is Lipschitz.*

It follows that δ, μ, *and* χ *are b-natural transformations in* (Met, B).

One crucial observation is that if the $H\Delta$-coalgebra m_σ obtained from an MDP m and a policy σ is bounded, then so is the determinized H-coalgebra $m_\sigma^\sharp$. The following proposition shows that our setup of b-structures ensures that this holds abstractly.

Proposition 5. *Let* λ *be a distributive law of monad* (T, η, μ) *over a functor* F *such that* T *is a strong b-endofunctor, and* $F\mu$ *and* λ *are b-natural transformations. Then* $\mathcal{B}$ *is closed under* $(-)^\sharp$, *i.e., if* $c \in \mathcal{B}$ *then* $c^\sharp \in \mathcal{B}$.

Proof. This follows instantly from $c^\sharp = F\mu_X \circ \lambda_{TX} \circ Tc$ (cf. Equation (10)). $\square$

Example 8. For our running example for MDPs where $F = H$, $T = \Delta$, and $\lambda = \chi$ is given by (12), we have the conclusion of Proposition 5 in the b-category (Met, B). Indeed, by Proposition 4, Δ is a strong b-endofunctor. By Lemma 6, χ is b-natural. Finally, by the second part of Lemmas 4 and 6 (2), $H\mu_X$ is Lipschitz for every X, and thus preserves bounded maps. Therefore, $H\mu$ is b-natural.

5.3 b-Corecursive Algebras (bcas)

As we explained in Sect. 4.3, the long-term value map LTV_σ is a certain coalgebra-to-algebra morphism, i.e., it is a solution to a set of recursive equations, but it is only uniquely defined if we restrict to bounded maps. The following notion of b-corecursive algebra categorifies this observation.

Definition 9. *Let* $(\mathsf{C}, \mathcal{B})$ *be a b-category,* F *an endofunctor on* C *(not necessarily a b-endofunctor), and* $\beta: FA \rightarrow A$ *an F-algebra. Then* β *is a b-corecursive algebra (bca) if for every F-coalgebra* $f: X \rightarrow FX$ *with* $f \in \mathcal{B}$, *there is a unique solution map* $f^\dagger \in \mathcal{B}$ *such that the diagram*

$$
\begin{array}{ccc}
X & \xrightarrow{\quad f \quad} & FX \\
{\scriptstyle f^\dagger}\Big\downarrow & & \Big\downarrow{\scriptstyle Ff^\dagger} \\
A & \xleftarrow{\quad \beta \quad} & FA
\end{array}
$$

commutes, or equivalently stated, such that $f^\dagger$ *is the fixed point of the operator* $\Phi_{f,\beta}: \mathsf{C}(X, A) \rightarrow \mathsf{C}(X, A)$, *defined for all* $j \in \mathsf{C}(X, A)$ *by* $\Phi_{f,\beta}(j) = \beta \circ Fj \circ f$.

We note that a *corecursive algebra* [6] is a bca with $\mathcal{B}$ the family of *all* morphisms in the underlying category.

Remark 3. A corecursive algebra is a special kind of *completely iterative algebra* (also called cias, see Milius [19]). With the obvious definition, the examples in this paper would be *b-cias*. Alas, we have not found any application of this fact.

The next lemma uses the *b*-category concepts to give conditions that ensure the operator $\Phi_{f,\beta}$ from Definition 9 restricts to $\mathcal{B}$-morphisms. This is the abstract analogue of showing that the Bellman operator maps bounded maps to bounded maps.

Lemma 7. *Let $(\mathsf{C}, \mathcal{B})$ be a b-category. If F is a b-endofunctor on $(\mathsf{C}, \mathcal{B})$, and $\beta\colon FA \to A$ is an F-algebra that preserves $\mathcal{B}$, it holds that for every F-coalgebra $f\colon X \to FX$ in $\mathcal{B}$, the operator $\Phi_{f,\beta}\colon \mathsf{C}(X, A) \to \mathsf{C}(X, A)$ from Definition 9 restricts to an operator $\Phi_{f,\beta}\colon \mathcal{B}(X, A) \to \mathcal{B}(X, A)$.*

Proof. Let $j \in \mathcal{B}(X, A)$. Since F is assumed to be an *b*-endofunctor, Fj preserves $\mathcal{B}$. Thus since $f \in \mathcal{B}$, $Fj \circ f \in \mathcal{B}$ as well. Finally, since β preserves $\mathcal{B}$, it follows that $\Phi_{f,\beta}(j) = \beta \circ Fj \circ f \in \mathcal{B}$. $\qquad\square$

The following result is the first step towards obtaining the long-term value map LTV_σ from the universal property of bcas.

Proposition 6. *The H-algebra $\alpha_\gamma\colon H\mathbb{R} \to \mathbb{R}$ is a bca in $(\mathsf{Met}, \mathcal{B})$.*

Proof. Fix a bounded $f\colon X \to HX$. Recall from Example 4 that the bounded function space $B(X, \mathbb{R})$ is a complete ordered metric space with the supremum metric. Since H is a *b*-endofunctor (Lemma 5) and α_γ preserves B (Example 7), the operator

$$\Phi = \Phi_{f,\alpha_\gamma}\colon B(X, \mathbb{R}) \to B(X, \mathbb{R})\colon \Phi(j) = \alpha_\gamma \circ Hj \circ f$$

is well-defined by Lemma 7.

We now show that Φ is a contractive map. So let $j, k \in B(X, \mathbb{R})$, and $x \in X$. We write $f = \langle f_1, f_2 \rangle\colon X \to \mathbb{R} \times X$. Then

$$d_{\mathbb{R}}(\Phi(j)(x), \Phi(k)(x)) \leq \gamma \cdot |j(f_2(x)) - k(f_2(x))| \leq \gamma \cdot d(j, k).$$

This holds for all $x \in X$. Since $0 \leq \gamma < 1$, it follows that Φ is contractive. By the Banach Fixpoint Theorem, Φ has a unique fixpoint. This proves that the operator Φ_{f,α_γ} has a unique *bounded* fixpoint, which is what we had to show. $\square$

The second step for obtaining the long-term value function LTV_σ from the universal property of bcas, is to show how to obtain a bca for $H\Delta$ from the bca α_γ for H. The next theorem shows that we can prove this result abstractly using *b*-structure.

We first note that given a *b*-structure $(\mathsf{C}, \mathcal{B})$ and a monad (T, η, μ), the category $\mathcal{EM}(T)$ has a *b*-structure consisting of the T-algebra morphisms φ such that $U\varphi \in \mathcal{B}$ in the base *b*-structure; we shall write the *b*-structure on $\mathcal{EM}(T)$ as $\mathcal{B}$ as well.

Theorem 3. *Let* $(C, \mathcal{B})$ *be a b-category,* F *a* C*-endofunctor,* (T, η, μ) *a monad on* C, *and* λ *a distributive law of* (T, η, μ) *over* F. *Assume further that* T *is a strong b-functor and that* λ *and* $F\mu$ *are b-natural in* $(C, \mathcal{B})$.

1. *If* $\beta \colon F_\lambda(A, \omega) \to (A, \omega)$ *is an* F_λ*-algebra in* $\mathcal{EM}(T)$ *such that the underlying* F*-algebra* $\beta \colon FA \to A$ *is a bca for* F *and* ω *preserves* $\mathcal{B}$, *then it holds that* $\beta \circ F\omega \colon FTA \to A$ *is a bca for* FT.
2. *Let the solution operation for the bca* $\beta \colon FA \to A$ *be denoted* $h \mapsto h^{\ddagger}$, *and the solution operation for the bca* $\beta \circ F\omega \colon FTA \to A$ *be denoted* $h \mapsto h^{\dagger}$. *Then for all* $g \colon X \to FTX$ *in* $\mathcal{B}$, *we have* $g^{\dagger} = (g^{\sharp})^{\ddagger} \circ \eta_X$ *and* $(g^{\sharp})^{\ddagger} = \omega \circ Tg^{\dagger}$.

Excluding the "b-considerations", Theorem 3 is formulated and proved in dual form (i.e., for comonads and recursive coalgebras) in [5, Theorem 19]. Our assumptions related to the b-structure ensure that the proof carries over to the case of bcas.

Using Theorem 3, we obtain the bca that will give us the long-term value.

Corollary 1. *The* $H\Delta$*-algebra* $\alpha = \alpha_\gamma \circ (\mathbb{R} \times \mathrm{E})$ *is a bca in* (Met, B).

Proof. This result follows from Theorem 3. We check the conditions. First, by Lemma 3, $\alpha_\gamma \colon H\chi(\mathbb{R}, \mathrm{E}) \to (\mathbb{R}, \mathrm{E})$ is a $H\chi$-algebra in $\mathcal{EM}(\Delta)$. Next, by Proposition 6, the underlying H-algebra α_γ is a bca for H, and in Example 7 we saw that E preserves $\mathcal{B}$. Finally, we saw in Example 8 that Δ is a strong b-endofunctor, and χ and $H\mu$ are b-natural. By Theorem 3, we have a bca structure for $H\Delta$ on $\mathbb{R}$, namely $\alpha_\gamma \circ H\mathrm{E} = \alpha_\gamma \circ (\mathbb{R} \times \mathrm{E})$. $\qquad\square$

Using that α_γ is a bca for H (Proposition 6), we obtain from the universal property of α_γ a unique bounded map $\ell'_\sigma \colon \Delta S \to \mathbb{R}$ that makes the diagram below on the left commute. Also, we obtain a map $\mathrm{LTV}'_\sigma \colon S \to \mathbb{R}$ from the universal property of $\alpha_\gamma \circ (\mathbb{R} \times \mathrm{E})$ as a bca for $H\Delta$ (Corollary 1). That is, LTV'_σ is the unique bounded map that makes the diagram below on the right commute.

$$
\begin{array}{ccc}
\Delta S \xrightarrow{\;m_\sigma^{\sharp}\;} \mathbb{R} \times \Delta S & \qquad & S \xrightarrow{\;m_\sigma = \langle u, t_\sigma \rangle\;} \mathbb{R} \times \Delta S \\
{\scriptstyle \ell'_\sigma}\downarrow \qquad \qquad \downarrow{\scriptstyle \mathbb{R} \times \ell'_\sigma} & \qquad & {\scriptstyle \mathrm{LTV}'_\sigma}\downarrow \qquad \qquad \downarrow{\scriptstyle \mathbb{R} \times \Delta(\mathrm{LTV}'_\sigma)} \\
\mathbb{R} \xleftarrow{\;\alpha_\gamma\;} \mathbb{R} \times \mathbb{R} & \qquad & \mathbb{R} \xleftarrow{\;\alpha_\gamma \circ (\mathbb{R} \times \mathrm{E})\;} \mathbb{R} \times \Delta\mathbb{R}
\end{array}
$$

Moreover, by Theorem 3 (2) we have that

$$\mathrm{LTV}'_\sigma = \ell'_\sigma \circ \delta_S \quad \text{and} \quad \ell'_\sigma = \mathrm{E} \circ \Delta\,\mathrm{LTV}'_\sigma. \tag{16}$$

In particular, LTV'_σ is the unique fixpoint of the operator

$$\Phi_\sigma = \Phi_{m_\sigma, \alpha_\gamma \circ (\mathbb{R} \times \mathrm{E})} \colon B(S, \mathbb{R}) \to B(S, \mathbb{R}) \qquad \Phi_\sigma(f) = \alpha_\gamma \circ \langle u, \mathrm{E} \circ (\Delta f) \circ t_\sigma \rangle. \tag{17}$$

It is not hard to see that, as expected, $\Phi_\sigma = \Psi_\sigma$ from (4) in Sect. 2 (note that $B(S, \mathbb{R}) = \mathbb{R}^S$ because S is finite). Hence, by unicity, $\mathrm{LTV}'_\sigma = \mathrm{LTV}_\sigma$. By the definition of ℓ_σ (cf. (2)) and the right-hand side of (16), we also see that $\ell'_\sigma = \ell_\sigma$. Therefore, the equations in (16) express that

$$\mathrm{LTV}_\sigma = \ell_\sigma \circ \delta_S \quad \text{and} \quad \ell_\sigma = \mathrm{E} \circ \Delta\,\mathrm{LTV}_\sigma. \tag{18}$$

In this way, we re-obtained LTV_σ and ℓ_σ using our categorical perspective. Note however that thanks to Corollary 1, in our novel approach we did not need to show that Φ_σ is contractive in order to get LTV_σ.

5.4   The Optimal Value Function V^*

We recall from Sect. 2 that the optimal value function V^* is the unique solution to Bellman's optimality equation, which we restate here for convenience:

$$V^*(s) = u(s) + \gamma \cdot \max_{a \in Act} \left\{ \sum_{s' \in S} t_a(s)(s') \cdot V^*(s') \right\}.$$

To say that V^* solves this is to say that the diagram below commutes:

$$
\begin{array}{ccc}
S & \xrightarrow{\;\;m = \langle u,t \rangle\;\;} & \mathbb{R} \times (\Delta S)^{Act} \\[2pt]
\scriptstyle V^* \downarrow & & \downarrow \scriptstyle \mathbb{R} \times (\Delta V^*)^{Act} \\[2pt]
\mathbb{R} & \xleftarrow[\;\alpha_\gamma \circ (\mathbb{R} \times \max_{Act} \circ\, \mathrm{E}^{Act})\;]{} & \mathbb{R} \times (\Delta\mathbb{R})^{Act}
\end{array}
$$

This diagram clearly looks like a bca diagram, and it is therefore natural to ask whether we can prove the existence of V^* by generalizing the results for LTV_σ. It turns out that many, but not all, do generalize. We give a brief overview.

The coalgebraic modeling is straightforward. Recall that an MDP is a $K\Delta$-coalgebra, where K is the functor $K = H \circ (-)^{Act}$ where $(-)^{Act}$ is the covariant hom-functor. There is also a distributive law ρ of Δ over K. It uses strength $str\colon \Delta \circ (-)^{Act} \Rightarrow (-)^{Act} \circ \Delta$; specifically, we have $\rho = \langle \mathrm{E} \circ \Delta\pi_1, str \circ \Delta\pi_2 \rangle$.

We can lift K to Met by viewing X^{Act} as an Act-fold product, i.e., we use the maximum metric. The metric version of K has the same nice properties as the metric version of H. For example, if f is Lipschitz, then so is Kf (generalizing the second part of Lemma 4), and $K\colon (\mathsf{Met}, B) \to (\mathsf{Met}, B)$ is a b-endofunctor (generalizing Proposition 3).

Moreover, we can show that the K-algebra $\alpha_\gamma \circ H \max_{Act}\colon \mathbb{R} \times \mathbb{R}^{Act} \to \mathbb{R}$ is a bca for K (generalizing Proposition 6). Part of the verification shows that the map $\max_{Act}\colon \mathbb{R}^{Act} \to \mathbb{R}$ preserves B; this uses the simple fact that for all sets A and all $h_1, h_2 \in \mathbb{R}^A$, it holds that $|\max_A h_1 - \max_A h_2| \leq \max_A |h_1 - h_2| = d_{\mathbb{R}^A}(h_1, h_2)$.

Things only go sour when we try to apply Theorem 3 to get a bca for $K\Delta$ from the bca $\beta = \alpha_\gamma \circ H \max_{Act}$ for K. The problem is that in order to do so, we need to show that β is an arrow in $\mathcal{EM}(\Delta)$, which entails that $\max_{Act}$ is an arrow in $\mathcal{EM}(\Delta)$, and this is not the case since, unlike α_γ, $\max_{Act}$ is not affine as it does not commute with convex linear combinations. Nevertheless, $\alpha_\gamma \circ (\mathbb{R} \times (\max_{Act} \circ E^{Act}))$ *is* a bca for $K\Delta$, since this is equivalent to the statement that the Bellman operator Φ^* (as a map on $B(S, \mathbb{R})$) has a unique fixpoint. The difference with the situation for H and LTV_σ is that we cannot use Theorem 3 to relate the bca structure for $K\Delta$ to the bca structure for K.

6 Extensions

We briefly discuss some extensions to our work.

6.1 Alternative Treatment of MDPs

In our definition of MDPs, rewards are associated with states. However, often MDPs are presented with rewards associated with transitions, i.e., an MDP is then of type $n: S \to (\mathbb{R} \times \Delta S)^{Act}$. The latter is an $H^{Act}\Delta$-coalgebra, where $H^{Act} = (-)^{Act} \circ H$. For the general results, not much changes. We again have a distributive law of H^{Act} over Δ, given by $\langle E \circ \Delta\pi_1, \Delta\pi_2 \rangle^{Act} \circ str$, and a policy σ yields an $H\Delta$-coalgebra given by $n_\sigma = \langle u_\sigma, t_\sigma \rangle = n \circ \sigma$. (Compare: $m_\sigma = \langle u, t_\sigma \rangle$.) So we again obtain ℓ_σ and LTV_σ from Proposition 6 together with Corollary 1. Also, (18) holds, just as before. The contractive operator characterizing LTV_σ in n_σ is defined as

$$\Phi_\sigma(f) = \alpha_\gamma \circ \langle u_\sigma, E \circ \Delta f \circ t_\sigma \rangle.$$

We adapt the definition of improved policies to the current setting by letting

$$\tau(s) = \mathrm{argmax}_{a \in Act}\, \alpha_\gamma(\pi_1(n(s)(a)), \ell_\sigma(\pi_2(n(s)(a)))).$$

We can prove the Policy Improvement Theorem. We expect that with similar adaptations, the results of Sect. 5.4 also go through as before.

6.2 Changing the Setting to Polish Metric Spaces

The main setting of this paper was the *b*-category (Met, B) of all metric spaces, where the hom-sets are those of Set and the B-morphisms are the bounded maps. Since it would be more satisfying to have a more "metric" category, we want to sketch how this can be done.

The first way is to restrict the morphisms between metric spaces to be non-expansive maps. We call the resulting category Met₁. Taking B to be the bounded, non-expansive maps, (Met₁, B) is a *b*-category. For our work, the problem with Met₁ is that when products are given the maximum metric, α_γ is not a morphism in Met₁. Using the sum metric on products, we get a *b*-category, and even bca structures for H, $H\Delta$, K and $K\Delta$. However, other technical problems arise that suggest that this is not a worthwhile approach.

A more fruitful direction is to work with the category of *Polish metric spaces* (complete separable spaces) and continuous functions, called PolMet. Let us write $\widehat{P}$: Met $\to$ Met for the endofunctor which takes a space M to the metric space of all Borel probability measures on M, using (for concreteness) the Kantorovich metric, defined using integrals instead of sums. The resulting topology is the weak topology. Giry [11] proved that $\widehat{P}$ restricts to an endofunctor P: PolMet $\to$ PolMet. For every Polish space X, the set ΔX of discrete probability distributions is a dense subset of $\widehat{P}X$ (see [29, Theorem 6.18]). For every map $k\colon X \to Y$ in Met, $\widehat{P}k$ and Δk both work the same way, by "pushing forward" a distribution. The appropriate version of E is $E(\mu) = \int x\,d\mu$. All of the results in Sects. 4 and 5 adapt to this setting, mutatis mutandis. The upshot is that we get a b-category (PolMet, B), where B is the class of bounded continuous functions. Furthermore, the policy improvement theorem can be done in that setting.

7   Conclusion

Our main goal has been to show that the value functions LTV_σ and V^* arise from a universal property of sorts, and to re-prove the correctness of policy improvement using a coinductive argument. The universal property was explained in terms of bca-structures, and for this we needed the notion of a b-category. The main examples led us to study boundedness preservation properties of the liftings of the stream functor and the distribution monad to metric spaces.

The coinductive analysis of policy improvement went by means of a new contraction coinduction principle. In essence, contraction coinduction allows one to infer qualitative relationships (e.g., policy improvement) without a detour into quantitative results. We would like to think this principle has many other uses.

We have a few comments on earlier work in the same general area.

Kozen and Ruozzi [24] surely had the intuition that aspects of the theory of MDPs should be understood coinductively. Their paper has a very interesting coinductive proof of the fact that the optimal policies in MDPs may be taken to be deterministic. They were not concerned with policy improvement, our target for coinduction. As for ourselves, we formulated contraction coinduction; this is an easy consequence of the metric coinduction principle from [17,24], and it seems to do the work one would want for inequalities as one finds in policy improvement.

One should go back to Shapley games and other infinite games to see if the metric coinduction principle from [17,24] could simplify the (subtle) positive results in the area. Also, the metric coinduction principle was used by Abramsky and Winschel [1] to establish a predicate coinduction principle. They use that result in connection with subgame perfect equilibria in infinite games such as the dollar auction. Pavlovic [22] shares some programmatic features with our work, even though the formal work appears different. There are connections to be made with all of these papers.

Denardo [7] is concerned with some of the same issues that we address. In some ways his work is more abstract than ours, as he does not assume a particular system type, and in some ways less. His work does not use categorical notions, so it does not directly compare with our work, but assumptions pertaining to contraction mappings and to order-preservation are prominent in the paper. Our contraction coinduction principle simplifies several of the proofs in [7].

A related point: Denardo assumes that (his version of) Φ maps bounded functions to bounded functions. Our notions of b-functor and b-preservation give us a compositional account of this fact. This was put to use in Proposition 6. On the other hand, to show that Φ is a contraction, our general machinery was not useful. So there certainly is more work to be done on that.

This paper emphasizes compositional reasoning about functions and functors. The classical theory of MDPs does not do this; it directly proves properties (such as boundedness) of composites viewed as monolithic entities, instead of deriving them from preservation properties of their constituents. So it neither needs nor uses the extra information that we obtained by working in a categorical setting. Indeed, most of our paper is devoted to this extra information. We hope that our work will be useful in settings beyond MDPs. We have some pilot results in this direction, but for lack of space these do not appear in this paper.

Acknowledgments. We would like to thank Tarmo Uustalu for pointing us to [5, Theorem 19], thereby improving the paper. We also thank Jasmine Blanchette, Wan Fokkink and Ana Sokolova for useful comments.

References

1. Abramsky, S., Winschel, V.: Coalgebraic analysis of subgame-perfect equilibria in infinite games without discounting. Math. Struct. Comput. Sci. **27**(5), 751–761 (2017)
2. Baldan, P., Bonchi, F., Kerstan, H., König, B.: Behavioral metrics via functor lifting. In: 34th International Conference on Foundation of Software Technology and Theoretical Computer Science, FSTTCS 2014, pp. 403–415 (2014)
3. Bartels, F.: On Generalised Coinduction and Probabilistic Specification Formats. Ph.D. thesis, Vrije Universiteit Amsterdam (2004)
4. Bellman, R.: Dynamic Programming, 1st edn. Princeton University Press, Princeton (1957)
5. Capretta, V., Uustalu, T., Vene, V.: Recursive coalgebras from comonads. Inf. Comp. **204**, 437–468 (2006)
6. Capretta, V., Uustalu, T., Vene, V.: Corecursive algebras: a study of general structured corecursion. In: Oliveira, M.V.M., Woodcock, J. (eds.) SBMF 2009. LNCS, vol. 5902, pp. 84–100. Springer, Heidelberg (2009). https://doi.org/10.1007/978-3-642-10452-7_7
7. Denardo, E.V.: Contraction mappings in the theory underlying dynamic programming. SIAM Rev. **9**(2), 165–177 (1967)
8. Desharnais, J., Edalat, A., Panangaden, P.: Bisimulation for labelled markov processes. Inf. Comput. **179**(2), 163–193 (2002)
9. Ferns, N., Panangaden, P., Precup, D.: Metrics for finite markov decision processes. In: Proceedings of the 20th Conference on Uncertainty in Artificial Intelligence, UAI 2004, pp. 162–169. AUAI Press, Arlington (2004). http://dl.acm.org/citation.cfm?id=1036843.1036863

10. Gibbs, A.L., Su, F.E.: On choosing and bounding probability metrics. Int. Stat. Rev./Revue Internationale de Statistique **70**(3), 419–435 (2002)
11. Giry, M.: A categorical approach to probability theory. In: Banaschewski, B. (ed.) Categorical Aspects of Topology and Analysis. LNM, vol. 915, pp. 68–85. Springer, Heidelberg (1982). https://doi.org/10.1007/BFb0092872
12. Howard, R.A.: Dynamic Programming and Markov Processes. The M.I.T. Press, Cambridge, MA (1960)
13. Jacobs, B.: Distributive laws for the coinductive solution of recursive equations. Inf. Comput. **204**(4), 561–587 (2006)
14. Jacobs, B., Silva, A., Sokolova, A.: Trace semantics via determinization. J. Comput. Syst. Sci. **81**(5), 859–879 (2015). 11th International Workshop on Coalgebraic Methods in Computer Science, CMCS 2012 (Selected Papers)
15. Johnstone, P.T.: Adjoint lifting theorems for categories of algebras. Bull. Lond. Math. Soc. **7**, 294–297 (1975)
16. Klin, B.: Bialgebras for structural operational semantics: an introduction. Theor. Comput. Sci. **412**(38), 5043–5069 (2011)
17. Kozen, D.: Coinductive proof principles for stochastic processes. Log. Methods Comput. Sci. **5**, 1–19 (2009)
18. Mac Lane, S.: Categories for the Working Mathematician. GTM, vol. 5. Springer, New York (1978). https://doi.org/10.1007/978-1-4757-4721-8
19. Milius, S.: Completely iterative algebras and completely iterative monads. Inf. Comput. **196**(1), 1–41 (2005)
20. Moore, A.W.: Markov Systems, Markov Decision Processes, and Dynamic Programming (2002). lecture slides available at https://www.autonlab.org/tutorials
21. Ó'Searcóid, M.: Metric Spaces. Springer, London (2006)
22. Pavlovic, D.: A semantical approach to equilibria and rationality. In: Kurz, A., Lenisa, M., Tarlecki, A. (eds.) CALCO 2009. LNCS, vol. 5728, pp. 317–334. Springer, Heidelberg (2009). https://doi.org/10.1007/978-3-642-03741-2_22
23. Puterman, M.L.: Markov Decision Processes: Discrete Stochastic Dynamic Programming. Wiley, Hoboken (2014)
24. Ruozzi, N., Kozen, D.: Applications of metric coinduction. Logical Methods in Computer Science, 5 (2009)
25. Rutten, J.: Universal coalgebra: a theory of systems. Theor. Comput. Sci. **249**(1), 3–80 (2000)
26. Silva, A., Bonchi, F., Bonsangue, M., Rutten, J.: Generalizing determinization from automata to coalgebras. Log. Methods Comput. Sci. **9**, 1–27 (2013)
27. Silva, A., Sokolova, A.: Sound and complete axiomatization of trace semantics for probabilistic systems. Electron. Notes Theor. Comput. Sci. **276**, 291–311 (2011). https://doi.org/10.1016/j.entcs.2011.09.027
28. Sokolova, A.: Probabilistic systems coalgebraically. Theor. Comput. Sci. **412**(38), 5095–5110 (2011). https://doi.org/10.1016/j.tcs.2011.05.008
29. Villani, C.: Optimal Transport, Grundlehren der Mathematischen Wissenschaften [Fundamental Principles of Mathematical Sciences], vol. 338. Springer, Heidelberg (2009)

(In)finite Trace Equivalence
of Probabilistic Transition Systems

Alexandre Goy[1,2] and Jurriaan Rot[2(✉)]

[1] MICS - CentraleSupélec, Gif-sur-Yvette, France
alexandre.goy@student.ecp.fr
[2] Radboud University, Nijmegen, The Netherlands
j.rot@cs.ru.nl

Abstract. We show how finite and infinite trace semantics of generative probabilistic transition systems arises through a determinisation construction. This enables the use of bisimulations (up-to) to prove equivalence. In particular, it follows that trace equivalence for finite probabilistic transition systems is decidable. Further, the determinisation construction applies to both discrete and continuous probabilistic systems.

1 Introduction

The theory of coalgebras encompasses a wide variety of probabilistic systems, and according notions of bisimulation and behavioural equivalence [18]. We focus on one of the most basic instances: (generative) probabilistic transition systems (PTS), consisting of a set of states X and for every state a probability distribution over next states and (explicit) termination. Formally, they are coalgebras of the form $\alpha \colon X \to \mathcal{D}(A \times X + 1)$, where A is a fixed set of transition labels, $\mathcal{D}$ the probability distribution functor and $1 = \{*\}$ a singleton, whose element we interpret as an extra 'accepting/termination' state (Sect. 3 for details).

There is a natural notion of finite trace semantics for such PTSs, assigning to every state a sub-probability distribution of words, as a quantitative analogue of acceptance of words in non-deterministic automata. The definition of *infinite* traces is more subtle: it requires assigning probability to *sets* of traces rather than individual traces (infinite traces often have probability zero), and to move to probability *measures*. It is shown in [13] how finite and infinite trace semantics arises by modelling PTSs as coalgebras in the Kleisli category of the Giry monad.

As such, the (in)finite traces semantics of PTSs is an instance of the general theory of trace semantics through Kleisli categories, as proposed in [10]. A fundamentally different way of obtaining trace semantics of coalgebras is through

The research leading to these results has received funding from the European Research Council (FP7/2007–2013, grant agreement nr. 320571), and the Erasmus+ program. It was carried out during the first author's master internship at Radboud University, reported in [9].

C. Cîrstea (Ed.): CMCS 2018, LNCS 11202, pp. 100–121, 2018.
https://doi.org/10.1007/978-3-030-00389-0_7

determinisation constructions, generalising the classical powerset construction of non-deterministic automata [12,16] but also encompassing many other examples. In particular, in [12,17] it is described how the *finite* traces of probabilistic transition systems arise through a certain determinisation construction, turning a PTS into a Moore automaton with sub-probability distributions as states. One of the advantages of determinisation is that it allows to use bisimulations (up-to) to prove trace equivalence. In particular, bisimulations up to congruence were used in Bonchi and Pous' HKC algorithm for non-deterministic automata [4] and in its extension to weighted automata [2].

In this paper, we show that the *finite and infinite* trace semantics of PTS, as in [13], arises through a determinisation construction (Sect. 4). The essential underlying idea that enables this approach, is that the (in)finite traces semantics in [13] is generated basically from two kinds of finite trace semantics: those that take into account termination/acceptance (as mentioned above), and those that do not (simply the probability of exhibiting a path in the PTS). In particular, for finite PTS, our determinisation construction yields an effective procedure for proving (in)finite trace equivalence using bisimulation up to congruence, using a variation of the HKC algorithm (Sect. 5). We finally show that the determinisation construction generalises to the setting of continuous PTS, working in the category of measurable spaces and with $\mathcal{D}$ replaced by the Giry monad (Sect. 6). While this generalises the discrete case, it is presented separately to make the discrete case accessible to a wider audience: the latter requires very little measure theory. We conclude with a discussion of related work (Sect. 7).

2 Preliminaries

Any finite set A can be called an *alphabet* and its elements *letters*. The set of words of length n with letters in A is denoted by A^n. By convention $A^0 = \{\varepsilon\}$ where ε is the empty word. The set of finite words over A is denoted by $A^* = \bigcup_{n \in \mathbb{N}} A^n$, the set of infinite words by $A^\omega = A^{\mathbb{N}}$ and the set of all (finite and infinite) words by $A^\infty = A^* \cup A^\omega$. A language L is a subset of $\mathcal{P}(A^*)$. It can be seen as a function $L \colon A^* \to \{0, 1\}$, by setting $L(w) = 1$ iff $w \in L$. The *language derivative* of L with respect to a letter a is defined by $L_a(w) = L(aw)$. The length of $w \in A^\infty$ is denoted by $|w| \in \mathbb{N} \cup \{\infty\}$. The concatenation function $c \colon A^* \times A^\infty \to A^\infty$ is denoted by juxtaposition $(c(u, v) = uv)$ and defined by $uv(n) = u(n)$ if $n < |u|$ and $uv(n) = v(n - |u|)$ if $|u| \le n < |u| + |v|$. It can be extended to languages $\mathcal{P}(A^*) \times \mathcal{P}(A^\infty) \to \mathcal{P}(A^\infty)$ by setting $LM = \{uv \mid u \in L, v \in M\}$. We sometimes abbreviate $\{w\}M$ by wM.

Coalgebras and Moore Automata. We recall the basic definition of coalgebras, see, e.g., [11,15] for details and examples. The only instances that we use in this paper are Moore automata (recalled below), probabilistic transition systems (Sect. 3) and measure-theoretic generalisations of both (Sect. 6). Let $\mathcal{C}$ be a category, and $F \colon \mathcal{C} \to \mathcal{C}$ a functor. An F-coalgebra consists of an object X and an arrow $\alpha \colon X \to FX$. Given coalgebras (X, α) and (Y, β), a coalgebra homomorphism is an arrow $f\colon X \to Y$ such that $\beta \circ f = Ff \circ \alpha$. Coalgebras and homomorphisms form a category $\mathsf{CoAlg}(F)$. A final object in $\mathsf{CoAlg}(F)$ is called *final coalgebra*; explicitly, a coalgebra (Ω, ω) is final if for every F-coalgebra (X, α) there is unique coalgebra homomorphism $\varphi\colon X \to \Omega$. We recall the notion of *bisimulation* only for Moore automata, below.

Let B be a set. Define the *machine functor* $F_B\colon \mathsf{Set} \to \mathsf{Set}$ by $F_B X = B \times X^A$ and $Ff = id_B \times f^A$. An F_B-coalgebra $\langle o, t \rangle\colon X \to B \times X^A$ is called a *Moore automaton* (with output in B). A relation $R \subseteq X \times X$ on the states of a Moore automaton $\langle o, t \rangle\colon X \to B \times X^A$ is a *bisimulation* if for all $(x, y) \in R$: $o(x) = o(y)$ and for all $a \in A$, $(t_a(x), t_a(y)) \in R$ (here, we used the classical notation $t_a(x)$ instead of writing $t(x)(a)$). We write $x \sim y$ if there exists a bisimulation R such that xRy, and in this case say that x and y are *bisimilar*. For every B, there exists a final F_B-coalgebra (Ω, ω) where $\Omega = B^{A^*}$. For an F_B-coalgebra (X, α), we write $\varphi_\alpha\colon X \to B^{A^*}$ or simply φ for the unique coalgebra morphism. We think of the elements of B^{A^*} as (weighted) languages, and of $\varphi(x)$ as the language of a state x. In particular, for $B = 2$, Moore automata are classical deterministic automata, and φ gives the usual language semantics. We have $\varphi(x) = \varphi(y)$ iff $x \sim y$, i.e., language equivalence coincides with bisimilarity.

Measure Theory. Let X be a set. A σ-*algebra* on X is a subset $\Sigma_X \subseteq \mathcal{P}(X)$ such that $\emptyset \in \Sigma_X$ and Σ_X is closed under complementation and countable union. Note that this implies that $X \in \Sigma_X$ and that Σ_X is closed under countable intersection and set difference. Given any subset $G \subseteq \mathcal{P}(X)$, there always exists a smallest σ-algebra containing G. Indeed, $\mathcal{P}(X)$ is a σ-algebra containing G, and the intersection of an arbitrary non-empty set of σ-algebras is itself a σ-algebra: just take the intersection of all σ-algebras containing G. We call it the σ-algebra *generated* by G and denote it by $\sigma_X(G)$. For example, $\mathcal{P}(X)$ is a σ-algebra on X. When working with real numbers $\mathbb{R}$, we will use the Borel σ-algebra $\mathcal{B}(\mathbb{R}) = \sigma_\mathbb{R}(\{(-\infty, x] \mid x \in \mathbb{R}\})$. We use $\mathcal{B}([0,1]) = \{B \cap [0,1] \mid B \in \mathcal{B}(\mathbb{R})\}$ as the canonical σ-algebra on $[0,1]$. If X is a set and Σ_X is a σ-algebra on X, the pair (X, Σ_X) is called a *measurable space*. We write X for (X, Σ_X) when the σ-algebra used is clear. A function $f\colon (X, \Sigma_X) \to (Y, \Sigma_Y)$ is *measurable* if for all $S_Y \in \Sigma_Y$, $f^{-1}(S_Y) \in \Sigma_X$. The composition of measurable functions is measurable. An (implicitly finite) *measure* is a map $m\colon \Sigma_X \to \mathbb{R}_+$ such that $m(\emptyset) = 0$ and $m\left(\bigcup_{n \in \mathbb{N}} A_n\right) = \sum_{n \in \mathbb{N}} m(A_n)$ if the union is disjoint (σ-additivity property). We write $\mathcal{M}(X)$ for the set of measures on a measurable space (X, Σ_X).

(Sub)distribution. The *distribution* functor $\mathcal{D}\colon \mathsf{Set} \to \mathsf{Set}$ is defined by $\mathcal{D}(X) = \{p\colon X \to [0,1] \mid \sum_{x \in X} p(x) = 1\}$ and, given a function $f\colon X \to Y$, $\mathcal{D}f(u)(y) = \sum_{x \in f^{-1}(\{y\})} u(x)$. The functor $\mathcal{D}$ extends to a monad, with the unit η given by the Kronecker delta $\eta_X(x) = \delta_x$ (i.e., $\eta(x)(y) = 1$ if $x = y$ and $\eta(x)(y) = 0$ otherwise), and the multiplication by $\mu_X(U)(y) = \sum_{u \in \mathcal{D}X} U(u) \cdot u(y)$. The *sub-distribution* functor $\mathcal{S}\colon \mathsf{Set} \to \mathsf{Set}$ is defined by $\mathcal{S}(X) = \{p\colon X \to [0,1] \mid \sum_{x \in X} p(x) \leq 1\}$. It extends to a monad in a similar way. There is a natural embedding of $\mathcal{D}$ in $\mathcal{S}$, which we denote by $\iota\colon \mathcal{D} \Rightarrow \mathcal{S}$.

3   Trace Semantics of Probabilistic Transition Systems

In this section, we recall PTSs and their (finite and infinite) trace semantics, following [13]. We start with the finite trace semantics.

Definition 3.1. *A* probabilistic transition system (PTS) *is a coalgebra for the functor* $\mathcal{D}(A \times \mathrm{Id} + 1)$, *i.e., a set* X *together with a map* $\alpha \colon X \to \mathcal{D}(A \times X + 1)$.

Definition 3.2. *Let* $\alpha \colon X \to \mathcal{D}(A \times X + 1)$ *be a PTS. The* finite trace semantics $[\![-]\!]_f \colon X \to \mathcal{S}(A^*)$ *is defined by the following equations.*

$$[\![x]\!]_f(\varepsilon) = \alpha(x)(*) \qquad [\![x]\!]_f(aw) = \sum_{y \in X} \alpha(x)(a,y) \cdot [\![y]\!]_f(w)$$

for all $x \in X$, $a \in A$, $w \in A^*$.

Consider as a first example the simple PTS below, where the element $*$ is represented as a distinguished double-circled state, and a transition is represented by an arrow labeled with its probability.

$$(1)$$

We have $[\![x]\!]_f(a^n) = \frac{1}{2^{n+1}}$ for all n. The trace semantics becomes more subtle if *infinite* words are also taken into account. Consider, for instance, the following PTS.

$$(2)$$

Intuitively both states accept any finite or infinite word w over $\{a,b\}$ with probability 0. However, the probability of 'starting with an a' in y or z is clearly different. This becomes apparent when we move to assigning probability to *sets* of traces, which is where we need a bit of measure theory. We therefore first define a suitable σ-algebra on the set A^∞ of finite and infinite words.

Definition 3.3. *Let* $S_\infty = \{\emptyset\} \cup \{\{w\} \mid w \in A^*\} \cup \{wA^\infty \mid w \in A^*\}$. *The* σ-algebra *of measurable sets of words is defined to be* $\Sigma_{A^\infty} = \sigma_{A^\infty}(S_\infty)$.

This σ-algebra is generated by a countable family of generators: the empty set, the singletons of finite words, and the cones, i.e., sets wA^∞ of words that have the finite word w as a prefix. This σ-algebra is very natural. Indeed, the usual measure-theoretical σ-algebra on $A^* \cup A^\omega$ would be the combination of the discrete σ-algebra $\mathcal{P}(A^*)$ and the product σ-algebra (see, e.g., [1], Definition 4.42) of all $\mathcal{P}(A)$ on A^ω. One can easily prove that this construction yields our Σ_{A^∞} too. In the sequel, this is the σ-algebra on A^∞ implicitly used. The following proposition establishes measurability for some useful sets.

Proposition 3.4. *The following sets of words are measurable:*

(i) The singleton $\{w\}$ for any $w \in A^\infty$;
(ii) any countable language;
(iii) any language of finite words;
(iv) $\emptyset$, A^*, A^ω, A^∞;
(v) the concatenation LS where $L \subseteq A^*$ and $S \in \Sigma_{A^\infty}$.

In the following, if m is a measure over A^∞ and $w \in A^\infty$, we will write $m(w)$ instead of $m(\{w\})$. We have the following key theorem, which follows easily from results in [13]:

Theorem 3.5. *Let* $m \colon S_\infty \to \mathbb{R}_+$ *be a map satisfying* $m(\emptyset) = 0$. *The two following conditions are equivalent.*

(i) *There exists a unique measure* $\tilde{m} \colon \Sigma_{A^\infty} \to \mathbb{R}_+$ *such that* $\tilde{m}_{|S_\infty} = m$.
(ii) *For all* $w \in A^*$, $m(wA^\infty) = m(w) + \sum_{a \in A} m(waA^\infty)$.

Proof. $(i) \Rightarrow (ii)$ The equation comes directly from the σ-additivity of $\tilde{m}$. $(ii) \Rightarrow$ (i) According to Lemma 3.18 of [13], (ii) is equivalent to the fact that m is a premeasure. Using Caratheodory's extension theorem (e.g., [14]), this pre-measure can be uniquely extended to a measure as in (i). $\square$

Recall that $\mathcal{M}(A^\infty)$ denotes the set of measures m on A^∞.

Definition 3.6. *Let* $\alpha \colon X \to \mathcal{D}(A \times X + 1)$ *be a PTS. The (finite and infinite) trace semantics* $[\![-]\!] \colon X \to \mathcal{M}(A^\infty)$ *is defined by the following equations.*

$$[\![x]\!](\varepsilon A^\infty) = 1 \qquad\qquad [\![x]\!](\varepsilon) = \alpha(x)(*)$$

$$[\![x]\!](awA^\infty) = \sum_{y \in X} \alpha(x)(a, y) \cdot [\![y]\!](wA^\infty) \quad [\![x]\!](aw) = \sum_{y \in X} \alpha(x)(a, y) \cdot [\![y]\!](w)$$

(These equations uniquely determine a measure by Theorem 3.5.)

Example 3.7. Consider the following PTS over the alphabet $A = \{a\}$.

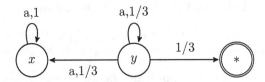

The semantics $[\![x]\!]$ is easy to compute for sets of words in S_∞ by induction: for every finite word w, $[\![x]\!](w) = 0$ and $[\![x]\!](wA^\infty) = 1$. Because $a^n A^\infty$ is a non-increasing sequence of measurable sets converging to $\{a^\omega\}$, properties of measures yield $[\![x]\!](a^\omega) = \lim_{n \to +\infty} [\![x]\!](a^n A^\infty) = 1$. Let us look at $[\![y]\!]$. Intuitively, the probability of performing n loops in state y and then moving to (and staying in) state x is $1/3^{n+1}$. Summing them for $n \in \mathbb{N} \cup \{0\}$ gives $1/2$, the probability of moving eventually to state x. Indeed, first observe that $[\![y]\!](\varepsilon) = 1/3$ and $[\![y]\!](\varepsilon A^\infty) = 1$. Let $n \in \mathbb{N} \cup \{0\}$, then:

$$[\![y]\!](a^{n+1}) = \frac{1}{3}[\![y]\!](a^n) + \frac{1}{3}[\![x]\!](a^n) = \frac{1}{3}[\![y]\!](a^n)$$

$$[\![y]\!](a^{n+1}A^\infty) = \frac{1}{3}[\![y]\!](a^n A^\infty) + \frac{1}{3}[\![x]\!](a^n A^\infty) = \frac{1}{3}[\![y]\!](a^n A^\infty) + \frac{1}{3}$$

One can then prove that $[\![y]\!](a^n) = 1/3^{n+1}$ and $[\![y]\!](a^\omega) = \lim_{n \to +\infty}[\![y]\!](a^n A^\infty) = \lim_{n \to +\infty}(1 + 3^{-n})/2 = 1/2$.

Example 3.8. Consider again the PTS in (2). We have $[\![y]\!](w) = [\![z]\!](w) = 0$ for all $w \in A^*$. However, $[\![y]\!](aA^\infty) = [\![y]\!](bA^\infty) = 1/2$ whereas $[\![z]\!](aA^\infty) = 3/4$ and $[\![z]\!](bA^\infty) = 1/4$. Hence $[\![y]\!] \neq [\![z]\!]$, as expected.

The above (in)finite trace semantics is essentially generated from two kinds of finite trace semantics: one for finite words w, and one for cones wA^∞, where $w \in A^*$. The probability of the latter is simply the probability the finite trace w without considering acceptance/termination, i.e., the probability of exhibiting the path w. This finite presentation is exploited in the determinisation construction in the next section, which essentially encodes both kinds of finite trace semantics simultaneously.

4   Determinisation

In this section, we show how the finite and infinite trace semantics of PTS (Definition 3.6) arises through a determinisation construction. This construction transforms any PTS into a certain kind of Moore machine with sub-probability distributions as states. The final coalgebra semantics of this Moore machine represents the trace semantics $[\![-]\!] \colon X \to \mathcal{M}(A^\infty)$ of the original PTS. The determinisation procedure is exploited in the next section to give an algorithm for computing (in)finite trace equivalence, based on bisimulations.

In Sect. 6, we consider a more general kind of PTS, with measurable sets as state spaces, which fully generalises the results and constructions of the current section. Most proofs in the current section are hence omitted. Moreover, it is explained in Sect. 6 that our approach is an instance of the abstract framework of coalgebraic determinisation based on distributive laws [12,16]. In the current section we mostly neglect this and present the concrete constructions.

Throughout this section, let $\alpha \colon X \to \mathcal{D}(A \times X + 1)$ be a PTS. Our approach to (in)finite traces resembles the determinisation construction of [12,17] for finite traces of PTSs. As explained below, there is one crucial addition for (in)finite traces: we make the total weight of sub-distributions in the determinised coalgebra observable, essentially to capture the probability of the 'cones' wA^∞. We will show that the resulting final coalgebra semantics factorises through the set $\mathcal{M}(A^\infty)$ of measures on words, recovering the trace semantics of Definition 3.6. The overall construction is as follows.

(i) Translate α into a coalgebra $\tilde{\alpha}\colon X \to [0,1] \times [0,1] \times (\mathcal{S}X)^A$.
(ii) Determinise it: define an $\alpha^{\#}\colon \mathcal{S}X \to [0,1] \times [0,1] \times (\mathcal{S}X)^A$ such that $\alpha^{\#} \circ \eta_X = \tilde{\alpha}$. Let $\varphi\colon \mathcal{S}X \to ([0,1] \times [0,1])^{A^*}$ be the unique map to the final coalgebra.
(iii) Factorise φ to get a coalgebra morphism $\mathcal{S}X \to \mathcal{M}(A^\infty)$, then precompose with η_X to get the desired trace semantics $X \to \mathcal{M}(A^\infty)$.

The construction is summed up in the following diagram. Below, we explain each of the steps in detail.

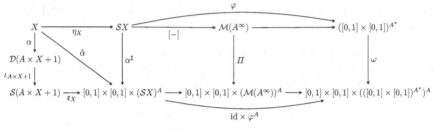

Remark 4.1. As mentioned above, the construction is quite close to the determinisation of finite traces [12,17]. There are two main differences: first, the latter determinises to a Moore automaton of the type $\mathcal{S}X \to [0,1] \times (\mathcal{S}X)^A$ (so with $[0,1]$ rather than $[0,1] \times [0,1]$). Second, the decomposition of φ here yields measures (to represent (in)finite trace semantics) rather than sub-probability distributions (to represent finite trace semantics).

(i) Translation: from α to $\tilde{\alpha}$. The first step, the definition of $\tilde{\alpha}$ from α, basically forgets certain information about probability distributions. The natural transformation $\mathfrak{e}$ is given on a component X by $\mathfrak{e}_X\colon \mathcal{S}(A \times X + 1) \to [0,1] \times [0,1] \times (\mathcal{S}X)^A$,

$$\mathfrak{e}_X(u) = \left\langle \sum_{z \in A \times X + 1} u(z), u(*), a \mapsto [y \mapsto u(a,y)] \right\rangle .$$

We have $\tilde{\alpha}(x) = \langle 1, \alpha(x)(*), a \mapsto [y \mapsto \alpha(x)(a,y)] \rangle$.

(ii) Determinisation. In the second step, we turn $\tilde{\alpha}$ into a Moore automaton over sub-distributions. Formally, the latter will be a coalgebra for the functor $F_{[0,1] \times [0,1]}\colon \mathbf{Set} \to \mathbf{Set}$; recall from Sect. 2 that this is defined by $F_{[0,1] \times [0,1]}X = [0,1] \times [0,1] \times X^A$. In the remainder of this section we abbreviate $F_{[0,1] \times [0,1]}$ by F. Notice that $\tilde{\alpha}$ is an FS-coalgebra. Any FS-coalgebra determinises to an F-coalgebra, but we spell it out here only for the necessary instance $\tilde{\alpha}$. For a concrete example, see the first part of Example 5.9 in the next section.

Definition 4.2. *The determinisation of the PTS $\alpha\colon X \to \mathcal{D}(A \times X + 1)$ is the Moore machine $\alpha^{\#}\colon \mathcal{S}X \to F\mathcal{S}X = [0,1] \times [0,1] \times (\mathcal{S}X)^A$, defined by:*

$$\alpha^{\#}(u) = \left\langle \sum_{x \in X} u(x), \sum_{x \in X} u(x) \cdot \alpha(x)(*), a \mapsto [y \mapsto \sum_{x \in X} u(x) \cdot \alpha(x)(a,y)] \right\rangle .$$

(iii) Factorisation of final coalgebra semantics. Since $\alpha^\sharp$ is an F-coalgebra, there exists a unique coalgebra morphism φ from $(SX, \alpha^\sharp)$ to the final coalgebra $(([0, 1] \times [0, 1])^{A^*}, \omega)$. This is not quite the right type: the (in)finite trace semantics $[\![-]\!]$ is a (probability) measure over words, hence, for each x, $[\![x]\!]$ should be an element of the set $\mathcal{M}(A^\infty)$ of measures over A^∞ (Sect. 2). In the last step (iii), we equip $\mathcal{M}(A^\infty)$ with an F-coalgebra structure Π which is final among F-coalgebras satisfying a certain property, satisfied by our determinisation $\alpha^\sharp$. This allows us to factor φ through a coalgebra homomorphism $[-]: S(X) \to \mathcal{M}(A^\infty)$. In the more general setting of Sect. 6 we show how the coalgebra structure on $\mathcal{M}(A^\infty)$ arises from the Giry monad and the final coalgebra of the Set endofunctor $X \mapsto A \times X + 1$. For now, we define it explicitly, which requires:

Definition 4.3 (Measure derivative). *Let m be a measure on A^∞ and $a \in A$. The map m_a defined by $m_a(S) = m(aS)$ for any $S \in \Sigma_{A^\infty}$ is a measure, called the* measure derivative *of m (with respect to a).*

It is easy to check that m_a as defined above is indeed a measure, so that the measure derivative is well-defined. Now, the coalgebra $\Pi: \mathcal{M}(A^\infty) \to [0, 1] \times [0, 1] \times (\mathcal{M}(A^\infty))^A$ is defined by $\Pi: m \mapsto \langle m(\varepsilon A^\infty), m(\varepsilon), a \mapsto m_a \rangle$. Since Π is an F-coalgebra, we obtain a coalgebra morphism to the final F-coalgebra ω.

Lemma 4.4. *The unique coalgebra morphism from Π to ω is injective.*

A proof is given in the more general setting of Lemma 6.8. The following crucial lemma states in which cases the factorisation is possible. It establishes the F-coalgebra Π as a final object in a certain subcategory of $\mathrm{CoAlg}(F)$.

Proposition 4.5. *Let $\beta = \langle \beta_\oplus, \beta_*, a \mapsto \tau_a \rangle: Y \to FY$ be an F-coalgebra. The two following conditions are equivalent:*

(i) There exists an F-coalgebra morphism $[-]$ from β to Π.
(ii) The equation $\beta_\oplus = \beta_* + \sum_{a \in A} \beta_\oplus \circ \tau_a$ holds.

In this case, this morphism is unique.

See Theorem 6.9 for a proof in the (more general) continuous setting.

Lemma 4.6. *The coalgebra $\alpha^\sharp: SX \to [0, 1] \times [0, 1] \times (SX)^A$ satisfies (ii) in Proposition 4.5.*

It is important to note that condition (ii) does not hold in general if the whole construction starts from a coalgebra of the form $\alpha: X \to S(A \times X + 1)$. The price to be paid for a PTS to be compatible enough to generate infinite trace semantics from the finite traces in a measure-theoretic way is to sum to 1, i.e., to use $\mathcal{D}$ and not $\mathcal{S}$.

The following result summarises the situation.

Corollary 4.7. *The morphism φ decomposes as a unique coalgebra morphism $[\![-]\!]\colon \mathcal{S}X \to \mathcal{M}(A^\infty)$ from $\alpha^\sharp$ to Π followed by an injective coalgebra morphism φ_Π from Π to ω, as shown in the following diagram.*

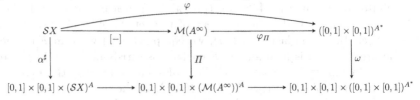

We thus obtain the semantics $[\![-]\!] \circ \eta_X \colon X \to \mathcal{M}(A^\infty)$ by precomposing with the unit of the monad $\mathcal{S}$. It coincides with the semantics $[\![-]\!]$ of Definition 3.6:

Theorem 4.8. *We have $[\![-]\!] = [\![-]\!] \circ \eta_X$.*

Theorem 4.8 is the main result of this section, stating that the (in)finite trace semantics is recovered by finality through a determinisation construction. Together with Lemma 4.4, it yields equivalence between the first two points below.

Corollary 4.9. *For any $x, y \in X$, the following are equivalent:*

1. *$[\![x]\!] = [\![y]\!]$,*
2. *$\varphi(\delta_x) = \varphi(\delta_y)$,*
3. *$\delta_x \sim \delta_y$,*

where $\sim$ is bisimilarity on the Moore automaton $\alpha^\sharp$, the determinisation of α (Definition 4.2).

The equivalence between 2. and 3. is standard, and was mentioned in Sect. 2. By the equivalence between 1. and 3., we can prove (in)finite trace equivalence by computing bisimulations, which is used in the next section.

5   Computing Trace Equivalence

The aim of this section is to give an algorithm that takes states $x, y \in X$ of a PTS and tells whether x and y are (in)finite trace equivalent (i.e., $[\![x]\!] = [\![y]\!]$) or not, based on the determinisation construction described in Sect. 4. Our algorithm is a variant of HKC, an algorithm for language equivalence of non-deterministic automata based on determinisation and bisimulation (up-to) techniques [4]. More specifically, we will use its generalisation to weighted automata given in [2].

Let $\alpha\colon X \to \mathcal{D}(A \times X + 1)$ be a *finite-state* PTS and $\alpha^\sharp$ its determinisation (Definition 4.2). By Corollary 4.9, to prove $[\![x]\!] = [\![y]\!]$ it suffices to show $\delta_x \sim \delta_y$, i.e., that there is a bisimulation $R \subseteq \mathcal{S}X \times \mathcal{S}X$ on the determinised Moore automaton such that $(\delta_x, \delta_y) \in R$. However, this task can be simplified using *bisimulation up-to techniques*, as explained next. In order to use the techniques from [2], we first move from sub-probability distributions to vector spaces. To this end, define $\mathbb{R}_\omega^X$ as the set of finitely supported functions $X \to \mathbb{R}$, i.e., $\mathbb{R}_\omega^X = \{u \colon X \to \mathbb{R} \mid u(x) \neq 0 \text{ for finitely many } x\}$. We define the Moore automaton $\overline{\alpha} = \langle \overline{\alpha}_\oplus, \overline{\alpha}_*, a \mapsto \overline{\alpha}_a \rangle \colon \mathbb{R}_\omega^X \to \mathbb{R} \times \mathbb{R} \times (\mathbb{R}_\omega^X)^A$ as follows on any $u \in \mathbb{R}_\omega^X$:

$$\overline{\alpha}_\oplus = \sum_{x \in X} u(x) \quad \overline{\alpha}_* = \sum_{x \in X} u(x) \cdot \alpha(x)(*) \quad \overline{\alpha}_a = \left[y \mapsto \sum_{x \in X} u(x) \cdot \alpha(x)(a, y) \right]$$
(3)

This is almost the same construction as in Definition 4.2, with sub-probability distributions replaced by vectors. (Note that this is well-defined since X is assumed to be finite; it would also suffice to assume that α is finitely branching.) It is easy to see that the embedding $i \colon SX \to \mathbb{R}_\omega^X$ is an injective $F_{[0,1] \times [0,1]}$-coalgebra morphism from $\alpha^\sharp$ to $\overline{\alpha}$. Together with Corollary 4.9, this yields:

Corollary 5.1. *For any $x, y \in X$: $[\![x]\!] = [\![y]\!]$ iff $\delta_x \sim \delta_y$, where $\sim$ is bisimilarity on the Moore automaton $\overline{\alpha}$.*

We now formulate bisimulation up to congruence, concretely for $\overline{\alpha}$.

Definition 5.2. *Let $R \subseteq \mathbb{R}_\omega^X \times \mathbb{R}_\omega^X$. Its congruence closure $c(R)$ is the least congruence that contains R, i.e., that satisfies*

$$\frac{(u, v) \in R}{(u, v) \in c(R)} \qquad \frac{}{(u, u) \in c(R)} \qquad \frac{(u, v) \in c(R)}{(v, u) \in c(R)} \qquad \frac{(u, v) \in c(R) \quad (v, w) \in c(R)}{(u, w) \in c(R)}$$

$$\frac{(r \cdot u, r \cdot v) \in c(R)}{(u, v) \in c(R)}(r \in \mathbb{R}) \qquad \frac{(u, u') \in c(R) \quad (v, v') \in c(R)}{(u + u', v + v') \in c(R)}$$

Definition 5.3. *Define $\overline{\alpha} \colon \mathbb{R}_\omega^X \to \mathbb{R} \times \mathbb{R} \times (\mathbb{R}_\omega^X)^A$ from a finite-state PTS α, as in Eq. (3). A relation $R \subseteq \mathbb{R}_\omega^X \times \mathbb{R}_\omega^X$ is a bisimulation up to congruence (on $\overline{\alpha}$) if for all $(u, v) \in R$:*

- *$\overline{\alpha}_\oplus(u) = \overline{\alpha}_\oplus(v)$, $\overline{\alpha}_*(u) = \overline{\alpha}_*(v)$, and*
- *$\forall a \in A$: $(\overline{\alpha}_a(u), \overline{\alpha}_a(v)) \in c(R)$.*

The following result states soundness of bisimulations up to congruence. This can either be proved from the abstract coalgebraic theory [3] or more directly using compatible functions, as in [2,4].

Theorem 5.4. *For any $u, v \in \mathbb{R}_\omega^X$: $u \sim v$ iff there is a bisimulation up to congruence R (on $\overline{\alpha}$) such that $(u, v) \in R$.*

Combined with Corollary 5.1, this means that to prove that $[\![x]\!] = [\![y]\!]$ for states x, y of a PTS, it suffices to show that there is a bisimulation up to congruence relating δ_x and δ_y. The following algorithm attempts to compute one given x, y.

$$\text{HKC}^\infty(x, y)$$

(1) $R := \emptyset$; $todo := \emptyset$
(2) insert (δ_x, δ_y) into $todo$
(3) while $todo$ is not empty do
    (3.1) extract (u, v) from $todo$
    (3.2) if $(u, v) \in c(R)$ then continue
    (3.3) if $\overline{\alpha}_\oplus(u) \neq \overline{\alpha}_\oplus(v)$ then return *false*
    (3.3') if $\overline{\alpha}_*(u) \neq \overline{\alpha}_*(v)$ then return *false*
    (3.4) for all $a \in A$, insert $(\overline{\alpha}_a(u), \overline{\alpha}_a(v))$ into $todo$
    (3.5) insert (u, v) into R
(4) return *true*

Theorem 5.5. *Whenever* $\text{HKC}^\infty(x, y)$ *terminates, it returns* true *iff* $[\![x]\!] = [\![y]\!]$.

Despite the fact that during the determinisation the state space always becomes infinite, the following results show that if the initial state space X is finite, then HKC^∞ does terminate.

Theorem 5.6 (see [6]). *Let* $\mathcal{R}$ *be a ring and* X *be a finite set. Let* $R \subseteq \mathcal{R}^X \times \mathcal{R}^X$ *be a relation and let* $(v, v') \in \mathcal{R}^X \times \mathcal{R}^X$ *be a pair of vectors. Let* $U_R = \{u - u' \mid (u, u') \in R\}$. *Then* $(v, v') \in c(R)$ *iff* $v - v' \in [U_R]$, *where* $[U_R]$ *is the submodule of* $\mathcal{R}^X$ *generated by* U_R.

Proposition 5.7. *If* X *is finite,* $\text{HKC}^\infty(x, y)$ *terminates for every* $x, y \in X$.

Example 5.8. To begin with, here is a very simple PTS which we use to demonstrate the need for bisimulation up to congruence over plain bisimulations.

A bisimulation on the determinised automaton containing (δ_x, δ_y) would require adding $(\delta_x/2^k, \delta_y/2^k)$ for all k to the relation. However, $\text{HKC}^\infty(x, y)$ (which computes a bisimulation up to congruence) stops after one step because it spots that $(\delta_x/2, \delta_y/2)$ is in the congruence closure of the relation $\{(\delta_x, \delta_y)\}$.

Example 5.9. Consider the PTS depicted on the left below. We will use HKC^∞ to check if the states x and z are (in)finite trace equivalent.

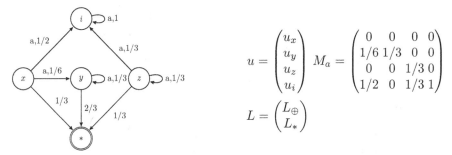

First, we compute part of the determinised automaton. To this end, observe that because X is finite, $\mathbb{R}_\omega^X = \mathbb{R}^X$ has a basis (e_x, e_y, e_z, e_i). An element $u \in \mathbb{R}_\omega^X$ is seen as a column vector $u_x e_x + u_y e_y + u_z e_z + u_i e_i$ in this basis. Moreover $\overline{\alpha}_\oplus$ and $\overline{\alpha}_*$ are linear forms that can be seen as the row vectors $L_\oplus = (1\,1\,1\,1)$ and $L_* = (1/3\,2/3\,1/3\,0)$, and $\overline{\alpha}_a$ is an endomorphism with a transition matrix M_a defined by $(M_a)_{j,k} = t_a(k)(j)$. This is depicted on the right above.

We represent here two parts of the determinised automaton. The first is the path beginning with the single state x; the second is the path beginning with the single state z. Each state here has two real outputs, obtained by matrix multiplication with $L_\oplus$ and L_*.

$$\begin{pmatrix}1\\0\\0\\0\end{pmatrix} \xrightarrow{a} \begin{pmatrix}0\\1/6\\0\\1/2\end{pmatrix} \xrightarrow{a} \begin{pmatrix}0\\1/18\\0\\1/2\end{pmatrix} \xrightarrow{a} \cdots \xrightarrow{a} \begin{pmatrix}0\\1/(2\times 3^n)\\0\\1/2\end{pmatrix} \xrightarrow{a} \cdots$$

$$\vdots \qquad \vdots \qquad \vdots \qquad \vdots$$

$$\begin{pmatrix}0\\0\\1\\0\end{pmatrix} \xrightarrow{a} \begin{pmatrix}0\\0\\1/3\\1/3\end{pmatrix} \xrightarrow{a} \begin{pmatrix}0\\0\\1/9\\4/9\end{pmatrix} \xrightarrow{a} \cdots \xrightarrow{a} \begin{pmatrix}0\\0\\1/3^n\\(1-3^{-n})/2\end{pmatrix} \xrightarrow{a} \cdots$$

Now, $\mathtt{HKC}^\infty(x, z)$ begins with $\mathtt{todo} = \{(\eta_X(x), \eta_X(z))\} = \{(e_x, e_z)\}$ and $\mathtt{R} = \emptyset$. It checks that $Le_x = Le_z$, etc. as shown in the following table. The check succeeds in loop 3 because $(u, v) \in c(R)$ according to Theorem 5.6:

$$\begin{pmatrix}0\\1/18\\0\\1/2\end{pmatrix} - \begin{pmatrix}0\\0\\1/9\\4/9\end{pmatrix} = \begin{pmatrix}0\\1/18\\-1/9\\1/18\end{pmatrix} = \frac{1}{3}\begin{pmatrix}0\\1/6\\-1/3\\1/6\end{pmatrix} = \frac{1}{3}\left(\begin{pmatrix}0\\1/6\\0\\1/2\end{pmatrix} - \begin{pmatrix}0\\0\\1/3\\1/3\end{pmatrix}\right)$$

Because $\mathtt{todo}$ is eventually empty, the algorithm returns $\mathtt{true}$. Indeed, if we compute directly the measures $[\![x]\!]$ and $[\![z]\!]$, we can see that $[\![x]\!](a^n) = 1/3^{n+1}$, $[\![x]\!](a^\omega) = 1/2$ and similarly for $[\![z]\!]$. Here the bisimulation up to congruence check is necessary for termination. The construction of a bisimulation up to equivalence (dashed + dotted lines on the determinised automaton picture) would take an infinite number of steps. But the construction of the bisimulation up to congruence (dashed lines) takes only 2 steps.

Step	(3.1)	(3.2)	(3.3)	(3.4)	(3.5)
Loop counter	(u,v) extracted from todo	Check $(u,v) \in c(R)$	Check $Lu = Lv$	$(M_a u, M_a v)$ added to todo	Cardinality of R
1	$(\begin{pmatrix}1\\0\\0\\0\end{pmatrix}, \begin{pmatrix}0\\0\\1\\0\end{pmatrix})$	Fail	$\begin{pmatrix}1\\1/3\end{pmatrix} = \begin{pmatrix}1\\1/3\end{pmatrix}$	$(\begin{pmatrix}0\\1/6\\0\\1/2\end{pmatrix}, \begin{pmatrix}0\\0\\1/3\\1/3\end{pmatrix})$	1
2	$(\begin{pmatrix}0\\1/6\\0\\1/2\end{pmatrix}, \begin{pmatrix}0\\0\\1/3\\1/3\end{pmatrix})$	Fail	$\begin{pmatrix}2/3\\1/9\end{pmatrix} = \begin{pmatrix}2/3\\1/9\end{pmatrix}$	$(\begin{pmatrix}0\\1/18\\0\\1/2\end{pmatrix}, \begin{pmatrix}0\\0\\1/9\\4/9\end{pmatrix})$	2
3	$(\begin{pmatrix}0\\1/18\\0\\1/2\end{pmatrix}, \begin{pmatrix}0\\0\\1/9\\4/9\end{pmatrix})$	Success	/	/	2
4	Empty	/	/	/	/

6　Continuous Systems

In this section, we generalise the determinisation construction for (in)finite trace semantics previously defined to the case of continuous PTS, defined later as coalgebras for the analogue of functor $\mathcal{D}(A \times - +1)$ in the category Meas (see [13] for examples of such PTSs). The underlying distributive law is brought to light, so that the origin of the determinisation process is better understood. The following table sums up the analogies and differences with the discrete case.

	Discrete case	General case
Category	Set	Meas
Usual operation	$\sum$	$\int$
Machine functor	$FX = [0,1] \times [0,1] \times X^A$	Measurable version of F
Probability monad	Distribution monad $\mathcal{D}$	Giry's monad $\mathbb{D}$
Determinisation monad	Sub-distribution monad $\mathcal{S}$	Sub-Giry's monad $\mathbb{S}$
PTS state space	Set X	Measurable space (X, Σ_X)
Determinised state	Finitely supported vector	Measure (≤ 1)
Transitions	Matrix $t_a \colon X \times X \to [0,1]$	Kernel $t_a \colon X \times \Sigma_X \to [0,1]$
Final F-coalgebra	ω	Measurable version of ω
Measure coalgebra	Π	Measurable version of Π
Pseudo-final morphism	$[-] \colon \mathcal{S}X \to \mathcal{M}(A^\infty)$	$[-] \colon \mathbb{S}X \to \mathbb{S}A^\infty$

In this section we work in the category Meas of measurable spaces and functions. It is easy to adapt F, but considering the monads we will need some additional measure-theoretic background.

Product. Given measurable spaces (X, Σ_X) and (Y, Σ_Y), we define a product σ-algebra on $X \times Y$ by $\Sigma_X \otimes \Sigma_Y = \sigma_{X \times Y}(\{S_X \times S_Y \mid S_X \in \Sigma_X, S_Y \in \Sigma_Y\})$. The product of measurable spaces is then defined by $(X, \Sigma_X) \otimes (Y, \Sigma_Y) = (X \times Y, \Sigma_X \otimes \Sigma_Y)$.

Sum. Given measurable spaces (X, Σ_X) and (Y, Σ_Y), we define a sum σ-algebra on the disjoint union $X + Y = \{(x, 0) \mid x \in X\} \cup \{(y, 1) \mid y \in Y\}$ by $\Sigma_X \oplus \Sigma_Y = \{S_X + S_Y \mid S_X \in \Sigma_X, S_Y \in \Sigma_Y\}$. The sum of measurable spaces is then defined by $(X, \Sigma_X) \oplus (Y, \Sigma_Y) = (X + Y, \Sigma_X \oplus \Sigma_Y)$.

Given measurable spaces X, Y and a measurable function $f \colon X \to Y$, define a new functor $\mathfrak{L} \colon \mathsf{Meas} \to \mathsf{Meas}$ by $\mathfrak{L}X = A \times X + 1$ along with its canonical σ-algebra $\Sigma_{\mathfrak{L}X} = \mathcal{P}(A) \otimes \Sigma_X \oplus \mathcal{P}(1)$, and $\mathfrak{L}f = id_A \times f + id_1$. Moreover, define $FX = [0,1] \times [0,1] \times X^A$ along with its σ-algebra $\mathcal{B}([0,1]) \otimes \mathcal{B}([0,1]) \otimes \bigotimes_{a \in A} \Sigma_X$ and $Ff = id_{[0,1]} \times id_{[0,1]} \times f^A$.

Integration. Let (X, Σ_X, m) be a measure space and $f \colon X \to \mathbb{R}$ be a measurable function. If $f(X) = \{\alpha_1, \ldots, \alpha_n\}$ for some $\alpha_1, \ldots, \alpha_n \in \mathbb{R}_+$, then f is called a *simple* function and its integral can be set as $\int_X f dm = \sum_{i=1}^n \alpha_i m(f^{-1}(\{\alpha_i\}))$. If f is non-negative, define $\int_X f dm = \sup \{\int_X g dm \mid g \leq f, g \text{ simple}\} \in [0, \infty]$. Finally, for any $f \colon X \to \mathbb{R}$, decompose $f = f^+ - f^-$ where $f^+ \geq 0$ and $f^- \geq 0$. If their integrals are not both ∞, define $\int_X f dm = \int_X f^+ dm - \int_X f^- dm$. If this is finite, we say that f is m-integrable. Furthermore, for any $S \in \Sigma_X$, the indicator function $\mathbf{1}_S$ is measurable and we define $\int_S f dm = \int_X \mathbf{1}_S f dm$.

Given a measurable function $g \colon X \to Y$ and measure $m \colon \Sigma_X \to \mathbb{R}_+$, the *pushforward measure* of m by g is $m \circ g^{-1}$. For any measurable $f \colon Y \to \mathbb{R}$, f is $m \circ g^{-1}$-integrable iff $f \circ g$ is m-integrable and in this case, $\int_Y f d(m \circ g^{-1}) = \int_X (f \circ g) dm$. Each positive measurable function $X \to \mathbb{R}_+$ is the pointwise limit of an increasing sequence of simple functions. To prove some property for every positive measurable function, one can prove it for simple functions (or for indicator functions, if it is preserved by linear combinations) and show it is preserved by limits. Many such proofs use the monotone convergence theorem (see [14]), which states that if $(f_n)_{n \in \mathbb{N}}$ is an increasing sequence of positive functions with pointwise limit f, then f is measurable and $\int_X f dm = \lim \int_X f_n dm$.

The Giry Monad. The Giry monad [8] provides a link between probability theory and category theory. In Meas, the Giry monad $(\mathbb{D}, \eta, \mu)$ is defined as follows. For any measurable space X, $\mathbb{D}X$ is the space of probability measures over (X, Σ_X), and $\Sigma_{\mathbb{D}X}$ is the σ-algebra generated by the functions $e_S^X \colon \mathbb{D}X \to [0,1]$ defined by $e_S^X(m) = m(S)$. For any measurable function $g \colon X \to Y$, $(\mathbb{D}g)(m) = m \circ g^{-1}$. The unit is defined by $\eta_X(x)(S) = \mathbf{1}_S(x)$ and the multiplication by $\mu_X(\Phi)(S) = \int_{\mathbb{D}X} e_S^X d\Phi$. Similarly, one defines the sub-Giry monad $(\mathbb{S}, \eta, \mu)$, with the only difference that $\mathbb{S}X$ is the space of sub-probability measures over (X, Σ_X). There is a natural embedding of $\mathbb{D}$ in $\mathbb{S}$, denoted by $\iota \colon \mathbb{D} \Rightarrow \mathbb{S}$.

6.1   Trace Semantics via Determinisation

The aim of this section is to define trace semantics for continuous PTS, i.e., coalgebras of the form $\alpha\colon X \to \mathbb{D}(A \times X + 1)$ where X is a measurable space. We proceed in the same way as for discrete systems.

(i) Transform α into a more convenient coalgebra $\tilde{\alpha}\colon X \to F\mathbb{S}X$.
(ii) Determinise $\tilde{\alpha}$ into an F-coalgebra $\alpha^{\sharp}\colon \mathbb{S}X \to F\mathbb{S}X$.
(iii) Factorise the final morphism : $\varphi_{\alpha^{\sharp}} = \varphi_{\Pi} \circ [-]$, then precompose with η_X.

The following diagram sums up the construction. Here $\Sigma_{([0,1]\times[0,1])^{A^*}}$ is the Σ-algebra generated by the functions $L \mapsto L(w)$.

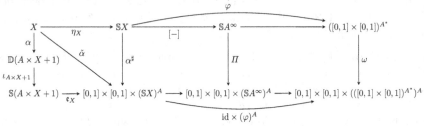

(i) Translation: from α to $\tilde{\alpha}$

Proposition 6.1. *For any measurable space X, the function $\mathfrak{e}_X\colon \mathbb{S}\mathfrak{L}X \to F\mathbb{S}X$ defined by $\mathfrak{e}_X(m) = \langle m(\mathfrak{L}X), m(1), a \mapsto [S \mapsto m(\{a\} \times S)]\rangle$ is measurable. Moreover, $\mathfrak{e}\colon \mathbb{S}\mathfrak{L} \Rightarrow F\mathbb{S}$ is a natural transformation.*

Now take $\tilde{\alpha} = \mathfrak{e}_X \circ \iota_{\mathfrak{L}X} \circ \alpha\colon X \to [0,1] \times [0,1] \times (\mathbb{S}X)^A$. Explicitly:

$$\tilde{\alpha}(x) = \langle \underbrace{\alpha(x)(\mathfrak{L}X)}_{1}, \alpha(x)(1), a \mapsto [S \mapsto \alpha(x)(\{a\} \times S)]\rangle$$

We decompose it as a pairing $\tilde{\alpha} = \langle \tilde{\alpha}_\oplus, \tilde{\alpha}_*, a \mapsto t_a\rangle$.

(ii) Determinisation. We recall some basic observations in abstract determinisation [12,16]. By *distributive law* here we mean the standard notion of distributive law of monad over functor (called EM-law in [12]).

Lemma 6.2. *Let C be a category, $F\colon C \to C$ be an endofunctor and (T, η, μ) be a monad on C. Let $f\colon X \to TFX$ be a TF-coalgebra and $h\colon TFTX \to FTX$ be an Eilenberg-Moore T-algebra. Then there exists a unique T-algebra morphism $f^{\sharp}\colon (TX, \mu_X) \to (FTX, h)$ such that $f = f^{\sharp} \circ \eta_X$.*

Lemma 6.3. *With the same notations as for Lemma 6.2, and given a distributive law $\lambda\colon TF \Rightarrow FT$, then $h = F\mu_X \circ \lambda_{TX}\colon TFTX \to FTX$ is an Eilenberg-Moore T-algebra.*

The next step is to define a distributive law $\lambda\colon \mathbb{S}F \Rightarrow F\mathbb{S}$ in order to apply Lemmas 6.2 and 6.3. In the following we write $id_{FX} = \langle \pi_X^\oplus, \pi_X^*, a \mapsto \pi_X^a \rangle$. Note that $\pi^\epsilon\colon F \Rightarrow [0,1]$ (for $\epsilon \in \{*, \oplus\}$) and $\pi^a\colon F \Rightarrow Id_{\mathbf{C}}$ (for $a \in A$) are natural transformations.

Lemma 6.4. *Let $g\colon \mathbb{S}([0,1]) \to [0,1]$ be defined by $g(m) = \int_{[0,1]} id_{[0,1]} dm$. Then g is measurable and an Eilenberg-Moore $\mathbb{S}$-algebra.*

For any object X of Meas, define $\lambda_X\colon \mathbb{S}FX \to F\mathbb{S}X$ by

$$\lambda_X = \langle g \circ \mathbb{S}\pi_X^\oplus, g \circ \mathbb{S}\pi_X^*, a \mapsto \mathbb{S}\pi_X^a \rangle$$

This is a measurable function because each component is measurable.

Proposition 6.5. $\lambda\colon \mathbb{S}F \Rightarrow F\mathbb{S}$ *is a distributive law.*

Let us compute the value of our resulting determinisation. Given $\tilde{\alpha}\colon X \to F\mathbb{S}X$, take $h = F\mu_X \circ \lambda_{\mathbb{S}X}$ (Lemma 6.3) and $\alpha^\sharp = h \circ \mathbb{S}\tilde{\alpha}$ (Lemma 6.2). We get

$$
\begin{aligned}
\alpha^\sharp &= h \circ \mathbb{S}\tilde{\alpha} \\
&= F\mu_X \circ \lambda_{\mathbb{S}X} \circ \mathbb{S}\tilde{\alpha} \\
&= F\mu_X \circ \langle g \circ \mathbb{S}(\pi_{\mathbb{S}X}^\oplus \circ \tilde{\alpha}), g \circ \mathbb{S}(\pi_{\mathbb{S}X}^* \circ \tilde{\alpha}), a \mapsto \mathbb{S}(\pi_{\mathbb{S}X}^a \circ \tilde{\alpha}) \rangle \\
&= \langle g \circ \mathbb{S}\tilde{\alpha}_\oplus, g \circ \mathbb{S}\tilde{\alpha}_*, a \mapsto \mu_X \circ \mathbb{S}t_a \rangle
\end{aligned}
$$

Let $m \in \mathbb{S}X$. This more explicit expression shows that the coalgebra that arises from the determinisation is natural in the sense that the components of $\alpha^\sharp$ are basically obtained by integrating the information provided by α.

$$
\begin{aligned}
\alpha^\sharp(m) &= \left\langle \int_X \tilde{\alpha}_\oplus dm, \int_X \tilde{\alpha}_* dm, a \mapsto \left[S \mapsto \int_X t_a(-)(S) dm \right] \right\rangle \\
&= \left\langle \int_X \alpha(-)(\mathcal{L}X) dm, \int_X \alpha(-)(1) dm, a \mapsto \left[S \mapsto \int_X \alpha(-)(\{a\} \times S) dm \right] \right\rangle
\end{aligned}
$$

(iii) Final coalgebra. This heavy determinisation part gives us an F-coalgebra $\alpha^\sharp$. There exists a final object in $\mathrm{CoAlg}(F)$:

Proposition 6.6. *Let $\Omega = ([0,1] \times [0,1])^{A^*}$ and Σ_Ω be the smallest σ-algebra that makes the functions $e_w\colon \Omega \to [0,1] \times [0,1]$ defined by $e_w(L) = L(w)$ measurable for every $w \in A^*$. Let $\omega\colon \Omega \to F\Omega$ be defined by $\omega(L) = \langle L(\varepsilon), a \mapsto L_a \rangle$. Then (Ω, ω) is a final F-coalgebra.*

Thus for any F-coalgebra β the final morphism towards ω, denoted φ_β, gives a canonical notion of semantics. What we want is something slightly more specific that takes into account the way $\alpha^\sharp$ was built to produce a probability measure in $\mathbb{S}A^\infty$. This is obtained via the coalgebra $\Pi\colon \mathbb{S}A^\infty \to F\mathbb{S}A^\infty$, built as follows.

Proposition 6.7. *Let $\pi\colon A^\infty \to \mathcal{L}A^\infty$ be defined by $\pi(\varepsilon) = *$ and $\pi(aw) = (a, w)$. This is a final $\mathcal{L}$-coalgebra.*

Let $\Pi = \mathfrak{e}_{A^\infty} \circ \mathbb{S}\pi$. One can check that with this definition, $\Pi : \mathbb{S}A^\infty \to F\mathbb{S}A^\infty$ has the same expression as the $\Pi : \mathcal{M}(A^\infty) \to F\mathcal{M}(A^\infty)$ of Sect. 4:

$$\Pi(m) = \langle m(\pi^{-1}(\mathcal{L}A^\infty)), m(\pi^{-1}(1)), a \mapsto [S \mapsto m(\pi^{-1}(\{a\} \times S))]\rangle$$
$$= \langle m(A^\infty), m(\varepsilon), a \mapsto m_a \rangle$$

The aim is now to factorise the semantics obtained via ω into semantics obtained via Π. The following result is a kind of completeness property for this operation.

Lemma 6.8. *The final morphism φ_Π from Π to ω is injective.*

Proof. For any $m, m' \in \mathbb{S}A^\infty$, in order to have $m = m'$, it is sufficient that $m_{|S_\infty} = m'_{|S_\infty}$ (see Theorem 3.5). By induction on w, we prove that for $m, m' \in \mathbb{S}A^\infty$ such that $\varphi_\Pi(m) = \varphi_\Pi(m')$, then $\langle m(wA^\infty), m(w)\rangle = \langle m'(wA^\infty), m'(w)\rangle$. First, $\langle m(\varepsilon A^\infty), m(\varepsilon)\rangle = \varphi_\Pi(m)(\varepsilon) = \varphi_\Pi(m')(\varepsilon) = \langle m'(\varepsilon A^\infty), m'(\varepsilon)\rangle$. Note that $\varphi_\Pi(m) = \varphi_\Pi(m')$ implies $\varphi_\Pi(m_a)(w) = \varphi_\Pi(m)(aw) = \varphi_\Pi(m')(aw) = \varphi_\Pi(m'_a)(w)$ so that $\varphi_\Pi(m_a) = \varphi_\Pi(m'_a)$. Use the induction hypothesis to obtain that $\langle m(awA^\infty), m(aw)\rangle = \langle m_a(wA^\infty), m_a(w)\rangle = \langle m'_a(wA^\infty), m'_a(w)\rangle = \langle m'(awA^\infty), m'(aw)\rangle$. This achieves the induction, so m and m' coincide on S_∞, hence $m = m'$. $\qquad\square$

The following proposition states precisely in which cases the factorisation can be done. This is a variant of Theorem 3.5 in which we really see that our system is making one step. This version is stronger than Proposition 4.5, because it also proves that the involved functions are measurable.

Theorem 6.9. *Let $\beta = \langle\beta_\oplus, \beta_*, a \mapsto \tau_a\rangle : Y \to FY$ be an F-coalgebra. The two following conditions are equivalent:*

(i) *There exists an F-coalgebra morphism $[-]$ from β to Π.*
(ii) *The equation $\beta_\oplus = \beta_* + \sum_{a \in A} \beta_\oplus \circ \tau_a$ holds.*

In this case, this morphism is unique.

For convenience we denote $e_S^{A^\infty} \circ [-]$ by $[-](S)$, and $\phi_a \circ [-]$ by $[-]_a$, where the measure derivative function $\phi_a : m \mapsto m_a$ is measurable as a component of Π.

Proof. $(i) \Rightarrow (ii)$. Suppose $[-]$ is a coalgebra morphism from β to Π. Commutation of the diagram yields $\langle\beta_\oplus, \beta_*, a \mapsto [-] \circ \tau_a\rangle = \langle[-](A^\infty), [-](\varepsilon), a \mapsto [-]_a\rangle$. Let $y \in Y$. Because $[y]$ is a measure, $\beta_\oplus(y) = [y](\varepsilon A^\infty) = [y](\varepsilon) + \sum_{a \in A}[y](aA^\infty)$. Thus $\beta_\oplus(y) = \beta_*(y) + \sum_{a \in A}[\tau_a(y)](A^\infty) = \beta_*(y) + \sum_{a \in A}(\beta_\oplus \circ \tau_a)(y)$.

Uniqueness. If $[-]'$ is another such morphism, we have $[-](A^\infty) = [-]'(A^\infty)$, $[-](\varepsilon) = [-]'(\varepsilon)$ and for any $a \in A$, $[-] \circ \tau_a = [-]_a$ and $[-]' \circ \tau_a = [-]'_a$. An immediate induction yields $[-]_{|S_\infty} = [-]'_{|S_\infty}$, thus $[-] = [-]'$ by Theorem 3.5.

$(ii) \Rightarrow (i)$ Assume that (ii) holds. Let us define $[-]$ on S_∞ by induction:

$$[y]_{|S_\infty}(\varepsilon A^\infty) = \beta_\oplus(y) \qquad\qquad [y]_{|S_\infty}(\varepsilon) = \beta_*(y)$$
$$[y]_{|S_\infty}(awA^\infty) = [\tau_a(y)]_{|S_\infty}(wA^\infty) \qquad [y]_{|S_\infty}(aw) = [\tau_a(y)]_{|S_\infty}(w)$$

We must prove that it can be extended to a measure, using Theorem 3.5. First, note that $[y]_{|S_\infty}(\varepsilon A^\infty) = \beta_\oplus(y) = \beta_*(y) + \sum_{a \in A}(\beta_\oplus \circ \tau_a)(y) = [y]_{|S_\infty}(\varepsilon) + \sum_{a \in A}[y]_{|S_\infty}(aA^\infty)$. If it is known that for all $y \in Y$, $[y]_{|S_\infty}(wA^\infty) = [y]_{|S_\infty}(w) + \sum_{a \in A}[y]_{|S_\infty}(waA^\infty)$ then for any $b \in A$ we obtain the equation $[y]_{|S_\infty}(bwA^\infty) = [\tau_b(y)]_{|S_\infty}(wA^\infty) = [\tau_b(y)]_{|S_\infty}(w) + \sum_{a \in A}[\tau_b(y)]_{|S_\infty}(waA^\infty) = [y]_{|S_\infty}(bw) + \sum_{a \in A}[y]_{|S_\infty}(bwaA^\infty)$. This proves the (ii) of Theorem 3.5. We denote by $[-]$ the extension of $[-]_{|S_\infty}$. We postpone the proof of the measurability of $[-]$; what is left is the commutation of the coalgebra diagram. The first line of the definition of $[-]_{|S_\infty}$ gives directly that $\beta_\oplus = [-](A^\infty)$ and $\beta_* = [-](\varepsilon)$. Let $a \in A$. For any $y \in Y$, according to the second line of the definition of $[-]_{|S_\infty}$, the measures $[\tau_a(y)]$ and $[y]_a$ coincide on S_∞, hence are equal according to Theorem 3.5, so $[-] \circ \tau_a = [-]_a$. This achieves the proof that the diagram commutes.

Measurability. It is not immediate to see why $[-] \colon Y \to \mathbb{S}A^\infty$ is a measurable function. What has to be shown is that for any $S \in \Sigma_{A^\infty}$, $[-](S)$ is measurable. This is true when $S \in S_\infty$. Indeed, $[-](\emptyset)$ is the zero function, which is measurable. For the rest we proceed by induction. Obviously $[-](\varepsilon A^\infty) = \beta_\oplus$ and $[-](\varepsilon) = \beta_*$ are measurable because β is. Furthermore, $[-](awA^\infty) = [-]_a(wA^\infty) = [-](wA^\infty) \circ \tau_a$ and $[-](aw) = [-]_a(w) = [-](w) \circ \tau_a$ are measurable by induction hypothesis and composition.

We need to introduce a widely known theorem of measure theory, namely the $\pi - \lambda$ theorem (see [1], Lemma 4.11). Let Z be a set. A set $P \subseteq \mathcal{P}(Z)$ is a π-*system* if it is non-empty and closed under finite intersections. A set $D \subseteq \mathcal{P}(Z)$ is a λ-*system* if it contains Z and is closed under difference (if $A, B \in D$ and $A \subseteq B$ then $B \setminus A \in D$) and countable *increasing* union. The $\pi - \lambda$ theorem states that given P a π-system, D a λ-system such that $P \subseteq D$, then $\sigma_Z(P) \subseteq D$.

Take $Z = A^\infty$, $P = S_\infty$ and $D = \{S \in \Sigma_{A^\infty} \mid [-](S) \text{ is measurable}\}$. It is easy to see that S_∞ is a π-system. Moreover, D is a λ-system. Indeed, $A^\infty \in D$ (see above), if $(S_n)_{n \in \mathbb{N}}$ is an increasing sequence of sets in D, then $[-](S_1 \setminus S_0) = [-](S_1) - [-](S_0)$ is measurable as a difference of measurable functions and $[-]\left(\bigcup_{n \in \mathbb{N}} S_n\right) = \lim_{n \to \infty}[-](S_n)$ is measurable as a pointwise limit of measurable functions. Finally, given the preceding paragraph, we have $S_\infty \subseteq D$. The $\pi - \lambda$ theorem therefore yields $\Sigma_{A^\infty} \subseteq D$. Thus $[-]$ is measurable. $\square$

An interpretation of the last proposition is that, in the subcategory of all F-coalgebras that satisfy the equation (ii), the final object is Π. If Theorem 6.9 holds, then note that $\varphi_\Pi \circ [-]$ is a coalgebra morphism from β into the final coalgebra ω. Hence by finality $\varphi_\Pi \circ [-] = \varphi_\beta$. Along with Lemma 6.8, this yields the following proposition, which is exactly the same as in Sect. 2.

Proposition 6.10. *Let $\beta \colon Y \to FY$ be an F-coalgebra for which Theorem 6.9 holds. Then for any $y, z \in Y$, $[y] = [z]$ iff $\varphi_\beta(y) = \varphi_\beta(z)$.*

Back to $\alpha \colon X \to \mathbb{D}\mathfrak{L}X$ we check that Theorem 6.9 holds for $\alpha^\sharp = \langle \alpha^\sharp_\oplus, \alpha^\sharp_*, a \mapsto \tau_a \rangle$. Note that because $\alpha(-)(\mathfrak{L}X) = 1$, we have for all $m \in \mathbb{S}X$ that $m(X) =$

$\int_X 1dm = \int_X \alpha(-)(\mathcal{L}X)dm = \alpha^\sharp_\oplus(m)$. This justifies the last equality:

$$\alpha^\sharp_\oplus(m) = \int_X \alpha(-)(\mathcal{L}X)dm = \int_X \left(\alpha(-)(1) + \sum_{a \in A} \alpha(-)(\{a\} \times X) \right) dm$$

$$= \int_X \alpha(-)(1)dm + \sum_{a \in A} \int_X \alpha(-)(\{a\} \times X)dm$$

$$= \alpha^\sharp_*(m) + \sum_{a \in A} \tau_a(m)(X) = \alpha^\sharp_*(m) + \sum_{a \in A} (\alpha^\sharp_\oplus \circ \tau_a)(m)$$

Conclusion. Any $\alpha\colon X \to \mathbb{D}\mathcal{L}X$ can be given a canonical trace semantics via a determinisation process. This is a function $[-]\colon \mathbb{S}X \to \mathbb{S}A^\infty$.

6.2   Correctness of the Resulting Trace Semantics

In [13], given a PTS $\alpha\colon X \to \mathbb{D}\mathcal{L}X$, the trace semantics $[\![-]\!]\colon X \to \mathbb{D}A^\infty$ (denoted by **tr** in [13]) is defined by

$$[\![x]\!](\varepsilon A^\infty) = \alpha(x)(\mathcal{L}X) \quad (= 1) \qquad\qquad [\![x]\!](\varepsilon) = \alpha(x)(1)$$

$$[\![x]\!](awA^\infty) = \int_X [\![-]\!](wA^\infty)dt_a(x) \qquad [\![x]\!](aw) = \int_X [\![-]\!](w)dt_a(x)$$

We will hereby prove that this semantics fits with ours, in the sense that the following diagram commutes.

$$\begin{array}{ccc}
X & \xrightarrow{\;[\![-]\!]\;} & \mathbb{D}A^\infty \\
{\scriptstyle \eta_X}\downarrow & & \downarrow{\scriptstyle \iota_{A^\infty}} \\
\mathbb{S}X & \xrightarrow[\;[-]\;]{} & \mathbb{S}A^\infty
\end{array} \qquad (4)$$

Lemma 6.11. *For any $m \in \mathbb{S}X$ and any $S \in S_\infty$, $[m](S) = \int_X([-]\circ\eta_X)(S)dm$.*

Proof. In this proof, $\int_X fdm$ may be denoted by $\int_{x \in X} f(x)m(dx)$. One can show using the monotone convergence theorem that for any measurable function $f\colon X \to [0,1]$,

$$\int_X fd\tau_a(m) = \int_{x \in X} \left(\int_X fdt_a(x) \right) m(dx)$$

Note further that $[\eta_X(x)](\varepsilon A^\infty) = (\alpha^\sharp_\oplus \circ \eta_X)(x) = \tilde{\alpha}_\oplus(x) = \alpha(x)(\mathcal{L}X)$ and in the same way $[\eta_X(x)](\varepsilon) = \alpha(x)(1)$. Now let us prove the lemma by induction, for all $m \in \mathbb{S}X$. First

$$[m](\varepsilon A^\infty) = \alpha^\sharp_\oplus(m) = \int_X \alpha(-)(\mathcal{L}X)dm = \int_X ([-] \circ \eta_X)(\varepsilon A^\infty)dm$$

$$[m](\varepsilon) = \alpha^\sharp_*(m) = \int_X \alpha(-)(1)dm = \int_X ([-] \circ \eta_X)(\varepsilon)dm$$

Assume the result is true for wA^∞ and w. Take $\diamond \in \{\{\varepsilon\}, A^\infty\}$.

$$[m](aw\diamond) = [\tau_a(m)](w\diamond) = \int_X ([-] \circ \eta_X)(w\diamond)d\tau_a(m) \quad \text{(induction hypothesis)}$$

$$= \int_{x \in X} \left(\int_X ([-] \circ \eta_X)(w\diamond)dt_a(x) \right) m(dx) \quad \text{(preliminary remark)}$$

$$= \int_{x \in X} [\tau_a(\eta_X(x))](w\diamond)m(dx) \quad \text{(definition of } \tau_a)$$

$$= \int_X ([-] \circ \eta_X)(aw\diamond)dm$$

$\square$

Using this last lemma and that $\tau_a \circ \eta_X = t_a$, we have for any $x \in X$:

$$[\eta_X(x)](\varepsilon A^\infty) = \alpha(x)(\mathcal{L}X)$$
$$[\eta_X(x)](\varepsilon) = \alpha(x)(1)$$
$$[\eta_X(x)](awA^\infty) = [(\tau_a \circ \eta_X)(x)](wA^\infty)$$
$$= [t_a(x)](wA^\infty) = \int_X [\eta_X(-)](wA^\infty)dt_a(x)$$
$$[\eta_X(x)](aw) = [(\tau_a \circ \eta_X)(x)](w) = [t_a(x)](w) = \int_X [\eta_X(-)](w)dt_a(x)$$

Thus, for any $x \in X$, $([-] \circ \eta_X)(x)$ and $(\iota_{A^\infty} \circ [\![-]\!])(x)$ are measures in $\mathbb{S}A^\infty$ that coincide on S_∞. Because of Theorem 3.5, they are equal. Consequently, the trace semantics we get via determinisation and Eilenberg-Moore algebras is the same as the Kleisli trace semantics of [13].

Theorem 6.12. *The diagram* (4) *commutes, i.e., the maps* $\iota_{A^\infty} \circ [\![-]\!]$ *and* $[-] \circ \eta_X$ *coincide.*

Finally, note that, in the event that $\alpha \colon X \to \mathbb{D}\mathcal{L}X$ can be seen as a discrete system, i.e., for all $x \in X$, $\alpha(x)$ is a convex countable sum of Dirac distributions, then the general semantics coincide with those obtained in Sect. 2.

7   Related Work

The (in)finite trace semantics of PTS discussed in this paper was presented coalgebraically in [13], through the Kleisli category of the (sub-)Giry monad. By using a determinisation construction, we obtain the same trace semantics, in a fundamentally different way. This determinisation construction is precisely what allows us to use bisimulations (up-to) to prove equivalence. Further, our determinisation construction can be presented separately for the discrete and continuous cases (the discrete case is much more basic), whereas in the Kleisli setting only the general continuous case can be presented (since discrete systems generate a probability measure). Other coalgebraic approaches to infinite traces (based on fixed points, e.g., [7,19]) do not use determinisation.

Our determinisation construction for (in)finite traces is strongly inspired by the one for finite traces in [12,17]. As explained in Sect. 4, the main technical difference is that the total probability mass of states in the determinised Moore automaton becomes observable, and that this yields a probability measure over sets of traces rather than a (sub)probability distribution over individual traces.

The above-mentioned equivalence between the determinisation and 'Kleisli' trace semantics for finite traces is a motivating example for the general comparison between coalgebraic determinisation and Kleisli traces in [12]. However, we do not know if those results can be applied here for at least one reason: the correspondence stated in [12] uses only one monad for both constructions, using, in case of finite traces, an extension natural transformation of the form $\mathfrak{e}\colon \mathbb{S}\mathcal{L} \Rightarrow F\mathbb{S}$ (actually, the discrete version). However, in our construction, we have to move from probability measures in the definition of PTSs (modeled by $\mathbb{D}$) to sub-probability measures in the determinised Moore automaton (modeled by $\mathbb{S}$). In contrast to the case of finite traces, we can not simply replace $\mathbb{D}$ by $\mathbb{S}$ in the definition of PTS, since the sums-to-1 condition is required for the condition (ii) of Theorem 6.9. One might try to nevertheless use only $\mathbb{S}$ as the monad, focusing on PTSs (involving $\mathbb{S}$) that satisfy the sums-to-one condition. But it is currently unclear to us how such a subclass fits into the framework of [12]; moreover, the Kleisli semantics for PTSs based on $\mathbb{S}$ is finite traces [13, Theorem 3.33]. Another idea is to use the isomorphism $\mathbb{D}(A \times X + 1) \simeq \mathbb{S}(A \times X)$, (via the map $m \mapsto m_{|\Sigma_A \otimes \Sigma_X}$) but this does not seem to solve the issue: the Kleisli semantics of PTS of the form $X \to \mathbb{S}(A \times X)$ is trivial [13, Theorem 3.33]. We leave a suitable extension of the abstract framework [12] for future work.

For the algorithm presented in Sect. 5, we embed convex combinations (in the transition structure of PTS) into vector spaces, in order to use a more general contextual closure, w.r.t. arbitrary linear combinations rather than only convex combinations. This guarantees termination of the algorithm based on bisimulation up to congruence. We do not know whether this move is really necessary: perhaps the contextual closure w.r.t. only convex combinations suffices. The recent [5] might be of use in answering this question.

This work was done primarily from a coalgebraic point of view. Actually, as pointed out by one of the reviewers, the determinization of a PTS involves to a standard construction in the theory of Markov chains and stochastic processes: the passage from a kernel to a stochastic operator. This perspective could be investigated further. Notably, one motivation for trying to do so is to study how the results of Sect. 5 could extend to (discrete approximations) of the measurable PTSs of Sect. 6.

Acknowledgments. We are grateful to Filippo Bonchi, Paul Levy, Damien Pous, Jan Rutten, Ana Sokolova and the anonymous reviewers for comments and suggestions.

References

1. Aliprantis, C.D., Border, K.C.: Infinite Dimensional Analysis: A Hitchhiker's Guide. Springer, Berlin (2006). https://doi.org/10.1007/3-540-29587-9
2. Bonchi, F., König, B., Küpper, S.: Up-to techniques for weighted systems. In: Legay, A., Margaria, T. (eds.) TACAS 2017. LNCS, vol. 10205, pp. 535–552. Springer, Heidelberg (2017). https://doi.org/10.1007/978-3-662-54577-5_31

3. Bonchi, F., Petrisan, D., Pous, D., Rot, J.: A general account of coinduction up-to. Acta Inf. **54**(2), 127–190 (2017). https://doi.org/10.1007/s00236-016-0271-4

4. Bonchi, F., Pous, D.: Checking NFA equivalence with bisimulations up to congruence. In: Principle of Programming Languages (POPL), Roma, Italy, pp. 457–468. ACM (2013). https://doi.org/10.1145/2429069.2429124. 16 p

5. Bonchi, F., Silva, A., Sokolova, A.: The power of convex algebras. In: Meyer, R., Nestmann, U. (eds.) 28th International Conference on Concurrency Theory (CONCUR 2017). Leibniz International Proceedings in Informatics (LIPIcs), vol. 85, pp. 23:1–23:18. Schloss Dagstuhl-Leibniz-Zentrum fuer Informatik, Dagstuhl, Germany (2017). https://doi.org/10.4230/LIPIcs.CONCUR.2017.23

6. Boreale, M.: Weighted bisimulation in linear algebraic form. In: Bravetti, M., Zavattaro, G. (eds.) CONCUR 2009. LNCS, vol. 5710, pp. 163–177. Springer, Heidelberg (2009). https://doi.org/10.1007/978-3-642-04081-8_12

7. Cîrstea, C.: From branching to linear time, coalgebraically. Fundam. Inform. **150**(3–4), 379–406 (2017). https://doi.org/10.3233/FI-2017-1474

8. Giry, M.: A categorical approach to probability theory. Lect. Notes Math. **915**, 68–85 (1982)

9. Goy, A.: Trace semantics via determinization for probabilistic transition systems. Master internship report, Radboud University (2017). arxiv:1802.09084

10. Hasuo, I., Jacobs, B., Sokolova, A.: Generic trace semantics via coinduction. Logical Methods Comput. Sci. 3(4) (2007). https://doi.org/10.2168/LMCS-3(4:11)

11. Jacobs, B.: Introduction to Coalgebra: Towards Mathematics of States and Observation, vol. 59. Cambridge University Press, Cambridge (2016). https://doi.org/10.1017/CBO9781316823187. Cambridge Tracts in Theoretical Computer Science

12. Jacobs, B., Silva, A., Sokolova, A.: Trace semantics via determinization. J. Comput. Syst. Sci. **81**(5), 859–879 (2015). https://doi.org/10.1016/j.jcss.2014.12.005. 11th International Workshop on Coalgebraic Methods in Computer Science, CMCS 2012 (Selected Papers)

13. Kerstan, H., König, B.: Coalgebraic trace semantics for continuous probabilistic transition systems. Logical Methods Comput. Sci. **9**(4) (2013). https://doi.org/10.2168/LMCS-9(4:16)2013

14. Klenke, A.: Probability Theory: A Comprehensive Course. Springer, London (2013). https://doi.org/10.1007/978-1-4471-5361-0

15. Rutten, J.: Universal coalgebra: a theory of systems. TCS **249**(1), 3–80 (2000)

16. Silva, A., Bonchi, F., Bonsangue, M.M., Rutten, J.J.M.M.: Generalizing determinization from automata to coalgebras. Log. Methods Comput. Sci. **9**(1) (2013). https://doi.org/10.2168/LMCS-9(1:9)2013

17. Silva, A., Sokolova, A.: Sound and complete axiomatization of trace semantics for probabilistic systems. Electr. Notes Theor. Comput. Sci. **276**, 291–311 (2011). https://doi.org/10.1016/j.entcs.2011.09.027. Twenty-seventh Conference on the Mathematical Foundations of Programming Semantics (MFPS XXVII)

18. Sokolova, A.: Probabilistic systems coalgebraically: a survey. Theor. Comput. Sci. **412**(38), 5095–5110 (2011). https://doi.org/10.1016/j.tcs.2011.05.008

19. Urabe, N., Hasuo, I.: Coalgebraic infinite traces and Kleisli simulations. In: Moss, L.S., Sobocinski, P. (eds.) CALCO 2015. LIPIcs, vol. 35, pp. 320–335. Schloss Dagstuhl - Leibniz-Zentrum fuer Informatik (2015). https://doi.org/10.4230/LIPIcs.CALCO.2015.320

Steps and Traces

Bart Jacobs[1], Paul Levy[2], and Jurriaan Rot[1]([✉])

[1] Institute for Computing and Information Sciences, Radboud Universiteit,
Nijmegen, The Netherlands
{bart@cs,jrot}@cs.ru.nl
[2] University of Birmingham, Birmingham, UK
P.B.Levy@cs.bham.ac.uk

Abstract. In the theory of coalgebras, trace semantics can be defined in various distinct ways, including through algebraic logics, the Kleisli category of a monad or its Eilenberg-Moore category. This paper elaborates two new unifying ideas: (1) coalgebraic trace semantics is naturally presented in terms of corecursive algebras, and (2) all three approaches arise as instances of the same abstract setting. Our perspective puts the different approaches under a common roof, and allows to derive conditions under which some of them coincide.

1 Introduction

Traces are used in the semantics of state-based systems as a way of recording the consecutive behaviour of a state in terms of sequences of observable (input and/or output) actions. Trace semantics leads to, for instance, the notion of trace equivalence, which expresses that two states cannot be distinguished by only looking at their iterated in/output behaviour.

For many years already, trace semantics is a central topic of interest in the coalgebra community — and not only there, of course. One of the key features of the area of coalgebra is that states and their coalgebras can be considered in different universes, formalised as categories. The break-through insight is that trace semantics for a system in universe A can often be obtained by switching to a different universe B. More explicitly, where the (ordinary) behaviour of the system can be described via a final coalgebra in universe A, the trace behaviour arises by finality in the different universe B. Typically, the alternative universe B is a category of algebraic logics, the Kleisli category, or the category of Eilenberg-Moore algebras, of a monad on universe A.

This paper elaborates two new unifying ideas.

1. We observe that the trace map from the state space of a coalgebra to a carrier of traces is in all three situations the unique 'coalgebra-to-algebra' map to a

J. Rot—The research leading to these results has received funding from the European Research Council under the European Union's Seventh Framework Programme (FP7/2007-2013)/ERC grant agreement nr. 320571.

C. Cîrstea (Ed.): CMCS 2018, LNCS 11202, pp. 122–143, 2018.
https://doi.org/10.1007/978-3-030-00389-0_8

corecursive algebra [6] of traces. This differs from earlier work which tries to describe traces as final coalgebras. For us it is quite natural to view languages as algebras, certainly when they consist of finite words/traces.

2. Next, these corecursive algebras, used as spaces of traces, all arise via a uniform construction, in a setting given by an adjunction together with a special natural transformation that we call a 'step'. We heavily rely on a basic result saying that in this situation, the (lifting of the) right adjoint preserves corecursive algebras, sending them from one universe to another. This is a known result [5], but its fundamental role in trace semantics has not be recognized before. For an arbitrary coalgebra there is then a unique map to the transferred corecursive algebra; this is the trace map that we are after.

The main contribution of this paper is the unifying step-based approach to coalgebraic trace semantics: it is shown that three existing flavours of trace semantics — logical, Eilenberg-Moore, Kleisli — are all instances of our approach. Moreover, comparison results are given relating two of these forms of trace semantics, namely logic-to-Eilenberg-Moore and logic-to-Kleisli. The other combinations involve subtleties which we do not fully grasp yet. Due to space limitations, we don't cover the whole field of coalgebraic trace semantics: we focus only on finite trace semantics, and also exclude at this stage the 'iteration' based approaches, e.g., in [8, 22, 25].

Outline. The paper is organised as follows. It starts in Sect. 1 with the abstract step-and-adjunction setting, and the relevant definitions and results for corecursive algebras. In the next three sections, it is explained how this setting gives rise to trace semantics, by presenting the above-mentioned three approaches to coalgebraic trace semantics in terms of steps and adjunctions: Eilenberg-Moore (Sect. 3), logical (Sect. 4) and Kleisli (Sect. 5). In each case, the relevant corecursive algebra is described. These sections are illustrated with several examples. The next section establishes a connection between the Eilenberg-Moore and the logical approach, and a connection between the Kleisli and logical approach (Sect. 6). In Sect. 7 we briefly show that our construction of corecursive algebras strengthens to a construction of completely iterative algebras. Finally, in Sect. 8 we provide some directions for future work.

Notation. In the context of an adjunction $F \dashv G$, we shall use overline notation $\overline{(-)}$ for adjoint transposition. The unit and counit of an adjunction are, as usual, written as η and ε.

For an endofunctor H, we write $\mathrm{Alg}(H)$ for its algebra category and $\mathrm{CoAlg}(H)$ for its coalgebra category. For a monad (T, η, μ) on $\mathbf{C}$, we write $\mathcal{EM}(T)$ for the Eilenberg-Moore category and $\mathcal{Kl}(T)$ for the Kleisli category.

We recall that any functor $S \colon \mathbf{Sets} \to \mathbf{Sets}$ has a unique strength st. We write $\mathrm{st} \colon S(X^A) \to S(X)^A$ for $\mathrm{st}(t)(a) = S(\mathrm{ev}_a)(t)$, where $\mathrm{ev}_a = \lambda f. f(a) \colon X^A \to X$.

2   Coalgebraic Semantics from a Step

This section is about the construction of corecursive algebras and their use for semantics. The notion of corecursive algebra, studied in [6,9] as the dual of Taylor's notion of recursive coalgebra [10], is defined as follows.

Definition 1. *Let H be an endofunctor on a category* **C**.

1. *A coalgebra-to-algebra morphism from a coalgebra $c\colon X \to H(X)$ to an algebra $a\colon H(A) \to A$ is a map $f\colon X \to A$ such that the diagram*

$$
\begin{array}{ccc}
X & \xrightarrow{\ f\ } & A \\
{\scriptstyle c}\downarrow & & \uparrow{\scriptstyle a} \\
H(X) & \xrightarrow[H(f)]{} & H(A)
\end{array}
$$

commutes. Equivalently: such a morphism is a fixpoint for the endofunction on the homset $\mathbf{C}(X,A)$ *sending f to the composite*

$$
X \xrightarrow{\ c\ } H(X) \xrightarrow{\ H(f)\ } H(A) \xrightarrow{\ a\ } A
$$

2. *An algebra $a\colon H(A) \to A$ is* corecursive *when for every coalgebra $c\colon H \to H(X)$ there is a unique coalgebra-to-algebra morphism $(X,c) \to (A,a)$.*

Here is some intuition.

- As explained in [14], the specification of a coalgebra-to-algebra morphism f is a "divide-and-conquer" algorithm. It says: to operate on an argument, first decompose it via the coalgebra c, then operate on each component via $H(f)$, then combine the results via the algebra a.
- For each final H-coalgebra $\zeta\colon A \xrightarrow{\cong} H(A)$, the inverse $\zeta^{-1}\colon H(A) \to A$ is a corecursive algebra. For most functors of interest, this final coalgebra gives semantics up to bisimilarity, which is finer than trace equivalence. So trace semantics requires a different corecursive algebra.

In all our examples, we use the same procedure for obtaining a corecursive algebra, which we shall now explain. Our basic setting consists of an adjunction, two endofunctors, and a natural transformation:

$$
H\!\!\left(\!\!\begin{array}{c}\ \end{array}\!\!\right)\!\mathbf{C} \underset{G}{\overset{F}{\underset{\longleftarrow}{\overset{\longrightarrow}{\perp}}}} \mathbf{D}\!\!\left(\!\!\begin{array}{c}\ \end{array}\!\!\right)\!L \qquad \text{with} \qquad HG \overset{\rho}{\Longrightarrow} GL \tag{1}
$$

The natural transformation $\rho\colon HG \Rightarrow GL$ will be called a *step*. Here H is the *behaviour functor*: we study H-coalgebras and give semantics for them in a corecursive H-algebra. This arrangement is well-known in the area of coalgebraic modal logic [3,7,20,24,27], but we shall see that its application is wider.

A step can be formulated in several equivalent ways [18,23].

Theorem 2. *In the situation* (1), *there are bijective correspondences between natural transformations* $\rho_1\colon HG \Rightarrow GL$, $\rho_2\colon FH \Rightarrow LF$, $\rho_3\colon FHG \Rightarrow L$ *and* $\rho_4\colon H \Rightarrow GLF$.

Moreover, if H and L happen to be monads, then ρ_1 is an $\mathcal{EM}$-law (map $HG \Rightarrow GL$ compatible with the monad structures) iff ρ_2 is a $\mathcal{K\ell}$-law (map $FH \Rightarrow LF$ compatible with the monad structures) iff ρ_4 is a monad map; and two further equivalent characterisations are respectively a lifting of G or an extension of F:

$$
\begin{array}{ccc}
\mathcal{EM}(H) \xleftarrow{\ \overline{G}\ } \mathcal{EM}(L) & \qquad & \mathcal{K\ell}(H) \xrightarrow{\ \overline{F}\ } \mathcal{K\ell}(L) \\
\downarrow \qquad \qquad \downarrow & & \uparrow \qquad \qquad \uparrow \\
\mathbf{C} \xleftarrow{\ \ G\ \ } \mathbf{D} & & \mathbf{C} \xrightarrow{\ \ F\ \ } \mathbf{D}
\end{array}
$$

Proof. We only mention the bijective correspondences: ρ_1 and ρ_3 correspond by adjoint transposition, and similarly for ρ_2 and ρ_4. Further, ρ_2 and ρ_3 are obtained from each other by:

$$
\rho_3 = \left(FHG \overset{\rho_2 G}{\Longrightarrow} LFG \overset{L\varepsilon}{\Longrightarrow} L \right)
$$

$$
\rho_2 = \left(FH \overset{FH\eta}{\Longrightarrow} FHGF \overset{\rho_3 F}{\Longrightarrow} LF \right).
$$

$\square$

It is common to refer to ρ_1 and ρ_2 as *mates*; the other two maps are their adjoint transposes. In diagrams we omit the subscript i in ρ_i and let the type determine which version of ρ is meant.

Further, in the remainder of this paper we drop the usual subscript of components of natural transformations.

Definition 3. *In the setting* (1), *the step natural transformation ρ gives rise to both:*

– *a lifting G_ρ of the right adjoint G, called the* step-induced algebra lifting:

$$
\begin{array}{cc}
\mathrm{Alg}(H) \xleftarrow{\ G_\rho\ } \mathrm{Alg}(L) & \qquad G_\rho\left(L(A) \overset{a}{\to} A \right) := \\
\downarrow \qquad \qquad \downarrow & \qquad \left(HG(A) \overset{\rho}{\to} GL(A) \overset{G(a)}{\longrightarrow} G(A) \right). \\
\mathbf{C} \xleftarrow{\ \ G\ \ } \mathbf{D} &
\end{array}
$$

– *dually, a lifting F^ρ of the left adjoint F, called the* step-induced coalgebra lifting:

$$
\begin{array}{cc}
\mathrm{CoAlg}(H) \xrightarrow{\ F^\rho\ } \mathrm{CoAlg}(L) & \qquad F^\rho\left(X \overset{c}{\to} H(X) \right) := \\
\downarrow \qquad \qquad \downarrow & \qquad \left(F(X) \overset{F(c)}{\longrightarrow} FH(X) \overset{\rho}{\to} LF(X) \right). \\
\mathbf{C} \xrightarrow{\ \ F\ \ } \mathbf{D} &
\end{array}
$$

Our approach relies on the following basic result.

Proposition 4 ([5]). *For each corecursive L-algebra a: $L(A) \to A$, the transferred H-algebra $G_\rho(A, a)\colon HG(A) \to G(A)$ is also corecursive. Explicitly, for any H-coalgebra (X, c), the unique coalgebra-to-algebra map $(X, c) \to G_\rho(A, a)$ is the adjoint transpose of the unique coalgebra-to-algebra map $F^\rho(X, c) \to (A, a)$.* □

Thus, by analogy with the familiar statement that *"right adjoints preserves limits"*, we have *"step-induced algebra liftings of right adjoints preserve corecursiveness"*. Now we give the complete construction for semantics of a coalgebra.

Theorem 5. *Suppose that L has a final coalgebra $\zeta\colon \Psi \overset{\cong}{\to} L(\Psi)$. Then for every H-coalgebra (X, c) there is a unique coalgebra-to-algebra map $c^\dagger$ as on the left below:*

$$
\begin{array}{ccc}
X \dashrightarrow^{c^\dagger} G(\Psi) & \qquad & F(X) \dashrightarrow^{\overline{c^\dagger}} \Psi \\
c\downarrow \quad \uparrow{G_\rho(\Psi,\zeta^{-1})} & & F^\rho(X,c)\downarrow \quad \uparrow{\zeta^{-1}} \\
H(X) \overset{H(c^\dagger)}{\dashrightarrow} HG(\Psi) & & LF(X) \overset{L(\overline{c^\dagger})}{\dashrightarrow} L(\Psi)
\end{array}
$$

The map $c^\dagger$ on the left can alternatively be characterized via its adjoint transpose $\overline{c^\dagger}$ on the right, which is the unique coalgebra-to-algebra morphism. The latter can also be seen as the unique map to the final coalgebra $\Psi \overset{\cong}{\to} L(\Psi)$. □

Note that Theorem 5 generalises final coalgebra semantics: taking in (1) $F = G = \mathrm{Id}_{\mathbf{C}}$ and $H = L$, the map $c^\dagger$ in the above theorem is the unique homomorphism to the final coalgebra. In the remainder of this paper we focus on instances where $c^\dagger$ captures traces, and we therefore refer to it as the *trace semantics* map.

3   Traces via Eilenberg-Moore

We recall the approach to trace semantics developed in [4,17,29], putting it in the framework of the previous section. The approach deals with coalgebras for the composite functor BT, where T is a monad that captures the 'branching' aspect. The following assumptions are required.

$$
\begin{array}{ccc}
\mathcal{EM}(T) & \overset{\overline{B}}{\longrightarrow} & \mathcal{EM}(T) \\
U\downarrow & & \downarrow U \qquad (2)\\
\mathbf{C} & \overset{B}{\longrightarrow} & \mathbf{C}
\end{array}
$$

1. An endofunctor $B\colon \mathbf{C} \to \mathbf{C}$ with a final coalgebra $\zeta\colon \Theta \overset{\cong}{\to} B(\Theta)$.
2. A monad (T, η^T, μ^T), with the standard adjunction $\mathcal{F} \dashv U$ between categories $\mathbf{C} \leftrightarrows \mathcal{EM}(T)$, where U is 'forget' and $\mathcal{F}$ is for 'free algebras'.
3. A lifting $\overline{B}$ of B, as in (2), or, equivalently, an $\mathcal{EM}$-law $\kappa\colon TB \Rightarrow BT$.

Example 6. To briefly illustrate these ingredients, we consider non-deterministic automata. These are BT-coalgebras with $B\colon \mathbf{Sets} \to \mathbf{Sets}$, $B(X) = 2 \times X^A$ with $2 = \{\bot, \top\}$ and T the finite powerset monad. The functor B has a final coalgebra carried by the set 2^{A^*} of languages. Further, $\mathcal{EM}(T)$ is the category of join semi-lattices (JSL). The lifting is defined by product in $\mathcal{EM}(T)$, using the JSL on 2 given by the usual ordering $\bot \leq \top$. By the end of this section, we revisit this example and obtain the usual language semantics.

These assumptions give rise to the following instance of our general setting (1):

$$BT \, \mathbf{C} \, \mathbf{C} \xrightleftharpoons[U]{\overset{\mathcal{F}}{\underset{\bot}{\rightleftarrows}}} \mathcal{EM}(T) \, \circlearrowright \overline{B} \qquad \text{with} \qquad \begin{array}{l} \rho \colon BTU \Longrightarrow U\overline{B} \ \text{ where} \\ \rho_{(X,a)} = \left(BTX \xrightarrow{Ba} BX \right) \end{array}$$

Actually — and equivalently, by Theorem 2 — the step ρ is most easily given in terms of $\rho_4 \colon BT \Rightarrow U\overline{B}\mathcal{F}$: since $\overline{B}$ lifts B, we have $U\overline{B}\mathcal{F} = BU\mathcal{F} = BT$, so that ρ_4 is then defined simply as the identity.

The following result is well-known, and is (in a small variation) due to [30].

Lemma 7. *There is a unique algebra structure $a \colon T(\Theta) \to \Theta$ making $((\Theta, a), \zeta)$ a $\overline{B}$-coalgebra. Moreover, this coalgebra is final.* □

We apply the step-induced algebra lifting $G_\rho \colon \mathrm{Alg}(\overline{B}) \to \mathrm{Alg}(BT)$ to the inverse of this final $\overline{B}$-coalgebra, obtaining a BT-algebra:

$$\left(BT(\Theta) \xrightarrow{\ell_{\mathrm{em}}} \Theta \right) := G_\rho((\Theta, a), \zeta^{-1}) = \left(BT(\Theta) \xrightarrow{B(a)} B(\Theta) \xrightarrow{\zeta^{-1}} \Theta \right).$$

By Theorem 5, this algebra is corecursive, giving us trace semantics of BT-coalgebras. More explicitly, given a coalgebra $c \colon X \to BT(X)$, the trace semantics is the unique map, written as em_c, making the following square commute.

$$\begin{array}{ccc} X & \dashrightarrow[\mathrm{em}_c] & \Theta \\ {\scriptstyle c}\downarrow & & \uparrow{\scriptstyle \ell_{\mathrm{em}}} \\ BT(X) & \underset{BT(\mathrm{em}_c)}{\dashrightarrow} & BT(\Theta) \end{array} \qquad (3)$$

The unique map em_c in (3) appears in the literature as a 'coiteration up-to' or 'unique solution' theorem [1]. Examples follow later in this section (Theorem 8, Example 9).

In [17,29], the above trace semantics of BT-coalgebras arises through 'determinisation', which we explain next. Given a coalgebra $c \colon X \to BT(X)$, one takes its adjoint transpose:

$$\frac{c \colon X \to BT(X) = BU\mathcal{F}(X) = U\overline{B}\mathcal{F}(X)}{\overline{c} \colon \mathcal{F}(X) \to \overline{B}\mathcal{F}(X)}$$

It follows from Theorem 2 and our definition of ρ that this transpose coincides with the application of the step-induced coalgebra lifting $\mathcal{F}^\rho \colon \mathrm{CoAlg}(BT) \to \mathrm{CoAlg}(\overline{B})$ from the previous section, i.e., $\mathcal{F}^\rho(X, c) = (\mathcal{F}(X), \overline{c})$. The

$$\begin{array}{ccc} T(X) & \dashrightarrow[\overline{\mathrm{em}}_c] & \Theta \\ {\scriptstyle \overline{c}}\downarrow & & \uparrow{\scriptstyle \zeta^{-1}} \\ BT(X) & \underset{B(\overline{\mathrm{em}}_c)}{\dashrightarrow} & B(\Theta) \end{array} \qquad (4)$$

functor $\mathcal{F}^\rho$ thus plays the role of determinisation, see [17]. By Theorem 5, the trace semantics em_c can equivalently be characterised in terms of $\mathcal{F}^\rho$, as the unique map $\overline{\mathrm{em}_c}$ making (4) commute. This is how the trace semantics via Eilenberg-Moore is presented in [17,29]: as the transpose $\mathrm{em}_c = \overline{\mathrm{em}_c} \circ \eta_X^T$.

We conclude this section by recalling a canonical construction of a distributive law [15] for a class of 'automata-like' examples.

Theorem 8. *Let Ω be a set, T a monad on* **Sets** *and $t\colon T(\Omega) \to \Omega$ an $\mathcal{EM}$-algebra. Let $B\colon$ **Sets** $\to$ **Sets**, $B(X) = \Omega \times X^A$, and $\kappa\colon TB \Rightarrow BT$ given by*

$$\kappa_X := \left(T(\Omega \times X^A) \xrightarrow{\langle T(\pi_1), T(\pi_2)\rangle} T(\Omega) \times T(X^A) \xrightarrow{t \times \mathrm{st}} \Omega \times T(X)^A \right).$$

Then κ is an $\mathcal{EM}$-law. Moreover, the final B-coalgebra (Ω^{A^*}, ζ) together with the algebra structure $T(\Omega^{A^*}) \xrightarrow{\mathrm{st}} T(\Omega)^{A^*} \xrightarrow{t^{A^*}} \Omega^{A^*}$ is a final $\overline{B}$-coalgebra. □

Example 9. By Theorem 8, we obtain an explicit description of the trace semantics arising from the corecursive algebra (3): for any $\langle o, f\rangle\colon X \to \Omega \times T(X)^A$, the trace semantics is the unique map em in

$$
\begin{array}{ccc}
X & \dashrightarrow^{\mathrm{em}} & \Omega^{A^*} \\
{\scriptstyle\langle o,f\rangle}\downarrow & & \uparrow{\scriptstyle\zeta^{-1}} \\
BT(X) \underset{BT(\mathrm{em})}{\dashrightarrow} BT(\Omega^{A^*}) \xrightarrow{\mathrm{st}} B(T(\Omega)^{A^*}) \xrightarrow{B(t^{A^*})} BT(\Omega^{A^*})
\end{array}
$$

We instantiate the trace semantics em for various choices of Ω, T and t. Given a coalgebra $\langle o, f\rangle\colon X \to \Omega \times T(X)^A$, we have $\mathrm{em}(x)(\varepsilon) = o(x)$ independently of these choices. The table below lists the inductive case $\mathrm{em}(x)(aw)$ respectively for non-deterministic automata (NDA) where branching is interpreted as usual (NDA-∃), NDA where branching is interpreted conjunctively (NDA-∀) and (reactive) probabilistic automata (PA). Here $\mathcal{P}_f$ is the finite powerset functor, and $\mathcal{D}_{\mathrm{fin}}$ the finitely supported distribution functor.

	T	Ω	$t\colon T(\Omega) \to \Omega$	$\mathrm{em}(x)(aw)$
NDA-∃	$\mathcal{P}_f$	$2 = \{\bot, \top\}$	$S \mapsto \bigvee S$	$\bigvee_{y \in f(x)(a)} \mathrm{em}(y)(w)$
NDA-∀	$\mathcal{P}_f$	$2 = \{\bot, \top\}$	$S \mapsto \bigwedge S$	$\bigwedge_{y \in f(x)(a)} \mathrm{em}(y)(w)$
PA	$\mathcal{D}_{\mathrm{fin}}$	$[0,1]$	$\varphi \mapsto \sum_{p \in [0,1]} p \cdot \varphi(p)$	$\sum_{y \in X} \mathrm{em}(y)(w) \cdot f(x)(a)(y)$

For other examples, and a concrete presentation of the associated determinisation constructions, see [17,29].

4   Traces via Logic

This section illustrates how the 'logical' approach to trace semantics of [21], started in [27], fits in our general framework. In essence, traces are built up from logical formulas, also called tests, which are evaluated for states. These tests are obtained via an initial algebra of a functor L. The approach works both for TB and BT-coalgebras (and could, in principle, be extended to more general combinations). We start by listing our assumptions in this section.

1. An adjunction $F \dashv G$ between categories $\mathbf{C} \leftrightarrows \mathbf{D}^{\mathrm{op}}$.
2. A functor T on $\mathbf{C}$ with a step $\tau \colon TG \Rightarrow G$.
3. A functor $B \colon \mathbf{C} \to \mathbf{C}$ and a functor $L \colon \mathbf{D} \to \mathbf{D}$ with a step $\delta \colon BG \Rightarrow GL$.
4. An initial algebra $\alpha \colon L(\Phi) \overset{\cong}{\to} \Phi$.

We deviate from the convention of writing ρ for 'step', since the above map τ gives rise to multiple steps δ_τ and δ^τ in (6) below, in the sense of Definition 2; here we use 'delta' instead of 'rho' notation since it is common in modal logic.

Example 10. We take $\mathbf{C} = \mathbf{D} = \mathbf{Sets}$, and F, G both the contravariant powerset functor 2^-. Non-deterministic automata are obtained either as BT-coalgebras with $B(X) = 2 \times X^A$ and T the finite powerset functor; or as TB-coalgebras, with $B(X) = A \times X + 1$ and T again the finite powerset functor. In both cases, L is given by $L(X) = A \times X + 1$. The map $\tau \colon T2^- \Rightarrow 2^-$ is defined by $\tau_X(S)(x) = \bigvee_{\varphi \in S} \varphi(x)$, and intuitively models the existential choice in the semantics of non-deterministic automata. The map ρ and the language semantics are defined later in this section.

The assumptions are close to the general step-and-adjunction setting (1). Here, we have an opposite category on the right, and instantiate H to TB or BT:

$$H \overset{\curvearrowright}{\mathbf{C}} \underset{G}{\overset{F}{\underset{\longleftarrow}{\underset{\perp}{\longrightarrow}}}} \mathbf{D}^{\mathrm{op}} \overset{\curvearrowleft}{L} \qquad \text{where } H = BT \text{ or } H = TB \qquad (5)$$

Notice that our assumptions already include a step δ (involving B, L) and a step τ, which we can compose to obtain steps for the TB respectively BT case:

$$\delta_\tau := \left(TBG \overset{T\delta}{\Longrightarrow} TGL \overset{\tau L}{\Longrightarrow} GL \right) \qquad \mathrm{CoAlg}(L) \overset{G_{\delta_\tau}}{\longrightarrow} \mathrm{Alg}(TB)$$

$$\delta^\tau := \left(BTG \overset{B\tau}{\Longrightarrow} BG \overset{\delta}{\Longrightarrow} GL \right) \qquad \mathrm{CoAlg}(L) \overset{G_{\delta^\tau}}{\longrightarrow} \mathrm{Alg}(BT)$$

$$(6)$$

Both δ^τ and δ_τ are steps, and hence give rise to step-induced algebra liftings G_{δ_τ} and G_{δ^τ} of G (Sect. 2). By Theorem 5, we obtain two corecursive algebras by applying these liftings to the inverse of the initial algebra, i.e., the (inverse of the) final coalgebra in $\mathbf{D}^{\mathrm{op}}$:

$$\ell_{\log} := \left(TBG(\Phi) \overset{\delta_\tau}{\longrightarrow} GL(\Phi) \overset{G(\alpha^{-1})}{\underset{\cong}{\longrightarrow}} G(\Phi) \right) ,$$

$$\ell^{\log} := \left(BTG(\Phi) \overset{\delta^\tau}{\longrightarrow} GL(\Phi) \overset{G(\alpha^{-1})}{\underset{\cong}{\longrightarrow}} G(\Phi) \right) .$$

$$(7)$$

These corecursive algebras define trace semantics for any TB-coalgebra (X, c) and BT-coalgebra (Y, d):

$$
\begin{array}{cc}
\begin{array}{ccc}
X & \overset{\log_c}{\dashrightarrow} & G(\Phi) \\
{\scriptstyle c}\downarrow & & \uparrow{\scriptstyle \ell_{\log}} \\
TB(X) & \underset{TB(\log_c)}{\dashrightarrow} & TBG(\Phi)
\end{array}
&
\begin{array}{ccc}
Y & \overset{\log_d}{\dashrightarrow} & G(\Phi) \\
{\scriptstyle d}\downarrow & & \uparrow{\scriptstyle \ell_{\log}} \\
BT(Y) & \underset{BT(\log_d)}{\dashrightarrow} & BTG(\Phi)
\end{array}
\end{array}
\tag{8}
$$

It is instructive to characterise this trace semantics in terms of the transpose and the step-induced coalgebra liftings F^{δ_τ} and F^{δ^τ}, showing how they arise as unique maps from an initial algebra:

$$
\begin{array}{cc}
\begin{array}{ccc}
F(X) & \overset{\overline{\log_c}}{\dashleftarrow} & \Phi \\
{\scriptstyle F^{\delta_\tau}(X,c)}\uparrow & & \downarrow{\scriptstyle \alpha^{-1}} \\
LF(X) & \underset{L(\overline{\log_c})}{\dashleftarrow} & L(\Phi)
\end{array}
&
\begin{array}{ccc}
F(Y) & \overset{\overline{\log_d}}{\dashleftarrow} & \Phi \\
{\scriptstyle F^{\delta^\tau}(Y,d)}\uparrow & & \downarrow{\scriptstyle \alpha^{-1}} \\
LF(Y) & \underset{L(\overline{\log_d})}{\dashleftarrow} & L(\Phi)
\end{array}
\end{array}
\tag{9}
$$

In the remainder of this section, we show two classes of examples of the logical trace semantics. With these descriptions we retrieve most of the examples from [21] in a smooth manner.

Proposition 11. *Let Ω be a set, $T\colon \mathbf{Sets} \to \mathbf{Sets}$ a functor and $t\colon T(\Omega) \to \Omega$ a map. Then the set of languages Ω^{A^*} carries a corecursive algebra for the functor $\Omega \times T(-)^A$. Given a coalgebra $\langle o, f\rangle\colon X \to \Omega \times T(X)^A$, the unique coalgebra-to-algebra morphism $\log\colon X \to \Omega^{A^*}$ satisfies*

$$
\log(x)(\varepsilon) = o(x) \qquad \log(x)(aw) = t\Big(T(\mathrm{ev}_w \circ \log)(f(x)(a))\Big)
$$

for all $x \in X$, $a \in A$ and $w \in A^*$.

Proof. We instantiate the assumptions in the beginning of this section by $\mathbf{C} = \mathbf{D} = \mathbf{Sets}$, $F = G = \Omega^-$, $B(X) = \Omega \times X^A$, $L(X) = A \times X + 1$ and T the functor from the statement. The initial L-algebra is $\alpha\colon A \times A^* + 1 \overset{\cong}{\to} A^*$. The map t extends to a modality $\tau\colon TG \Rightarrow G$, given on components by

$$
\tau_X := \Big(T(\Omega^X) \overset{\mathrm{st}}{\longrightarrow} T(\Omega)^X \overset{t^X}{\longrightarrow} \Omega^X \Big).
$$

The logic $\delta\colon BG \Rightarrow GL$ is given by the isomorphism $\Omega \times (\Omega^-)^A \cong \Omega^{(A \times -)+1}$. Instantiating (7) we obtain the corecursive BT-algebra

$$
\Omega \times T(\Omega^{A^*})^A \overset{\mathrm{id} \times (\mathrm{st})^A}{\longrightarrow} \Omega \times (T(\Omega)^{A^*})^A \overset{\mathrm{id} \times (t^{A^*})^A}{\longrightarrow} \Omega \times (\Omega^{A^*})^A \overset{\Omega^{\alpha^{-1}} \circ \delta}{\longrightarrow} \Omega^{A^*} .
$$

The concrete description of $\log$ follows by spelling out the coalgebra-to-algebra diagram that characterises it. $\qquad\square$

Example 12. We instantiate the trace semantics $\log$ from Proposition 11 for various choices of Ω, T and t. Similar to the instances in Example 9, we consider a coalgebra $\langle o, f \rangle \colon X \to \Omega \times T(X)^A$, and we always have $\log(x)(\varepsilon) = o(x)$. The cases of non-deterministic automata (NDA-$\exists$, NDA-$\forall$) and probabilistic automata (PA) are the same as in Example 9. However, in contrast to the Eilenberg-Moore approach and other approaches to trace semantics, a monad structure on T is not required here. This is convenient as it also allows to treat alternating automata (AA), where $T = \mathcal{P}_f \mathcal{P}_f$; it is unclear whether T carries a suitable monad structure in that case.

	T	Ω	$t \colon T(\Omega) \to \Omega$	$\log(x)(aw)$
NDA-$\exists$	$\mathcal{P}_f$	$2 = \{\bot, \top\}$	$S \mapsto \bigvee S$	$\bigvee_{y \in f(x)(a)} \log(y)(w)$
NDA-$\forall$	$\mathcal{P}_f$	$2 = \{\bot, \top\}$	$S \mapsto \bigwedge S$	$\bigwedge_{y \in f(x)(a)} \log(y)(w)$
PA	$\mathcal{D}_{\text{fin}}$	$[0,1]$	$\varphi \mapsto \sum_{p \in [0,1]} p \cdot \varphi(p)$	$\sum_{y \in X} \log(y)(w) \cdot f(x)(a)(y)$
AA	$\mathcal{P}_f \mathcal{P}_f$	$2 = \{\bot, \top\}$	$S \mapsto \bigvee_{T \in S} \bigwedge_{b \in T} b$	$\bigvee_{T \in f(x)(a)} \bigwedge_{y \in T} \log(y)(w)$

We also describe a logic for polynomial functors constructed from a signature. Here, we model a signature by a functor $\Sigma \colon \mathbb{N} \to \mathbf{Sets}$, where $\mathbb{N}$ is the discrete category of natural numbers. This gives rise to a functor $H_\Sigma \colon \mathbf{Sets} \to \mathbf{Sets}$ as usual by $H_\Sigma(X) = \coprod_{n \in \mathbb{N}} \Sigma(n) \times X^n$. We abuse notation and write $\sigma(x_1, \ldots, x_n)$ instead of $(\sigma, x_1, \ldots x_n)$. The initial algebra of H_Σ consists of closed terms (or finite node-labelled trees) over the signature.

Proposition 13. *Let Ω be a meet semi-lattice with top element $\top$ as well as a bottom element $\bot$, let $T \colon \mathbf{Sets} \to \mathbf{Sets}$ be a functor, and $t \colon T(\Omega) \to \Omega$ a map. Let Φ be the initial H_Σ-algebra. The set Ω^Φ of 'tree' languages carries a corecursive algebra for the functor $T H_\Sigma$. Given a coalgebra $c \colon X \to T H_\Sigma(X)$, the unique coalgebra-to-algebra map $\log \colon X \to \Omega^\Phi$ is given by*

$$\log(x)(\sigma(t_1, \ldots, t_n)) = t(T(m) \circ c(x)), \text{ where}$$

$$m = \left(t \mapsto \begin{cases} \bigwedge_i \log(x_i)(t_i) & \text{if } \exists x_1 \ldots x_n. t = \sigma(x_1, \ldots, x_n) \\ \bot & \text{otherwise} \end{cases} \right) \colon H_\Sigma(X) \to \Omega$$

for all $x \in X$ and $\sigma(t_1, \ldots, t_n) \in \Phi$.

Proof. We use $\mathbf{C} = \mathbf{D} = \mathbf{Sets}$, $F = G = \Omega^-$, $B = L = H_\Sigma$. The map t extends to a modality $\tau \colon TG \Rightarrow G$ as in the proof of Proposition 11. The logic $\delta \colon H_\Sigma \Omega^- \Rightarrow \Omega^{H_\Sigma(-)}$ is:

$$\delta_X(\sigma(\phi_1, \ldots, \phi_n))(t) = \begin{cases} \bigwedge_i \phi_i(x_i) & \text{if } \exists x_1 \ldots x_n. t = \sigma(x_1, \ldots, x_n) \\ \bot & \text{otherwise} \end{cases}$$

The corecursive algebra $\ell_{\log}$ is then given by:

$$TH_\Sigma(\Omega^\Phi) \xrightarrow{T(\delta)} T(\Omega^{H_\Sigma(\Phi)}) \xrightarrow{\text{st}} T(\Omega)^{H_\Sigma(\Phi)} \xrightarrow{t^{H_\Sigma(\Phi)}} \Omega^{H_\Sigma(\Phi)} \xrightarrow{\cong} \Omega^\Phi .$$

The explicit characterisation of log is a straightforward computation.   □

Example 14. Given a signature Σ, a coalgebra $f \colon X \to \mathcal{P}_f H_\Sigma(X)$ is a (top-down) *tree automaton*. With $\Omega = \{\bot, \top\}$ and $t(S) = \bigvee S$, Proposition 13 gives:

$$\log(x)(\sigma(t_1,\ldots,t_n)) = \top \text{ iff } \exists x_1 \ldots x_n.\sigma(x_1,\ldots,x_n) \in f(x) \wedge \bigwedge_{1 \leq i \leq n} \log(x_i)(t_i)$$

for every state $x \in X$ and tree $\sigma(t_1,\ldots,t_n)$. This is the standard semantics of tree automata. It is easily adapted to *weighted* tree automata, see [21].

In both Examples 14 and 12, the step-induced coalgebra lifting F_{δ^τ} (respectively F_{δ_τ}) of the underlying logic corresponds to reverse determinisation, see [21,28] for details. In particular, in Example 14 it maps a top-down tree automaton to the corresponding bottom-up tree automaton.

5   Traces via Kleisli

In this section we briefly recall the 'Kleisli approach' to trace semantics [12], and cast it in our abstract framework. It applies to coalgebras for a composite functor TB, where T is a monad modelling the type of branching. For example,

$$\begin{array}{ccc} \mathcal{K}\ell(T) & \xrightarrow{\overline{B}} & \mathcal{K}\ell(T) \\ J\uparrow & & \uparrow J \\ \mathbf{C} & \xrightarrow{B} & \mathbf{C} \end{array} \quad (10)$$

a coalgebra $X \to \mathcal{P}(A \times X + S)$ has an associated map $X \to \mathcal{P}(A^* \times S)$ that sends a state $x \in X$ to the set of its complete traces. (Taking $S = 1$, this is the usual language semantics of a nondeterministic automaton.) To fit this to our framework, the monad T is $\mathcal{P}$ and the functor B is $(A \times -) + S$. In general, the following assumptions are required.

1. An endofunctor $B \colon \mathbf{C} \to \mathbf{C}$ with an initial algebra $\beta \colon B(\Psi) \xrightarrow{\cong} \Psi$.
2. A monad (T, η^T, μ^T), with the standard adjunction $J \dashv U$ between categories $\mathbf{C} \leftrightarrows \mathcal{K}\ell(T)$, where $J(X) = X$ and $U(Y) = T(Y)$.
3. An extension $\overline{B}$ of B, as in (10), or, equivalently, a $\mathcal{K}\ell$-law $\lambda \colon BT \Rightarrow TB$.
4. $(\Psi, J(\beta^{-1}))$ is a final $\overline{B}$-coalgebra.

In the case that B is the functor $(A \times -) + S$, its initial algebra is carried by $A^* \times S$, and the canonical $\mathcal{K}\ell$-law is given at X by

$$[T\mathsf{inl} \circ st_{A,X}, T\mathsf{inr} \circ \eta_S^T] \colon A \times TX + S \to T(A \times X + S)$$

A central observation for the Kleisli approach to traces is that the fourth assumption holds under certain order enrichment requirements on $\mathcal{K}\ell(T)$, see [12]. In particular, these hold when T is the powerset monad, the (discrete) sub-distribution monad or the lift monad.

The above assumptions give rise to the following instance of our setting (1):

$$TB \subset \mathbf{C} \underset{U}{\overset{J}{\underset{\perp}{\rightleftarrows}}} \mathcal{K\ell}(T) \supset \overline{B} \quad \text{with} \quad \begin{array}{l} \rho \colon TBU \Longrightarrow U\overline{B} \text{ where } \rho_X = \\ (TBTX \xrightarrow{T(\lambda)} T^2BX \xrightarrow{\mu^T} TBX) \end{array}$$

Similar to the $\mathcal{EM}$-case in Sect. 3, the map of adjunctions is most easily given in terms of $\rho_4 \colon TB \Rightarrow U\overline{B}J$ as the identity, using that $\overline{B}$ extends B.

We apply the step-induced algebra lift $G_\rho \colon \mathrm{Alg}(\overline{B}) \to \mathrm{Alg}(TB)$ to the inverse of the final $\overline{B}$-coalgebra, and call it ℓ_{kl}:

$$\begin{aligned} \left(TBT(\Psi) \xrightarrow{\ell_{\mathrm{kl}}} T(\Psi)\right) &:= G_\rho(\Psi, J(\beta^{-1})^{-1}) \\ &= G_\rho(\Psi, J(\beta)) \\ &= \left(TBT(\Psi) \xrightarrow{T(\lambda)} T^2B(\Psi) \xrightarrow{\mu^T} TB(\Psi) \xrightarrow{T(\beta)} T(\psi)\right). \end{aligned}$$

By Theorem 5, this algebra is corecursive, i.e., for every coalgebra $c \colon X \to TB(X)$, there is a unique map kl_c as below:

$$\begin{array}{ccc} X & \dashrightarrow^{\mathrm{kl}_c} & T(\Psi) \\ c \downarrow & & \uparrow \ell_{\mathrm{kl}} \\ TB(X) & \xrightarrow[TB(\mathrm{kl}_c)]{} & TBT(\Psi) \end{array}$$

The trace semantics is exactly as in [12], to which we refer for examples.

6 Comparison

The presentation of trace semantics in terms of corecursive algebras allows us compare the different approaches by constructing algebra morphisms between them. In this section, we compare the Eilenberg-Moore against the logical app-roach, and the Kleisli against the logical approach as well. For a comparison between Kleisli and Eilenberg-Moore we refer to [17]. The latter is not in terms of corecursive algebras; we leave such a reformulation for future work. In [21], logical traces are also compared to determinisation constructions. But the tech-nique is different, with the primary difference that no corecursive algebras are used there.

6.1 Eilenberg-Moore and Logic

To compare the Eilenberg-Moore approach to the logical approach, we combine their assumptions. This amounts to an adjunction $F \dashv G$, endofunctors B, L and a monad T as follows:

$$L \subset \mathbf{D}^{\mathrm{op}} \underset{G}{\overset{F}{\underset{\perp}{\rightleftarrows}}} \mathbf{C} \overset{BT}{\underset{U}{\overset{\mathcal{F}}{\underset{\perp}{\rightleftarrows}}}} \mathcal{EM}(T) \supset \overline{B}$$

together with:

1. A final B-coalgebra $\zeta\colon \Theta \overset{\cong}{\Rightarrow} B(\Theta)$.
2. An $\mathcal{EM}$-law $\kappa\colon TB \Rightarrow BT$, or equivalently, a lifting $\overline{B}$ of B.
3. An initial algebra $\alpha\colon L(\Phi) \overset{\cong}{\Rightarrow} \Phi$.
4. A step $\delta\colon BG \Rightarrow GL$.
5. A step $\tau\colon TG \Rightarrow G$, whose components are $\mathcal{EM}$-algebras (a *monad action*).

The map τ is an assumption of the logical approach, but the compatibility with the monad structure was not assumed before (in the logical approach, T is not assumed to be a monad). We note that τ being a monad action is the same thing as τ being an $\mathcal{EM}$-law (involving the monad T on the left and the identity monad on the right). Therefore, by Theorem 2:

Lemma 15. *The following are equivalent:*

1. *a monad action $\tau_1\colon TG \Rightarrow G$;*
2. *a map $\tau_2\colon F \Rightarrow FT$, satisfying the obvious dual equations;*
3. *a monad morphism $\tau_4\colon T \Rightarrow GF$;*
4. *an extension $\widehat{F}\colon \mathcal{K}\ell(T) \to \mathbf{D}^{\mathrm{op}}\ (= \mathcal{K}\ell(\mathrm{Id}))$ of F.*
5. *a lifting $\widehat{G}\colon \mathbf{D}^{\mathrm{op}} \to \mathcal{EM}(T)$ of G.*

Such monad actions and the corresponding liftings are used, e.g., in [11,13,16] where $\widehat{F}$ is called Pred. We turn back to the comparison between the Eilenberg-Moore and logical approach. First, observe that since $\delta\colon BG \Rightarrow GL$ is a step, it induces a corecursive B-algebra $BG(\Phi) \overset{\delta}{\to} GL(\Phi) \overset{G(\alpha^{-1})}{\longrightarrow} G(\Phi)$. Hence, we obtain a unique map e as in the following diagram:

$$
\begin{array}{ccc}
\Theta & \xrightarrow{\quad\quad \mathsf{e} \quad\quad} & G(\Phi) \\
{\scriptstyle\zeta}\downarrow & & \uparrow{\scriptstyle G(\alpha^{-1})} \\
B(\Theta) & \xrightarrow[B(\mathsf{e})]{} BG(\Phi) \xrightarrow[\delta]{} & GL(\Phi)
\end{array}
\tag{11}
$$

This is a map from the carrier of the corecursive algebra ℓ_{em} (from the Eilenberg-Moore approach) to the carrier of the corecursive algebra ℓ^{log} (from the logical approach). Note that, by the above diagram, it is a B-algebra morphism, whereas ℓ_{em} and ℓ^{log} are BT-algebras. The following is a sufficient condition under which the map e is a BT-algebra morphism from ℓ_{em} to ℓ^{log}, which implies that the logical trace semantics factors through the Eilenberg-Moore trace semantics (Theorem 17).

Lemma 16. *The distributive law κ commutes with the logics in (6), as in:*

$$
\begin{array}{ccc}
TBG & \xrightarrow{\quad \kappa G \quad} & BTG \\
& {\scriptstyle\delta_\tau}\searrow \quad \swarrow {\scriptstyle\delta^\tau} & \\
& GL &
\end{array}
\tag{12}
$$

iff there is a natural transformation $\varrho\colon \overline{B}\widehat{G} \Rightarrow \widehat{G}L$ such that $U(\varrho) = \delta$ — where the functor $\widehat{G}\colon \mathbf{D}^{\mathrm{op}} \to \mathcal{EM}(T)$ is the lifting corresponding to τ (Lemma 15).

Proof. The existence of such a ϱ amounts to the property that each component $\delta_X : BG(X) \to GL(X)$ is a T-algebra homomorphism from $\overline{B}\widehat{G}(X)$ to $\widehat{G}L(X)$, i.e., the following diagram commutes:

$$
\begin{array}{ccc}
TBGX & \xrightarrow{\;T\delta\;} & TGLX \\
{\scriptstyle \kappa G}\downarrow & & \downarrow{\scriptstyle \tau L} \\
BTGX & & \\
{\scriptstyle B\tau}\downarrow & & \\
BGX & \xrightarrow{\;\;\delta\;\;} & GLX
\end{array}
$$

This corresponds exactly to (12). $\qquad\qquad\qquad\qquad\qquad\qquad\qquad\qquad\square$

Theorem 17. *If the equivalent conditions in Lemma 16 hold, then the map* e *defined in* (11) *is an algebra morphism from* ℓ_{em} *to* ℓ^{log}, *as on the left below.*

In that case, for any coalgebra $X \xrightarrow{c} BT(X)$ *the triangle on the right commutes.*

Proof. We use that $\ell_{em} = \zeta^{-1} \circ B(a) : BT(\Theta) \to \Theta$, where $((\Theta, a), \zeta)$ is the final $\overline{B}$-coalgebra, see Sect. 3. We need to prove that the outside of the following diagram commutes.

$$
\begin{array}{ccccccc}
 & & & \ell_{em} & & & \\
BT(\Theta) & \xrightarrow{B(a)} & B(\Theta) & \xrightarrow[\cong]{\;\;\zeta^{-1}\;\;} & & & \Theta \\
{\scriptstyle BT(e)}\downarrow & & {\scriptstyle B(e)}\downarrow & & & & \downarrow{\scriptstyle e} \\
BTG(\Phi) & \xrightarrow[B(\tau_1)]{} & BG(\Phi) & \xrightarrow{\;\;\delta\;\;} GL(\Phi) & \xrightarrow[G(\alpha^{-1})]{\cong} & G(\Phi) \\
 & & & \ell^{log} & & &
\end{array}
$$

The rectangle on the right commutes by definition of e. For the square on the left, it suffices to show $e \circ a = \tau_1 \circ T(e)$; this is equivalent to $F(a) \circ \overline{e} = \tau_2 \circ \overline{e}$ in:

$$
\Phi \xrightarrow{\;\overline{e} = F(e) \circ \epsilon\;} F(\Theta) \xrightarrow[\tau_2]{\;\;F(a)\;\;} FT(\Theta)
$$

Indeed, by transposing we have on the one hand:

$$
\overline{e \circ a} = F(a \circ e) \circ \epsilon_{\Phi} = F(a) \circ F(e) \circ \epsilon_{\Phi} = F(a) \circ \overline{e}
$$

And on the other hand, using that $\tau_2 = F(\tau_1 \circ T(\eta)) \circ \epsilon$,

$$
\begin{aligned}
\tau_2 \circ \overline{e} &= F(\tau_1 \circ T(\eta)) \circ \epsilon \circ F(e) \circ \epsilon \\
&= F(\tau_1 \circ T(\eta)) \circ FG(F(e) \circ \epsilon) \circ \epsilon \\
&= F\big(G(F(e) \circ \epsilon) \circ \tau_1 \circ T(\eta)\big) \circ \epsilon \\
&= F\big(\tau_1 \circ TG(F(e) \circ \epsilon) \circ T(\eta)\big) \circ \epsilon \\
&= F\big(\tau_1 \circ T(G(\epsilon) \circ GF(e) \circ \eta)\big) \circ \epsilon = F\big(\tau_1 \circ T(e)\big) \circ \epsilon = \overline{\tau_1 \circ T(e)}.
\end{aligned}
$$

By transposing the maps in (11), it follows that $\overline{e} \colon \Phi \to F(\Theta)$ is the unique morphism from the initial L-algebra to $F(\zeta) \circ \delta_2 \colon LF(\Theta) \to F(\Theta)$. Hence, for the desired equality $F(a) \circ \overline{e} = \tau_2 \circ \overline{e}$, it suffices to prove that $F(a)$ and τ_2 are both algebra homomorphisms from $F(\zeta) \circ \delta_2$ to a common algebra, which in turn follows from commutativity of the following diagram.

$$
\begin{array}{ccccc}
LF(\Theta) & \xrightarrow{\ L(\tau_2)\ } & LFT(\Theta) & \xleftarrow{\ LF(a)\ } & LF(\Theta) \\
& & \downarrow{\scriptstyle \delta_2 T} & & \downarrow{\scriptstyle \delta_2} \\
\downarrow{\scriptstyle \delta_2} & & FBT(\Theta) & \xleftarrow{\ FB(a)\ } & FB(\Theta) \\
& & \downarrow{\scriptstyle F\kappa} & & \\
FB(\Theta) & \xrightarrow{\ \tau_2 B\ } & FTB(\Theta) & & \downarrow{\scriptstyle F(\zeta)} \\
{\scriptstyle F(\zeta)}\downarrow & & \downarrow{\scriptstyle FT(\zeta)} & & \\
F(\Theta) & \xrightarrow[\ \tau_2\ ]{} & FT(\Theta) & \xleftarrow[\ F(a)\ ]{} & F(\Theta)
\end{array}
$$

Using the translation $(-)_1 \leftrightarrow (-)_2$ (of Theorem 2), one shows that the upper-left rectangle is equivalent to the assumption (12). To see this, we use that $(\delta^\tau)_2 = (\delta_1 \circ B\tau_1)_2 = \delta_2 T \circ L\tau_2$ and $(\delta_\tau)_2 = (\tau_1 L \circ T\delta_1)_2 = \tau_2 B \circ \delta_2$ (as stated, e.g., in [21]); moreover, it is easy to check that $(\delta_1 \circ B\tau_1 \circ \kappa G)_2 = F\kappa \circ (\delta_1 \circ B\tau_1)_2$. The lower-right rectangle commutes since $((\Theta, a), \zeta)$ is a $\overline{B}$-coalgebra. The other two squares commute by naturality.

For the second part of the theorem, let $c \colon X \to BT(X)$ be a coalgebra. Since e is an algebra morphism, the equation $e \circ \mathrm{em}_c = \log_c$ follows by uniqueness of morphisms from c to the corecursive algebra on $G(\Phi)$. $\qquad\square$

The equality $e \circ \mathrm{em}_c = \log_c$ means that equivalence wrt Eilenberg-Moore trace semantics implies equivalence wrt the logical trace semantics. The converse is, of course, true if e is monic. For that, it is sufficient if $\delta \colon BG \Rightarrow GL$ is *expressive*. Here expressiveness is the property that for any B-coalgebra, the unique coalgebra-to-algebra morphism to the corecursive algebra on $G(\Phi)$ factors as a B-coalgebra homomorphism followed by a mono. This holds in particular if the components $\delta_A \colon BG(A) \to GL(A)$ are all monic (in **C**) [20].

Lemma 18. *If $\delta \colon BG \Rightarrow GL$ is expressive, then e is monic. Moreover, if δ is an isomorphism, then e is an iso as well.*

Proof. Expressivity of δ means that we have $\mathsf{e} = m \circ h$ for some coalgebra homomorphism h and mono m. By finality of ζ there is a B-coalgebra morphism h' such that $h' \circ h = \mathrm{id}$. It follows that h is monic (in **C**), so that $m \circ h = \mathsf{e}$ is monic too. For the second claim, if δ is an isomorphism, then $G(\alpha^{-1}) \circ \delta: BG(\varPhi) \to G(\varPhi)$ is an invertible corecursive B-algebra, which implies it is a final coalgebra (see [5, Proposition 7], which states the dual). It then follows from (11) that e is a coalgebra morphism from one final B-coalgebra to another, which means it is an isomorphism.   □

Previously, we have seen both a class of examples of the Eilenberg-Moore approach (Theorem 8), and the logical approach (Proposition 11). Both arise from the same data: a monad T (just a functor in the logical approach) and an $\mathcal{EM}$-algebra t. We thus obtain, for these automata-like examples, both a logical trace semantics and a matching 'Eilenberg-Moore' semantics, where the latter essentially amounts to a determinisation procedure. The underlying distributive laws satisfy (12) by construction, so that the two approaches coincide (as already seen in the concrete examples).

Theorem 19. *Let $\varOmega$ be a set, $T: $ **Sets** $\to$ **Sets** *a monad and* $t: T(\varOmega) \to \varOmega$ *an $\mathcal{EM}$-algebra. The $\mathcal{EM}$-law κ of Theorem 8, together with δ, τ as defined in the proof of Proposition 11, satisfies (12). For any coalgebra $c: X \to \varOmega \times T(X)^A$, the map $\log_c$ coincides (up to isomorphism) with the map em_c.*

Proof. To prove (12), *i.e.*, $\delta^\tau \circ \kappa = \delta_\tau$, we first compute, following (6),

$$(\delta^\tau)_X = \delta_X \circ (\mathrm{id} \times \tau_X^A) = \delta_X \circ (\mathrm{id} \times (t^X \circ \mathrm{st})^A) : \varOmega \times (T(\varOmega^X))^A \to \varOmega^{A \times X + 1}$$
$$(\delta_\tau)_X = \tau_{A \times X + 1} \circ T(\delta_X) = t^{A \times X + 1} \circ \mathrm{st} \circ T(\delta_X) : T(\varOmega \times (\varOmega^X)^A) \to \varOmega^{A \times X + 1}$$

Hence, we need to show that

$$\delta_X \circ (\mathrm{id} \times (t^X \circ \mathrm{st})^A) \circ (t \times \mathrm{st}) \circ \langle T(\pi_1), T(\pi_2) \rangle = t^{A \times X + 1} \circ \mathrm{st} \circ T(\delta_X) \quad (13)$$

for every set X. To this end, let $S \in T(\varOmega \times (\varOmega^X)^A)$ and $t \in (A \times X + 1)$. We first spell out the right-hand side:

$$
\begin{aligned}
&(t^{A \times X + 1} \circ \mathrm{st} \circ T(\delta_X)(S))(t) \\
&= t((\mathrm{st} \circ T(\delta_X)(S))(t)) \\
&= t(T(\mathrm{ev}_t \circ \delta_X)(S)) \\
&= \begin{cases} t(T(\pi_1)(S)) & \text{if } t = * \in 1 \\ t(T(\mathrm{ev}_x \circ \mathrm{ev}_a \circ \pi_2)(S)) & \text{if } t = (a, x) \in A \times X \end{cases}
\end{aligned}
$$

In the last step, we used the definition of δ:

$$\mathrm{ev}_* \circ \delta_X(\omega, f) = \delta_X(\omega, f)(*) = \omega = \pi_1(\omega, f),$$
$$\mathrm{ev}_{(a,x)} \circ \delta_X(\omega, f) = \delta_X(\omega, f)(a, x) = f(a)(x) = \mathrm{ev}_x \circ \mathrm{ev}_a \circ \pi_2(\omega, f).$$

For the left-hand side of (13), distinguish cases $* \in 1$ and $(a, x) \in A \times X$.

$$(\delta_X \circ (\mathrm{id} \times (t^X \circ \mathrm{st})^A) \circ (t \times \mathrm{st}) \circ \langle T(\pi_1), T(\pi_2) \rangle(S))(*)$$
$$= \pi_1(\mathrm{id} \times (t^X \circ \mathrm{st})^A) \circ (t \times \mathrm{st}) \circ \langle T(\pi_1), T(\pi_2) \rangle(S))$$
$$= t(T(\pi_1)(S))$$

which matches the right-hand side of (13). For $(a, x) \in A \times X$, we have:

$$(\delta_X \circ (\mathrm{id} \times (t^X \circ \mathrm{st})^A) \circ (t \times \mathrm{st}) \circ \langle T(\pi_1), T(\pi_2) \rangle(S))(a, x)$$
$$= (((t^X \circ \mathrm{st})^A \circ \mathrm{st})(T(\pi_2)(S)))(a)(x)$$
$$= (((t^X)^A \circ \mathrm{st}^A \circ \mathrm{st})(T(\pi_2)(S)))(a)(x)$$
$$= (t^X \circ \mathrm{st}(\mathrm{st}(T(\pi_2)(S))(a)))(x)$$
$$= (t^X \circ \mathrm{st}(T(\mathrm{ev}_a)(T(\pi_2)(S))))(x)$$
$$= (t^X \circ \mathrm{st}(T(\mathrm{ev}_a \circ \pi_2)(S)))(x)$$
$$= t(\mathrm{st}(T(\mathrm{ev}_a \circ \pi_2)(S))(x))$$
$$= t(T(\mathrm{ev}_x) \circ T(\mathrm{ev}_a \circ \pi_2)(S))$$
$$= t(T(\mathrm{ev}_x \circ \mathrm{ev}_a \circ \pi_2)(S))$$

which also matches the right-hand side, hence we obtain (13) as desired.

Since (12) is satisfied, it follows from Theorem 17 that $e \circ \mathrm{em}_c = \log_c$. Since δ is an iso, e is an iso as well by Lemma 18.   □

6.2   Kleisli and Logic

To compare the Kleisli approach to the logical approach, we combine their assumptions. This amounts to an adjunction $F \dashv G$, endofunctors B, L and a monad T as follows:

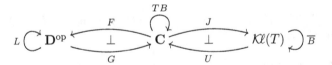

together with:

1. An initial algebra $\beta \colon B(\Psi) \overset{\cong}{\Rightarrow} \Psi$.
2. A $\mathcal{K}\ell$-law $\lambda \colon BT \Rightarrow TB$, or equivalently, an extension $\overline{B}$ of B.
3. $(\Psi, J(\beta^{-1}))$ is a final $\overline{B}$-coalgebra.
4. An initial algebra $\alpha \colon L(\Phi) \overset{\cong}{\Rightarrow} \Phi$.
5. A step $\delta \colon BG \Rightarrow GL$.
6. A step $\tau \colon TG \Rightarrow G$, whose components are $\mathcal{EM}$-algebras (a *monad action*).

Again, we assume τ to be compatible with the monad, satisfying the equivalent conditions in Lemma 15. Since δ is a step, we obtain the following unique coalgebra-to-algebra morphism k from the initial B-algebra:

$$
\begin{array}{ccc}
\Psi & \xrightarrow{\quad k \quad} & G(\Phi) \\
{\scriptstyle \beta^{-1}}\downarrow & & \uparrow{\scriptstyle G(\alpha^{-1})} \\
B(\Psi) & \xrightarrow[\quad B(k) \quad]{} BG(\Phi) \xrightarrow{\ \delta\ } & GL(\Phi)
\end{array}
\tag{14}
$$

Since τ is a monad action, for every X, $G(X)$ carries an Eilenberg-Moore algebra τ_X. Thus we can take the adjoint transpose $\overline{k} = \tau_\Phi \circ T(k)\colon T(\Psi) \to G(\Phi)$. We have the following analogue of Theorem 17.

Lemma 20. *The distributive law λ commutes with the logics in (6), as in:*

$$
\begin{array}{ccc}
BTG & \xrightarrow{\quad \lambda G \quad} & TBG \\
& {\scriptstyle \delta^\tau}\searrow \quad \swarrow{\scriptstyle \delta_\tau} & \\
& GL &
\end{array}
\tag{15}
$$

iff there is a natural transformation $\varrho\colon L\widehat{F} \Rightarrow \widehat{F}B$ given by $\varrho J = \delta$ — where the functor $\widehat{F}\colon \mathcal{K}\ell(T) \to \mathbf{D}^{\mathrm{op}}$ is the extension corresponding to τ (Lemma 15).

Proof. The condition $\varrho J = \delta$ simply means that $\varrho_X = \delta_X$ for every object X in **C**. Naturality of ϱ amounts to commutativity of the outside of the diagram below, for every map $f\colon X \to T(Y)$.

$$
\begin{array}{ccc}
LF(Y) \xrightarrow{\ \delta\ } FB(Y) \xrightarrow{\ \tau B\ } FTB(Y) \\
{\scriptstyle L\tau}\downarrow \qquad\qquad\qquad\qquad \downarrow{\scriptstyle F\lambda} \\
LFT(Y) \xrightarrow{\qquad\qquad\ \delta T\ \qquad\qquad} FBT(Y) \\
{\scriptstyle LF(f)}\downarrow \qquad\qquad\qquad\qquad \downarrow{\scriptstyle FB(f)} \\
LF(X) \xrightarrow{\qquad\qquad\ \delta\ \qquad\qquad} FB(X)
\end{array}
$$

The lower rectangle commutes by naturality, the upper is equivalent to (15). Hence, (15) implies naturality. Conversely, if ϱ is natural, then the upper rectangle commutes for each Y by taking $f = \mathrm{id}_{TY}$ (the identity map in **C**). $\square$

Theorem 21. *If the equivalent conditions in Lemma 20 hold, then the map $\overline{k} = \tau_\Phi \circ T(k)\colon T(\Psi) \to G(\Phi)$ is an algebra morphism from ℓ_{kl} to ℓ_{log}, as on the left below.*

$$
\begin{array}{ccc}
TBT(\Psi) & \xrightarrow{\ TB(\overline{k})\ } & TBG(\Phi) \\
{\scriptstyle \ell_{\mathrm{kl}}}\downarrow & & \downarrow{\scriptstyle \ell_{\mathrm{log}}} \\
T(\Psi) & \xrightarrow{\quad \overline{k}\quad } & G(\Phi)
\end{array}
\qquad\qquad
\begin{array}{ccc}
& X & \\
{\scriptstyle \mathrm{kl}_c}\swarrow & & \searrow{\scriptstyle \mathrm{log}_c} \\
T(\Psi) & \xrightarrow{\quad \overline{k}\quad } & G(\Phi)
\end{array}
$$

In that case, for any coalgebra $c\colon X \to TB(X)$ there is a commuting triangle as on the right above.

Proof. Consider the following diagram.

$$
\begin{array}{c}
\text{diagram}
\end{array}
$$

Everything commutes: the upper right rectangle by assumption (15), the rightmost square in the middle row since τ is an action, the outer shapes by definition of ℓ_{kl} and $\ell_{\log}$, the lower left rectangle by (14) and the rest by naturality. □

The above result gives a sufficient condition under which 'Kleisli' trace equivalence implies logical trace equivalence. However, contrary to the case of traces in Eilenberg-Moore, in Lemma 18, we currently do not have a converse. If δ has monic components, then it is easy to use corecursiveness to define a map from $\ell_{\log}$ to ℓ_{kl}, but this surprisingly is not sufficient to show $\overline{k}$ to be monic, as confirmed by Example 22 below. In the comparison between Eilenberg-Moore and Kleisli traces [17], a similar difficulty arises: it is unclear under what conditions the map from the final coalgebra in Kleisli to the final coalgebra in Eilenberg-Moore obtained there is mono (and hence, if Eilenberg-Moore trace equivalence implies Kleisli trace equivalence).

Example 22. We give an example where $\delta\colon BG \Rightarrow GL$ is monic and (15) commutes, but where nevertheless logical equivalence is stronger than 'Kleisli' trace equivalence. Let $\mathbf{C} = \mathbf{D} = \mathbf{Sets}$, $F = G = 2^-$, $B = L = (A \times -) + 1$, $T = \mathcal{P}$, $\tau\colon \mathcal{P}2^- \Rightarrow 2^-$ given by union as before, and define the step δ by $\delta_X(a, \varphi)(t) = \top$ iff $\exists x.t = (a, x) \wedge \varphi(x)$, and $\delta_X(*)(t) = \top$ (the latter differs from the step in Proposition 13). Notice that δ indeed has monic components.

Let $\lambda\colon BT \Rightarrow TB$ be the distributive law from [12], given by $\lambda_X(a, S) = \{(a, x) \mid x \in S\}$ and $\lambda(*) = \{*\}$. Then (15) is satisfied:

$$
\begin{array}{ccc}
A \times \mathcal{P}(2^X) + 1 & \xrightarrow{\ \ \lambda\ \ } & \mathcal{P}(A \times 2^X + 1) \\
{\scriptstyle \mathrm{id} \times \tau + 1}\downarrow & & \downarrow{\scriptstyle \mathcal{P}(\delta)} \\
A \times 2^X + 1 & \xrightarrow{\ \delta\ } 2^{A \times X + 1} \xleftarrow{\ \tau\ } & \mathcal{P}(2^{A \times X + 1})
\end{array}
$$

It is straightforward to check that this commutes. However, given a coalgebra $f\colon X \to TB(X)$, the induced logical semantics $\log\colon X \to 2^{A^*}$ is: $\log(x)(w) = \top$ iff $* \in f(x)$ or $\exists a \in A, v \in A^*, y \in X.w = av \wedge (a, y) \in f(x) \wedge \log(y)(v) = \top$. In particular, this means that if $* \in f(x)$ and $* \in f(y)$ for some states x, y, then they are trace equivalent. This differs from the Kleisli semantics, which amounts to the usual language semantics of non-deterministic automata [12].

Cîrstea [8] compares logical traces to a 'path-based semantics', which resembles the Kleisli approach (as well as [22]) but does not require a final $\overline{B}$-coalgebra. In particular, given a commutative monad T on **Sets** and a signature Σ, she considers a canonical distributive law $\lambda\colon H_\Sigma T \Rightarrow T H_\Sigma$, which coincides with the one in [12]. Cîrstea shows that, with $\Omega = T(1)$, $t = \mu_1\colon TT(1) \to T(1)$ and δ from the proof of Proposition 13 (assuming $T1$ to have enough structure to define that logic), the triangle (15) commutes (see [8, Lemma 5.12]).

7   Completely Iterative Algebras

In this paper, we constructed several corecursive algebras. We briefly show that they all satisfy the following stronger property [26].

Definition 23. *For an endofunctor H on $\mathbf{C}$, an H-algebra $a\colon HA \to A$ is completely iterative when $[\mathrm{id}, a]$ is a corecursive $A + H$-coalgebra. Explicitly: when for every $c\colon X \to A + HX$ there is a unique $f\colon X \to A$ such that the following diagram commutes.*

$$
\begin{array}{ccc}
X & \xrightarrow{\quad f \quad} & A \\
{\scriptstyle c}\downarrow & & \downarrow{\scriptstyle [\mathrm{id}, a]} \\
A + HX & \xrightarrow[\;A+Hf\;]{} & A + HA
\end{array}
$$

Following [14, 26], we have two ways of constructing such algebras.

Proposition 24. *1. If $\zeta\colon A \to HA$ is a final H-coalgebra, then (A, ζ^{-1}) is completely iterative.*
2. Given a step as in Sect. 2, the functor G_ρ preserves complete iterativity.

We may thus say: *"step-induced algebra liftings of right adjoints preserve complete iterativity"*. Consequently, by analogy with Theorem 5, if L has a final coalgebra (Ψ, ζ) then $G_\rho(A, \zeta^{-1})$ is completely iterative. For our examples, this may be seen as a trace semantics for a coalgebra c that may sometimes stop following the behaviour functor and instead provide semantics directly.

8   Future Work

The main contribution of this paper is a general treatment of trace semantics via corecursive algebras, constructed through an adjunction and a step, covering the 'Eilenberg-Moore', 'Kleisli' and 'logic' approaches to trace semantics. It is expected that our framework also works for other examples, such as the 'quasi-liftings' in [2], but this is left for future work. In [19], several examples of adjunctions are discussed in the context of automata theory, some of them the same as the adjunctions here, but with the aim of lifting them to categories of coalgebras, under the condition that what we call the step is an iso. In our case, it usually is not an iso, since the behaviour functor is a composite TB or BT;

however, it remains interesting to study cases in which such adjunction liftings appear, as used for instance in the aforementioned paper and [21,28]. Further, our treatment in Sect. 3 (Eilenberg-Moore) assumes a monad to construct the corecursive algebra, but it was shown by Bartels [1] that this algebra is also corecursive when the underlying category has countable coproducts (and dropping the monad assumption). We currently do not know whether this fits our abstract approach. Finally, the Eilenberg-Moore/logic and Kleisli/logic comparisons (Sect. 6) seem to share certain aspects (the conditions look very similar), but so far we have been unable to derive a general perspective on such comparisons that covers both, and possibly also the Eilenberg-Moore/Kleisli comparison of [17].

Acknowledgement. We are grateful to the anonymous referees for various comments and suggestions.

References

1. Bartels, F.: Generalised coinduction. Math. Struct. Comput. Sci. **13**(2), 321–348 (2003)
2. Bonchi, F., Silva, A., Sokolova, A.: The power of convex algebras. In: Meyer, R., Nestmann, U. (eds.) 28th International Conference on Concurrency Theory, CONCUR 2017, LIPIcs. vol. 85, pp. 23:1–23:18. Schloss Dagstuhl - Leibniz-Zentrum fuer Informatik (2017)
3. Bonsangue, M.M., Kurz, A.: Duality for logics of transition systems. In: Sassone, V. (ed.) FoSSaCS 2005. LNCS, vol. 3441, pp. 455–469. Springer, Heidelberg (2005). https://doi.org/10.1007/978-3-540-31982-5_29
4. Bonsangue, M., Milius, S., Silva, A.: Sound and complete axiomatizations of coalgebraic language equivalence. ACM Trans. Comput. Log. **14**(1), 7:1–7:52 (2013)
5. Capretta, V., Uustalu, T., Vene, V.: Recursive coalgebras from comonads. Inf. Comput. **204**(4), 437–468 (2006)
6. Capretta, V., Uustalu, T., Vene, V.: Corecursive algebras: a study of general structured corecursion. In: Oliveira, M.V.M., Woodcock, J. (eds.) SBMF 2009. LNCS, vol. 5902, pp. 84–100. Springer, Heidelberg (2009). https://doi.org/10.1007/978-3-642-10452-7_7
7. Chen, L.-T., Jung, A.: On a categorical framework for coalgebraic modal logic. Electron. Notes Theor. Comput. Sci. **308**, 109–128 (2014)
8. Cîrstea, C.: A coalgebraic approach to quantitative linear time logics. CoRR, abs/1612.07844 (2016)
9. Eppendahl, A.: Coalgebra-to-algebra morphisms. Electron. Notes Theor. Comput. Sci. **29**, 42–49 (1999)
10. Girard, J.Y., Lafont, Y., Taylor, P.: Proofs and Types. Cambridge Tracts in Theoretical Computer Science 7. Cambridge University Press, Cambridge (1988)
11. Hasuo, I.: Generic weakest precondition semantics from monads enriched with order. Theor. Comput. Sci. **604**, 2–29 (2015)
12. Hasuo, I., Jacobs, B., Sokolova, A.: Generic trace semantics via coinduction. Log. Methods Comput. Sci. **3**(4) (2007)
13. Hino, W., Kobayashi, H., Hasuo, I., Jacobs, B.: Healthiness from duality. In: Logic in Computer Science. Computer Science Press, IEEE (2016)

14. Hinze, R., Wu, N., Gibbons, J.: Conjugate hylomorphisms - or: the mother of all structured recursion schemes. In: Rajamani, S.K., Walker, D. (eds.) Proceedings of the Symposium on Principles of Programming Languages, POPL 2015, pp. 527–538. ACM (2015)

15. Jacobs, B.: A bialgebraic review of deterministic automata, regular expressions and languages. In: Futatsugi, K., Jouannaud, J.-P., Meseguer, J. (eds.) Algebra, Meaning, and Computation. LNCS, vol. 4060, pp. 375–404. Springer, Heidelberg (2006). https://doi.org/10.1007/11780274_20

16. Jacobs, B.: A recipe for state and effect triangles. Log. Methods Comput. Sci. **13**(2), (2017). https://lmcs.episciences.org/3660

17. Jacobs, B., Silva, A., Sokolova, A.: Trace semantics via determinization. J. Comput. Syst. Sci. **81**(5), 859–879 (2015)

18. Kelly, G.M., Street, R.: Review of the elements of 2-categories. In: Kelly, G.M. (ed.) Category Seminar. LNM, vol. 420, pp. 75–103. Springer, Heidelberg (1974). https://doi.org/10.1007/BFb0063101

19. Kerstan, H., König, B., Westerbaan, B.: Lifting adjunctions to coalgebras to (re)discover automata constructions. In: Bonsangue, M.M. (ed.) CMCS 2014 2014. LNCS, vol. 8446, pp. 168–188. Springer, Heidelberg (2014). https://doi.org/10.1007/978-3-662-44124-4_10

20. Klin, B.: Coalgebraic modal logic beyond sets. In: Fiore, M. (ed.) Mathematical Foundations of Programming Semantics, ENTCS. vol. 173. Elsevier, Amsterdam (2007)

21. Klin, B., Rot, J.: Coalgebraic trace semantics via forgetful logics. Log. Methods Comput. Sci. **12**(4:10) (2016). https://lmcs.episciences.org/2622

22. Kurz, A., Milius, S., Pattinson, D., Schröder, L.: Simplified coalgebraic trace equivalence. In: De Nicola, R., Hennicker, R. (eds.) Software, Services, and Systems. LNCS, vol. 8950, pp. 75–90. Springer, Cham (2015). https://doi.org/10.1007/978-3-319-15545-6_8

23. Leinster, T.: Higher Operads, Higher Categories. London Mathematical Society Lecture Notes, vol. 298. Cambridge University Press, Cambridge (2004)

24. Levy, P.: Final coalgebras from corecursive algebras. In: Moss, L., Sobocinski, P. (eds.) Conference on Algebra and Coalgebra in Computer Science (CALCO 2015), LIPIcs. vol. 35, pp. 221–237. Schloss Dagstuhl (2015)

25. Milius, S., Pattinson, D., Schröder, L.: Generic trace semantics and graded monads. In: 6th Conference on Algebra and Coalgebra in Computer Science, CALCO 2015, Nijmegen, The Netherlands, 24–26 June 2015, pp. 253–269 (2015)

26. Milius, S.: Completely iterative algebras and completely iterative monads. Inf. Comput. **196**(1), 1–41 (2005)

27. Pavlovic, D., Mislove, M., Worrell, J.B.: Testing semantics: connecting processes and process logics. In: Johnson, M., Vene, V. (eds.) AMAST 2006. LNCS, vol. 4019, pp. 308–322. Springer, Heidelberg (2006). https://doi.org/10.1007/11784180_24

28. Rot, J.: Coalgebraic minimization of automata by initiality and finality. Electron. Notes Theor. Comput. Sci. **325**, 253–276 (2016)

29. Silva, A., Bonchi, F., Bonsangue, M., Rutten, J.: Generalizing determinization from automata to coalgebras. Log. Methods Comput. Sci. **9**(1) (2013)

30. Turi, D., Plotkin, G.D.: Towards a mathematical operational semantics. In: Proceedings of the 12th Annual IEEE Symposium on Logic in Computer Science, Warsaw, Poland, 29 June–2 July 1997, pp. 280–291. IEEE Computer Society (1997)

On Algebras with Effectful Iteration

Stefan Milius[2][(✉)], Jiří Adámek[1], and Henning Urbat[2]

[1] Czech Technical University in Prague, Prague, Czech Republic
[2] Friedrich-Alexander-Universität Erlangen-Nürnberg, Erlangen, Germany
stefan.milius@fau.de

Abstract. For every finitary monad T on sets and every endofunctor F on the category of T-algebras we introduce the concept of an ffg-Elgot algebra for F, that is, an algebra admitting coherent solutions for finite systems of recursive equations with effects represented by the monad T. The goal of this paper is to study the existence and construction of free ffg-Elgot algebras. To this end, we investigate the locally ffg fixed point φF, the colimit of all F-coalgebras with free finitely generated carrier, which is shown to be the initial ffg-Elgot algebra. This is the technical foundation for our main result: the category of ffg-Elgot algebras is monadic over the category of T-algebras.

1 Introduction

Terminal coalgebras yield a fully abstract domain of behavior for a given type of state-based systems whose transition type is described by an endofunctor F. Often one is mainly interested in the study of the semantics of *finite* coalgebras; for instance, regular languages are the behaviors of finite deterministic automata, while the terminal coalgebra of the corresponding functor is formed by *all* formal languages. For endofunctors on sets, the *rational fixed point* introduced by Adámek et al. [2] yields a fully abstract domain of behavior for finite coalgebras. However, in recent years there has been a lot of interest in studying coalgebras over more general categories than sets. In particular, categories of algebras for a (finitary) monad T on sets are a paradigmatic setting; they are used, for instance, in the generalized determinization framework of Silva et al. [30] and yield *coalgebraic language equivalence* [9] as a semantic equivalence of systems with a side effect provided by the monad T. In the category $\mathscr{C}$ of T-algebras, several notions of 'finite' object are natural to consider, and each of those yields an ensuing notion of 'finite' coalgebra: free objects on finitely many generators (*ffg* objects) yield precisely the coalgebras that are the target of generalized determinization; finitely presentable (*fp*) objects are the ones that can be presented by finitely many generators and relations and yield the rational fixed point; and finitely generated (*fg*) objects, i.e. those presented by finitely many generators

---

S. Milius—Supported by Deutsche Forschungsgemeinschaft (DFG) under project MI 717/5-1.

C. Cîrstea (Ed.): CMCS 2018, LNCS 11202, pp. 144–166, 2018.
https://doi.org/10.1007/978-3-030-00389-0_9

(but possibly infinitely many relations). Taking the colimits of all coalgebras with ffg, fp, and fg carriers, respectively, yields three coalgebras φF, ϱF and ϑF which, under suitable assumptions on F, are all fixed points of F [2,24,34]. Our present paper is devoted to studying the fixed point φF, which we call the *locally ffg fixed point* of F. For a finitary endofunctor F preserving surjective and non-empty injective morphisms in $\mathscr{C}$, the three fixed points are related (to the terminal coalgebra νF) as shown in the picture below:

$$\varphi F \twoheadrightarrow \varrho F \twoheadrightarrow \vartheta F \rightarrowtail \nu F, \tag{1.1}$$

where $\twoheadrightarrow$ denotes a quotient coalgebra and $\rightarrowtail$ a subcoalgebra. The three right-hand fixed points are characterized by a universal property both as a coalgebra and (inverting their coalgebra structure) as an algebra [2,19,24]; see [34] for one uniform proof. We recall this in more detail in Sect. 2.4.

The main contribution of this paper is a new characterization of the locally ffg fixed point φF by a universal property as an algebra. As already observed by Urbat [34], as a coalgebra, φF does not satisfy the expected finality property. A simple initiality property of φF as an algebra was recently established by Milius [21]. Here we go a step further and introduce the notion of an *ffg-Elgot algebra* (Sect. 4), which is an algebra for F equipped with an operation that allows to take solutions of *effectful iterative equations* (see Remark 4.5) subject to two natural axioms. These axioms are inspired by and closely related to the axioms of (ordinary) Elgot algebras [1], which we recall in Sect. 3. We then prove that φF is the initial ffg-Elgot algebra (Theorem 4.11).

In addition, we study the construction of *free* ffg-Elgot algebras. In the case of ordinary Elgot algebras, it was shown in [1] that the parametrized rational fixed point $\varrho(F(-)+Y)$ is a free Elgot algebra on Y. In addition, the category of Elgot algebras is the Eilenberg-Moore category for the corresponding monad on $\mathscr{C}$. In the present paper, we first prove that free ffg-Elgot algebras exist on every object Y of $\mathscr{C}$. But is it true that the free ffg-Elgot algebra on Y is $\varphi(F(-)+Y)$? We do not know the answer for arbitrary objects Y, but if Y is a free T-algebra (on a possibly infinite set of generators), the answer is affirmative (Theorem 4.15).

Finally, we prove that the category of ffg-Elgot algebras is monadic over $\mathscr{C}$, i.e. ffg-Elgot algebras are precisely the Eilenberg-Moore algebras for the monad that assigns to a given object Y of $\mathscr{C}$ its free ffg-Elgot algebra (Theorem 4.16). Full proofs of all results presented here can be found in [18].

2   Preliminaries

2.1   Varieties and 'Finite' Algebras

Throughout the paper we will work with a (finitary, many-sorted) variety $\mathscr{C}$ of algebras. Equivalently, $\mathscr{C}$ is the category of Eilenberg-Moore algebras for a finitary monad T on the category Set^S of S-sorted sets [6]. We will speak about objects of $\mathscr{C}$ (rather than algebras for T) and reserve the word 'algebra' for algebras for an endofunctor on $\mathscr{C}$. All the usual categories of algebraic structures and their homomorphisms are varieties: monoids, (semi-)groups, rings, vector spaces over a fixed field, modules for a (semi-)ring, positive convex algebras, join-semilattices, Boolean algebras, distributive lattices, and many others. In each case, the corresponding monad T assigns to a set the free object on it, e.g. $TX = X^*$ for monoids, the finite power-set monad $T = \mathcal{P}_f$ for join-semilattices, and the subdistribution monad $\mathcal{D}$ for positive convex algebras, etc.

As mentioned in the introduction, every variety $\mathscr{C}$ of algebras comes with three natural notions of 'finite' objects, each of which admits a neat category-theoretic characterization (see [6]):

Finitely presentable objects (fp objects, for short) can be presented by finitely many generators and relations. An object X is fp iff the covariant hom-functor $\mathscr{C}(X, -): \mathscr{C} \to Set$ is *finitary*, i.e. it preserves filtered colimits.[1] We denote by $\mathscr{C}_{\mathsf{fp}}$ the full subcategory of $\mathscr{C}$ given by all fp objects. In our proofs we will use the well-known fact that every object X is the filtered colimit of the canonical diagram $\mathscr{C}_{\mathsf{fp}}/X \to \mathscr{C}$, i.e. objects in the diagram scheme are morphisms $P \to X$ in $\mathscr{C}$ with P fp.

Finitely generated objects (fg objects, for short) are presented by finitely many generators but, possibly, infinitely many relations. An object X is fg iff $\mathscr{C}(X, -)$ preserves filtered colimits with monic connecting morphisms. Hence, every fp object is fg but not conversely. In fact, the fg objects are precisely the (regular) quotients of the fp objects [6, Proposition 5.22].

Free finitely generated objects (ffg objects, for short) are the objects (TX_0, μ_{X_0}) where X_0 is a finite S-sorted set (i.e. the coproduct of all components X_s, $s \in S$ is finite). An object X is a split quotient of an ffg object iff $\mathscr{C}(X, -)$ preserves *sifted* colimits [6, Corollary 5.14]. Recall from [6] that sifted colimits are more general than filtered colimits: a sifted colimit is a colimit of a diagram $D: \mathcal{D} \to \mathscr{C}$ whose diagram scheme $\mathcal{D}$ is a sifted category, which means that finite products commute with colimits over $\mathcal{D}$ in Set. For instance, every filtered category and every category with finite coproducts is sifted, see [6, Example 2.16].

The category $\mathscr{C}$ is cocomplete and the forgetful functor $\mathscr{C} \to Set^S$ preserves and reflects sifted colimits, that is, sifted colimits in $\mathscr{C}$ are formed on the level of underlying sets [6, Proposition 2.5].

A finitely cocomplete category has sifted colimits if and only if it has filtered colimits and reflexive coequalizers, and, moreover a functor preserves sifted colimits if and only if it preserves filtered colimits and reflexive coequalizers [5].

We denote by $\mathscr{C}_{\mathsf{ffg}}$ the full subcategory of ffg objects of $\mathscr{C}$. Analogously to the corresponding result for fp objects, every object X is a sifted colimit of the canonical diagram $\mathscr{C}_{\mathsf{ffg}}/X \to \mathscr{C}$; this follows from [6, Proposition 5.17].

2.2   Relation Between the Object Classes

We already mentioned that every fp object is fg (but not conversely, in general). Clearly, every ffg object is fg, but not conversely in general. So, in general, we have full embeddings

---

[1] These are colimits of diagrams $D: \mathcal{D} \to \mathscr{C}$ where $\mathcal{D}$ is *filtered*, i.e. every finite subcategory $\mathcal{D}_0 \hookrightarrow \mathcal{D}$ has a cocone in $\mathcal{D}$.

$$\mathscr{C}_{\mathsf{ffg}} \overset{\neq}{\hookrightarrow} \mathscr{C}_{\mathsf{fp}} \overset{\neq}{\hookrightarrow} \mathscr{C}_{\mathsf{fg}}.$$

In rare cases, all three object classes coincide; e.g. in Set (considered as a variety) and the category of vector spaces over a field.

In addition to those examples, the equation $\mathscr{C}_{\mathsf{fg}} = \mathscr{C}_{\mathsf{fp}}$ holds true, for example, for all locally finite varieties (i.e. where ffg objects are carried by finite sets), for positive convex algebras [31], commutative monoids [14,29], abelian groups, and more generally, in any category of (semi-)modules for a semiring $ that is *Noetherian* in the sense of Ésik and Maletti [12], i.e. every subsemimodule of an fg semimodule is fg itself. For example, the following semirings are Noetherian: every finite semiring, every field, every principal ideal domain such as the ring of integers and therefore every finitely generated commutative ring by Hilbert's Basis Theorem. The tropical semiring $(\mathbb{N} \cup \{\infty\}, \min, +, \infty, 0)$ is not Noetherian [11]. The usual semiring of natural numbers is also not Noetherian, but for the category of $\mathbb{N}$-semimodules (= commutative monoids), $\mathscr{C}_{\mathsf{fp}} = \mathscr{C}_{\mathsf{fg}}$ still holds.

2.3   Functors and Liftings

We will consider coalgebras for functors F on the variety $\mathscr{C}$. In many cases F is a *lifting* of a set functor, i.e. we have functor $F_0 \colon \mathsf{Set}^S \to \mathsf{Set}^S$ such that $F_0 \cdot U = U \cdot F$, where $U \colon \mathscr{C} \to \mathsf{Set}^S$ is the forgetful functor. It is well-known [7,15] that liftings of a given functor F_0 on Set^S to $\mathscr{C}$, the variety given by the monad T, are in bijective correspondence with distributive laws of the monad T over the functor F_0. It was observed by Turi and Plotkin [28] that a final coalgebra for F_0 lifts to a final coalgebra for the lifting F, and this is then the final *bialgebra* for the corresponding distributive law.

Coalgebras for lifted functors are significant for us because the targets of *finite* coalgebras X under *generalized determinization* [30] are precisely those coalgebras for the lifting F carried by ffg objects TX. In more detail, generalized determinization is the process of turning a given coalgebra $c \colon X \to F_0 TX$ in Set^S into a coalgebra for the lifting F: one uses the freeness of TX and the fact that FTX is a T-algebra to extend c to a T-algebra homomorphism $c^* \colon TX \to FTX$. The *coalgebraic language semantics* [9] of (X, c) is then the final semantics of c^* in $\mathscr{C}$. The classical instance of this is the language semantics of non-deterministic automata considered as coalgebras $X \to \{0,1\} \times (\mathcal{P}_f X)^{\Sigma}$; here the generalized determinization with $T = \mathcal{P}_f$ and $F = \{0,1\} \times X^{\Sigma}$ on Set is the well-known subset construction from automata theory.

2.4   Four Fixed Points

Let us now consider a finitary endofunctor $F \colon \mathscr{C} \to \mathscr{C}$ on our variety. Then we know that F has a terminal coalgebra [4], which we denote by νF. Its coalgebra structure $\nu F \to F(\nu F)$ is an isomorphism by Lambek's lemma [17], and so νF is a fixed point of F.

There are three more fixed points of F obtained from 'finite' coalgebras, where 'finite' can mean each of the three notions discussed in the previous subsection. More precisely, we consider the full subcategories of the category $\mathsf{Coalg}\, F$ given by those coalgebras with fp, fg, and ffg carriers, respectively and denote them as shown below:

$$\mathsf{Coalg}_{\mathrm{ffg}}\, F \hookrightarrow \mathsf{Coalg}_{\mathrm{fg}}\, F \hookrightarrow \mathsf{Coalg}_{\mathrm{fp}}\, F \hookrightarrow \mathsf{Coalg}\, F.$$

Since all three categories $\mathsf{Coalg}_x\, F$ (for $x = \mathrm{fp}, \mathrm{fg}$ or ffg) are essentially small, we can form coalgebras as the colimits of the above inclusions as follows:

$$\varphi F = \mathrm{colim}(\mathsf{Coalg}_{\mathrm{ffg}}\, F \hookrightarrow \mathsf{Coalg}\, F),$$
$$\vartheta F = \mathrm{colim}(\mathsf{Coalg}_{\mathrm{fg}}\, F \hookrightarrow \mathsf{Coalg}\, F),$$
$$\varrho F = \mathrm{colim}(\mathsf{Coalg}_{\mathrm{fp}}\, F \hookrightarrow \mathsf{Coalg}\, F).$$

Note that the latter two colimits are filtered; in fact, $\mathsf{Coalg}_{\mathrm{fg}}\, F$ and $\mathsf{Coalg}_{\mathrm{fp}}\, F$ are clearly closed under finite colimits in $\mathsf{Coalg}\, F$, whence they are filtered categories. The first colimit is a sifted colimit since its diagram scheme $\mathsf{Coalg}_{\mathrm{ffg}}\, F$ is closed under finite coproducts [22, Lemma 3.7]. In what follows, the objects of $\mathsf{Coalg}_{\mathrm{ffg}}\, F$ are called *ffg-coalgebras*.

We now discuss the three above coalgebras in more detail.

The rational fixed point is the coalgebra ϱF; that this is a fixed point was proved by Adámek et al. [2]. In addition, ϱF is characterized by a universal property both as a coalgebra and as an algebra: (a) as a coalgebra, ϱF is the terminal *locally finitely presentable* (lfp) coalgebra, where a coalgebra is called lfp if it is a filtered colimit of a diagram formed by coalgebras from $\mathsf{Coalg}_{\mathrm{fp}}\, F$ [20]; and (b) as an algebra, ϱF is the initial iterative algebra for F. An *iterative algebra* is an F-algebra $a\colon FA \to A$ such that every *fp-equation*, i.e. a morphism $e\colon X \to FX + A$ with X fp, has a unique *solution* in A. The latter means that there exists a unique morphism $e^\dagger$ such that the following square commutes:

$$
\begin{array}{ccc}
X & \xrightarrow{\;\; e^\dagger \;\;} & A \\
\downarrow{\scriptstyle e} & & \uparrow{\scriptstyle [a,A]} \\
FX + A & \xrightarrow[\;Fe^\dagger + A\;]{} & FA + A
\end{array}
\tag{2.1}
$$

(Note that in a diagram we usually denote identity morphisms simply by the (co)domain object.) This notion is a categorical generalization of iterative Σ-algebras for a single-sorted signature Σ originally introduced by Nelson [27]; see also Tiuryn [33] for a closely related concept.

The locally finite fixed point is the coalgebra ϑF; this coalgebra was recently introduced and studied by Milius et al. [24] for a finitary and mono-preserving functor F. It was proved to be a fixed point of F and characterized by two universal properties analogous to the rational fixed point: (a) as a coalgebra, ϑF is the terminal *locally finitely generated* (lfg) coalgebra, where a coalgebra is called lfg if it is a colimit of a directed diagram of coalgebras in $\mathsf{Coalg}_{\mathrm{fg}}\, F$; and (b) as an algebra, ϑF is the initial fg-iterative algebra for F, where fg-iterative is simply the variation of iterative where the domain object of $e\colon X \to FX + A$ is required to be fg in lieu of fp. Moreover, ϑF always is a subcoalgebra of νF [24, Theorem 3.10] and thus fully abstract w.r.t. behavioral equivalence.

The locally ffg fixed point is the coalgebra φF. Recently, Urbat [34] has proved that φF is indeed a fixed point of F, provided that F preserves sifted colimits. Actually, he defined φF as the colimit of all F-coalgebras whose carrier is a split quotient of an ffg object. However, this is the same colimit as the one we use above . Moreover, loc. cit. provides a general framework that allows to prove that all four coalgebras ϱF, ϱF, ϑF and νF are fixed points by one uniform proof. Also, a uniform proof of the universal properties of ϱF, ϑF and νF is given.

Somewhat surprisingly, the coalgebra φF fails to have the finality property w.r.t. to coalgebras in $\mathsf{Coalg}_{\mathrm{ffg}}\, F$: Urbat [34, Example 4.12] gives an example of a coalgebra for the identity functor on the category $\mathscr{C}$ of algebras with one unary operation (and no equations) that admits two coalgebra homomorphisms into φF; see Example 2.2 below. This also shows that φF cannot have a universal property as some kind of iterative algebra (i.e. where solutions are unique).

Relations between the Fixed Points. Recall that a *quotient* of a coalgebra is represented by a coalgebra homomorphism carried by a regular epimorphism (= surjective algebra morphism) in $\mathscr{C}$. Suppose we have a finitary functor F on $\mathscr{C}$ preserving surjective and non-empty injective morphisms.[2] Then the subcoalgebra ϑF of νF is a quotient of ϱF, which in turn is a quotient of φF [22,23]; see (1.1). Whenever, $\mathscr{C}_{\mathrm{fp}} = \mathscr{C}_{\mathrm{fg}}$, we clearly have $\mathsf{Coalg}_{\mathrm{fp}}\, F = \mathsf{Coalg}_{\mathrm{fg}}\, F$ and hence $\varrho F \cong \vartheta F$ (i.e. ϱF is fully abstract w.r.t. behavioral equivalence), and if $\mathscr{C}_{\mathrm{fp}} = \mathscr{C}_{\mathrm{fg}} = \mathscr{C}_{\mathrm{ffg}}$ then those two coincide with φF as well. Moreover, Milius [22] introduced the notion of a *proper* functor (generalizing the notion of a proper semiring of Ésik and Maletti [11]) and proved that a functor F is proper if and only if the three fixed points coincide, i.e. the picture (1.1) collapses to $\varphi F \cong \varrho F \cong \vartheta F \hookrightarrow \nu F$. Loc. cit. also shows that on a variety $\mathscr{C}$ where fg objects are closed under taking kernel pairs, every endofunctor mapping kernel pairs to weak pullbacks in Set is proper [22, Proposition 3.18].[3]

Instances of the three fixed points have mostly been considered for proper functors (where the three are the same, e.g. for functors on Set), or else on algebraic

---

[2] These are mild assumptions; e.g. if $\mathscr{C}$ is single-sorted and F a lifting of a set functor, then the conditions are fulfilled.

[3] Note that these conditions are fulfilled in particular by every locally finite variety and every category of semirings for a Noetherian semiring and any lifted endofunctor whose underlying Set functor preserves weak pullbacks.

categories where $\mathscr{C}_{\mathsf{fp}} = \mathscr{C}_{\mathsf{fg}}$ (where the rational and locally finite fixed points coincide). For example, regular languages for the automaton functor $2 \times (-)^{\Sigma}$ on Set; rational formal power series for the functor $\$ \times (-)^{\Sigma}$ on $\$$-semimodules (whenever $\$$ is a proper semiring the three fixed points coincide); rational (a.k.a. regular) Σ-trees for the polynomial functor on Set associated to the signature Σ; eventually periodic and rational streams for the functor $k \times (-)$ on Set and vector spaces over the field k, respectively; the behaviors of probabilistic automata modelled as coalgebras for $[0, 1] \times (-)^{\Sigma}$ on the category of positive convex algebras (that this functor is proper was recently proved by Sokolova and Woracek [32]); finally, (deterministic) context-free languages and constructively $\$$-algebraic formal power-series (the weighted counterpart of context-free languages) [24]. Note that the last two examples are instances of the locally finite fixed point ϑF, but a description of φF and ϱF is unknown.

Remark 2.1. The rational and locally finite fixed points are defined and studied more generally than in the present setting, namely for finitary functors F on a locally finitely presentable category $\mathscr{C}$ (see Adámek and Rosický [4] for an introduction to locally presentable categories). The following are instances of ϱF and ϑF for F on a locally finitely presentable category $\mathscr{C}$: (a) Courcelle's algebraic trees [10] as proved in [24]; (b) rational λ-trees (modulo α-equivalence) for a functor on the category of presheaves over finite sets [3] or for a related functor on the category of nominal sets [26]; more generally, (c) rational trees over an arbitrary binding signature (see Fiore et al. [13]) as proved in [25]. Again, (a) is an instance of the locally finite fixed point ϑF but a description of the rational fixed point is unknown. In the setting of general locally finitely presentable categories, there is no analogy to φF, of course.

We now present a new example where only φF is interesting but the other three fixed points are trivial.

Example 2.2. We consider the monad T on Set whose algebras are the algebras with one unary operation u (with no equation):

$$TX = \mathbb{N} \times X \quad \text{with} \quad u(n, x) = (n + 1, x).$$

The functor F is the identity functor Id on the category $\mathscr{C} = \mathsf{Set}^T$. The final coalgebra for Id is (lifted from Set and therefore is) the trivial algebra on 1 with id_1 as coalgebra structure. Since 1 is clearly finitely presented by one generator x and the relation $u(x) = x$, both of the diagrams $\mathsf{Coalg}_{\mathsf{fp}}$ Id and $\mathsf{Coalg}_{\mathsf{fg}}$ Id have a terminal object which is then their colimit, whence $\varrho\mathsf{Id} \cong \vartheta\mathsf{Id} \cong 1$.

However, $\varphi\mathsf{Id}$ is non-trivial and interesting: an ffg-coalgebra $TX \xrightarrow{\gamma} TX$ may be viewed (by restricting it to its generators in X) as obtained by generalized determinization of an FT-coalgebra with $F = \mathsf{Id}$ on Set, i.e. a map $X \xrightarrow{\langle o, \delta \rangle} \mathbb{N} \times X$ that we call *stream coalgebra*. Given a state $x \in X$, we call the sequence of natural numbers

$$(o(x), o(\delta(x)), o(\delta^2(x)), \dots)$$

the *stream generated by* x. Since X is finite, this stream is eventually periodic, i.e. of the form $s = s_0 s_1^\omega$ for finite lists s_0 and s_1 of natural numbers. (Here $(-)^\omega$ means infinite iteration.) Two eventually periodic streams $s = s_0 s_1^\omega$ and $t = t_0 t_1^\omega$ with $s_1 = (s_{1,0}, \ldots, s_{1,p-1})$ and $t_1 = (t_{1,0}, \ldots, t_{1,q-1})$ are called *equivalent* if one has

$$q \cdot \sum_{i<p} s_{1,i} = p \cdot \sum_{j<q} t_{1,j}, \tag{2.2}$$

i.e. the entries of the two lists s_1^q and t_1^p of length $p \cdot q$ have the same sum. For instance, the streams

$$s = (1, 2, 7, 4)(1, 3, 2)^\omega = (1, 2, 7, 4, 1, 3, 2, 1, 3, 2, 1, 3, 2, \ldots)$$

and

$$t = (5, 6)(0, 4)^\omega = (5, 6, 0, 4, 0, 4, 0, 4, 0, 4, \ldots)$$

are equivalent. Note that the above notion of equivalence is well-defined, i.e. not depending on the choice of the finite lists s_0, s_1 and t_0, t_1 in the representation of s and t. In fact, given alternative representations $s = \bar{s}_0 \bar{s}_1^\omega$ and $t = \bar{t}_0 \bar{t}_1^\omega$ with $\bar{s}_1 = (\bar{s}_{1,0}, \ldots, \bar{s}_{\bar{p}-1})$ and $\bar{t}_1 = (\bar{t}_{1,0}, \ldots, \bar{t}_{1,\bar{q}-1})$, the lists $s_1^{\bar{q}}$ and $\bar{s}_1^p$ are equal up to cyclic shift, as are the lists $t_1^{\bar{q}}$ and $\bar{t}_1^q$. Therefore from (2.2) it follows that

$$\bar{q} \cdot q \cdot p \cdot \sum_{i<\bar{p}} \bar{s}_{1,i} = \bar{q} \cdot q \cdot \bar{p} \cdot \sum_{i<p} s_{1,i} = \bar{q} \cdot \bar{p} \cdot p \cdot \sum_{j<q} t_{1,j} = \bar{p} \cdot p \cdot q \cdot \sum_{j<\bar{q}} \bar{t}_{1,j}.$$

Dividing by $p \cdot q$ yields

$$\bar{q} \cdot \sum_{i<\bar{p}} \bar{s}_{1,i} = \bar{p} \cdot \sum_{j<\bar{q}} \bar{t}_{1,j},$$

as required.

Lemma 2.3. *(a) The coalgebra $\varphi\mathsf{Id}$ is carried by the set of equivalence classes of eventually periodic streams. The unary operation and the coalgebra structure are both given by* id: $\varphi\mathsf{Id} \to \varphi\mathsf{Id}$. *(b) For any* Id-*coalgebra* (TX, γ_X) *with X finite, the colimit injection* $\gamma_X^{\#}: TX \to \varphi\mathsf{Id}$ *maps* $(m, x) \in TX$ *to the equivalence class of the stream generated by x.*

Proof. (1) We first show that the morphisms $(-)^{\#}$ form a cocone. Given an ffg-coalgebra (TX, γ_X) and elements $(m, x), (n, y) \in TX$ with $\gamma_X(m, x) = (n, y)$, the stream generated by y is the tail of the stream generated by x, and thus the two streams are equivalent. This shows that $\gamma_X^{\#}$ is a coalgebra homomorphism.

To show that the morphisms $(-)^{\#}$ form a compatible family, suppose that $h: (TX, \gamma_X) \to (TY, \gamma_Y)$ is a homomorphism of ffg-coalgebras, and let $(m, x) \in TX$ and $(n, y) \in TY$ with $h(m, x) = (n, y)$ be given. We need to show that the streams generated by x and y are equivalent. Denote by

$$(m_j, x_j) := \gamma_X^j(m, x) \quad \text{and} \quad (n_j, y_j) := \gamma_Y^j(n, y) \qquad (j = 0, 1, 2, \ldots) \tag{2.3}$$

the states reached from (m, x) and (n, y) after j steps. Since h is a coalgebra homomorphism, one has $h(m_j, x_j) = (n_j, y_j)$ for all j. Since X is finite, there exist natural numbers $k \geq 0$ and $p > 0$ with $x_k = x_{k+p}$. Then the eventually periodic stream generated by x is given by

$$(m_1 - m_0, m_2 - m_1, \ldots, m_k - m_{k-1})(m_{k+1} - m_k, \ldots, m_{k+p} - m_{k+p-1})^\omega$$

Since $h(m_k, x_k) = (n_k, y_k)$ and $h(m_{k+p}, x_{k+p}) = (n_{k+p}, y_{k+p})$, one has $y_k = y_{k+p}$, which implies that y generates the stream

$$(n_1 - n_0, n_2 - n_1, \ldots, n_k - n_{k-1})(n_{k+1} - n_k, \ldots, n_{k+p} - n_{k+p-1})^\omega$$

To show that the streams generated by x and y are equivalent, it suffices to verify that $m_{k+p} - m_k = n_{k+p} - n_k$, as this entails that

$$p \cdot \sum_{i<p} m_{k+i+1} - m_{k+i} = p \cdot (m_{k+p} - m_k) = p \cdot (n_{k+p} - n_k)$$

$$= p \cdot \sum_{i<p} n_{k+i+1} - n_{k+i}.$$

To prove the desired equation, we compute

$$\begin{aligned} (n_{k+p}, y_{k+p}) &= h(m_{k+p}, x_{k+p}) \\ &= h(m_{k+p}, x_k) \\ &= h(m_{k+p} - m_k + m_k, x_k) \\ &= (m_{k+p} - m_k + n_k, y_k) \end{aligned}$$

where the last equality uses that $h(m_k, x_k) = (n_k, y_k)$ and that h is a morphism in $\mathscr{C}$. This implies $n_{k+p} = m_{k+p} - m_k + n_k$.

(2) We prove that the cocone $(-)^\#$ is a colimit cocone. Since sifted colimits in $\mathsf{Coalg}\,\mathsf{Id}$ are formed as in $\mathscr{C}$ and thus as in Set, it suffices to show that (i) the morphisms $\gamma_X^\#$ are jointly surjective and (ii) given ffg-coalgebras (TX, γ_X) and (TY, γ_Y) and two states $(m, x) \in TX$ and $(n, y) \in TY$ merged by $\gamma_X^\#$ and $\gamma_Y^\#$, there exists a zig-zag in $\mathsf{Coalg}_{\mathrm{ffg}}\,\mathsf{Id}$ connecting the two states. Statement (i) is clear because finite stream coalgebras generate precisely the eventually periodic streams. For (ii), we adapt the argument of the first part of our proof and continue to use the notation (2.3). Since X and Y are finite, there exist natural numbers $k \geq 0$ and $p > 0$ with $x_k = x_{k+p}$ and $y_k = y_{k+p}$. As the streams generated by x and y are equivalent, one has $m_{k+p} - m_k = n_{k+p} - n_k$. Consider the ffg-coalgebra (TZ, γ_Z) with $Z = \{z_0, z_1, \ldots, z_{k+p-1}\}$, and γ_Z defined by

$$\gamma_Z(z_j) = (0, z_{j+1}) \ (j < k + p - 1) \quad \text{and} \quad \gamma_Z(z_{k+p-1}) = (m_{k+p} - m_k, z_k).$$

Form the morphisms $g: TZ \to TX$ and $h: TZ \to TX$ given by

$$g(z_j) = (m_j, x_j) \quad \text{and} \quad h(z_j) = (n_j, y_j) \qquad (j < k + p).$$

Then g and h are coalgebra homomorphisms. Indeed, for $j < k + p - 1$ we have

$$
\begin{aligned}
g(\gamma_Z(z_j)) &= g(0, z_{j+1}) & &(\text{def. } \gamma_Z) \\
&= (m_{j+1}, x_{j+1}) & &(\text{def. } g) \\
&= \gamma_X(m_j, x_j) & &(\text{def. } m_{j+1}, x_{j+1}) \\
&= \gamma_X(g(z_j)) & &(\text{def. } g)
\end{aligned}
$$

and moreover

$$
\begin{aligned}
g(\gamma_Z(z_{k+p-1})) &= g(m_{k+p} - m_k, z_k) & &(\text{def. } \gamma_Z) \\
&= (m_{k+p} - m_k + m_k, x_k) & &(\text{def. } g) \\
&= (m_{k+p}, x_{k+p}) & & \\
&= \gamma_X(m_{k+p-1}, x_{k+p-1}) & &(\text{def. } m_{k+p}, x_{k+1}) \\
&= \gamma_X(g(z_{k+p-1})) & &(\text{def. } g)
\end{aligned}
$$

and analogously for h. Thus we have constructed a zig-zag

$$
(TX, \gamma_X) \xleftarrow{\;g\;} (TZ, \gamma_Z) \xrightarrow{\;h\;} (TY, \gamma_Y)
$$

in $\mathsf{Coalg}_{\mathrm{ffg}}\,\mathrm{Id}$ connecting (m, x) and (n, y), as required.   □

Observe that every non-empty ffg-coalgebra (TX, γ_X) admits infinitely many coalgebra homomorphisms into $\varphi\mathrm{Id}$, for instance, any constant map into $\varphi\mathrm{Id}$ is one. This shows that, in general, the coalgebra φF is not final w.r.t. the coalgebras in $\mathsf{Coalg}_{\mathrm{ffg}}\,F$.

3   Recap: Elgot Algebras

In this section we briefly recall the notion of an Elgot algebra [1] and some key results to contrast this with our subsequent development of ffg-Elgot algebras in Sect. 4. Throughout this section we assume the endofunctor $F \colon \mathscr{C} \to \mathscr{C}$ to be finitary.

Definition 3.1. *An* fp-equation *is a morphism*

$$
e \colon X \to FX + A,
$$

where X is an fp object (of variables) and A an arbitrary object of parameters.
    Suppose that A carries the structure of an F-algebra $a \colon FA \to A$. Then a solution *of e in A is a morphism $e^\dagger \colon X \to A$ such that the square (2.1) commutes.*

Notation 3.2. We use the following notation for fp-equations:

(1) Given an fp-equation $e \colon X \to FX + A$ and a morphism $h \colon A \to B$ we have an fp-equation

$$
h \bullet e = \left(X \xrightarrow{e} FX + A \xrightarrow{FX + h} FX + B \right).
$$

(2) Given a pair of fp-equations $e\colon X \to FX{+}Y$ and $f\colon Y \to FY{+}Z$ we combine them into the following fp-equation

$$e \bullet f = \left(X + Y \xrightarrow{[e,\mathrm{inr}]} FX + Y \xrightarrow{FX+f} FX + FY + Z \xrightarrow{\mathrm{can}+Z} F(X+Y) + Z \right),$$

where $\mathrm{can} = [F\mathrm{inl}, F\mathrm{inr}]\colon FX + FY \to F(X+Y)$ denotes the canonical morphism.

Definition 3.3. *An Elgot algebra is a triple $(A, a, \dagger)$ where (A, a) is an F-algebra and $\dagger$ is an operation*

$$\frac{e\colon X \to FX + A}{e^{\dagger}\colon X \to A}$$

assigning to every fp-equation in A a solution subject to the following two conditions:

(1) Weak Functoriality. Given a pair of equations $e\colon X \to FX{+}Z$, $f\colon Y \to FY{+}Z$, where Z is an fp object, and a coalgebra homomorphism $m\colon X \to Y$ for $F(-)+Z$, then for every morphism $h\colon Z \to A$ we have $(h \bullet f)^{\dagger}{\cdot}m = (h \bullet e)^{\dagger}$:

$$
\begin{array}{ccc}
X & \xrightarrow{\;e\;} & FX + Z \\
{\scriptstyle m}\downarrow & & \downarrow{\scriptstyle Fm+Z} \\
Y & \xrightarrow[f]{} & FY + Z
\end{array}
\quad\Longrightarrow\quad
\begin{array}{c}
X \;\;\xrightarrow{(h\bullet e)^{\dagger}} \\
{\scriptstyle m}\downarrow \;\searrow\;\; A \\
Y \;\;\xrightarrow[(h\bullet f)^{\dagger}]{}
\end{array}
\quad \text{for all } h\colon Z \to A.
$$

(2) Compositionality. For every pair of fp-equations $e\colon X \to FX{+}Y$ and $f\colon Y \to FY + A$ we have

$$(e \bullet f)^{\dagger} \cdot \mathrm{inl} = (f^{\dagger} \bullet e)^{\dagger}\colon X \to A.$$

Remark 3.4. Later we will need the following properties of $\bullet$ and $\blacksquare$:

(1) $t \bullet (s \bullet e) = (t \cdot s) \bullet e$ for every $e\colon X \to FX + A$, $s\colon A \to B$ and $t\colon B \to C$;
(2) $s \bullet (e \blacksquare f) = e \blacksquare (s \bullet f)$ for every $e\colon X \to FX + Y$, $f\colon Y \to FY + A$ and $s\colon A \to B$;
(3) $(e \blacksquare f) \blacksquare g = (\mathrm{inl} \bullet e) \blacksquare (f \blacksquare g)$ for every $e\colon X \to FX + Y$, $f\colon Y \to FY + Z$ and $g\colon Z \to FZ + V$.

For the proof of the first two see [1, Remark 4.6]. The remaining one is easy to prove by considering the three coproduct components of $X + Y + Z$ separately, we leave this as an easy exercise for the reader.

Note that, in lieu of weak functoriality, $\dagger$ previously [1] was required to satisfy (full) functoriality, i.e. given fp-equations $e\colon X \to FX + A$, $f\colon Y \to FY + A$ and a coalgebra homomorphism $m\colon (X, e) \to (Y, f)$ we have $f^{\dagger} \cdot m = e^{\dagger}\colon X \to A$. However, this makes no difference:

Lemma 3.5. *Functoriality and weak functoriality are equivalent properties of $\dagger$.*

Proof. Functoriality clearly implies Weak Functoriality. In order to prove the converse, let $e: X \to FX + A$, $f: Y \to FY + A$ be fp-equations, and let $m: (X, e) \to (Y, f)$ be a coalgebra morphism. Write A is the filtered colimit of its canonical diagram $\mathscr{C}_{\mathsf{fp}}/A$ (cf. Sect. 2.1). The functor $FX + (-)$ preserves filtered colimits, and so $FX + A$ is the filtered colimit of the diagram formed by all morphisms $FX + h: FX + Z \to FX + A$. Since X is fp, the morphism $e: X \to FX + A$ factors through one of these morphisms, i.e. there exists a morphism $h: Z \to A$ with Z fp and $e': X \to FX + Z$ such that $e = h \bullet e'$:

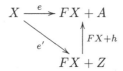

Similarly, we have a factorization of $f: Y \to FY + A$, and by filteredness of the diagram $\mathscr{C}_{\mathsf{fp}}/A \to \mathscr{C}$ we can assume the same $h: Z \to A$ is used. Thus a morphsm $f': Y \to FY + Z$ is given such that $h \bullet f' = (FY + h) \cdot f' = f$. We do not claim that m is a coalgebra homomorphism from (X, e') to (Y, f'). However, the corresponding equation holds when postcomposed by the colimit injection $FY + h$:

$$(FX + h) \cdot (Fm + Z) \cdot e' = (Fm + A) \cdot (FX + h) \cdot e'$$
$$= (Fm + A) \cdot e$$
$$= f \cdot m$$
$$= (FY + h) \cdot f' \cdot m.$$

Therefore there exists a morphism $h: Z' \to A$ with Z' fp and a connecting morphism $z: Z \to Z'$ in $\mathscr{C}_{\mathsf{fp}}/A$, i.e. z satisfies $h' \cdot z = h$, such that $FY + z$ merges $(Fm + Z) \cdot e'$ and $f' \cdot m$. It follows that m is a coalgebra homomorphism from $z \bullet e'$ to $z \bullet f'$:

$$
\begin{array}{ccccc}
& & z \bullet e' & & \\
X & \xrightarrow{e'} & FX + Z & \xrightarrow{FX+z} & FX + Z' \\
\downarrow{m} & & \downarrow{Fm+Z} & & \downarrow{Fm+Z'} \\
Y & \xrightarrow{f'} & FY + Z & \xrightarrow{FY+z} & FY + Z' \\
& & z \bullet f' & &
\end{array}
$$

Indeed, the left-hand square commutes when postcomposed with $FY + z$; thus, since the upper and lower parts as well as the right-hand square commute, so does the outside, as desired. By weak functoriality, we thus conclude

$$f^{\dagger} \cdot m = (h \bullet f')^{\dagger} \cdot m = ((h' \cdot z) \bullet f')^{\dagger} \cdot m = (h' \bullet (z \bullet f'))^{\dagger} \cdot m$$
$$= (h' \bullet (z \bullet e'))^{\dagger} = ((h' \cdot z) \bullet e')^{\dagger} = (h \bullet e')^{\dagger} = e^{\dagger}. \qquad \square$$

Example 3.6. Let us recall a few examples of Elgot algebras [1].

(1) Iterative F-algebras (cf. Sect. 2.4): the operation † assigning to every equation its unique solution satisfies Compositionality and (Weak) Functoriality, see [1, 2.15–1.19]. It follows that ϱF, ϑF and νF are Elgot algebras.

(2) Cpo enrichable algebras. Recall that a *complete partial order* (*cpo*, for short) is a partially ordered set having joins of ω-chains. Cpos form a category CPO together with the *continuous* functions, i.e. functions preserving joins of ω-chains. Let $F_0 \colon \mathsf{Set} \to \mathsf{Set}$ be a functor having a *locally continuous* lifting $F \colon \mathsf{CPO} \to \mathsf{CPO}$, i.e. a lifting such that the hom mappings $\mathsf{CPO}(X, Y) \to \mathsf{CPO}(FX, FY)$ are continuous. For example, every polynomial functor F_Σ associated to the signature Σ has a lifting to CPO.

Suppose further that $a \colon FA \to A$ is an algebra where A is a CPO with a least element $\bot$ and a is continuous. Then A is an Elgot algebra w.r.t. the operation † assigning the least solution. More precisely, given an fp-equation $e \colon X \to FX + A$ (in Set) consider X as a cpo with discrete order and let $e^\dagger \colon X \to A$ be the least fixed point of the continuous function

$$h \mapsto [a, A] \cdot (Fh + A) \cdot e$$

on the cpo of continuous functions from X to A. For details see [1, 3.5–3.8].

(3) CMS enrichable algebras. A related example is based on *complete metric spaces*, i.e. metric spaces in which every Cauchy sequence has a limit. Here one considers the category CMS of complete metric spaces with distances in $[0, 1]$ and non-expanding maps, i.e. maps $f \colon X \to Y$ such that for every $x, x' \in X$ one has $d_Y(fx, fx') \le d_X(x, x')$. Let $F_0 \colon \mathsf{Set} \to \mathsf{Set}$ have a *locally contracting* lifting to CMS, i.e. a lifting $F \colon \mathsf{CMS} \to \mathsf{CMS}$ such that there exists some $\varepsilon < 1$ such that for all $f, g \colon X \to Y$ in CMS one has

$$d_{X,Y}(f, g) \le \varepsilon d_{FX,FY}(Ff, Fg),$$

where $d_{X,Y}$ denotes the sup-metric on $\mathsf{CMS}(X, Y)$. Again, polynomial set functors have locally contracting liftings to CMS.

Now suppose that $a \colon FA \to A$ is a non-empty algebra such that A carries a complete metric space and a is a non-expanding map. Then A is iterative, whence an Elgot algebra. In fact, for every equation $e \colon X \to FX + A$ consider X as a discrete metric space (i.e. all distances are 1) and consider the ε-contracting function

$$h \mapsto [a, A] \cdot (Fh + A) \cdot e$$

on $\mathsf{CMS}(X, A)$. Then, by Banach's fixed point theorem, this function has a unique fixed point, viz. a unique solution of e. For details see [1, 2.8–2.11].

(4) As a concrete instance of the previous point one can obtain fractals as solutions of equations. For example, let A be the set of closed subsets of the unit interval $[0, 1]$ equipped with the following binary operation:

$$(C, C') \mapsto \frac{1}{3}C \cup \left(\frac{1}{3}C' + \frac{2}{3}\right),$$

where $\frac{1}{3}C = \{\frac{1}{3}c \mid c \in C\}$ etc. Then A is an algebra for $F_0 X = X \times X$ on Set, and this F_0 has the locally contracting lifting $F(X, d) = (X \times X, \frac{1}{3} d_{\max})$, where $d_{\max}$ denotes the usual maximum metric on the cartesian product. One sees that A is an algebra for F when equipped with the so-called Hausdorff metric. Hence, it is an Elgot algebra. For example, let $X = \{x\}$ and let $e \colon X \to FX + A$ be given by $e(x) = (x, x)$. Then $e^{\dagger}(x)$ is the well-known Cantor set.

The rational fixed point ϱF is, besides being the initial iterative F-algebra, also an initial Elgot algebra. Moreover, for every object Y, the rational fixed point $\varrho(F(-) + Y)$ is a free iterative algebra on Y. Thus, the object assignment $R \colon Y \mapsto \varrho(F(-) + Y)$ yields a monad on $\mathscr{C}$, and one obtains the following

Theorem 3.7. ([1]). *The category of Eilenberg-Moore algebras for R is isomorphic to the category of Elgot algebras for F.*

Thus, in particular, $\varrho(F(-) + Y)$ is not only a free iterative algebra but also a free Elgot algebra on Y.

4 FFG-Elgot Algebras

The rest of our paper is devoted to studying the fixed point φF, the colimit of all ffg-coalgebras for F, in its own right and establish a universal property of it as an algebra.

Assumption 4.1. *Throughout the rest of the paper we assume that $\mathscr{C}$ is a variety of algebras and that $F \colon \mathscr{C} \to \mathscr{C}$ is an endofunctor preserving sifted colimits.*

Example 4.2.(1) For the monad T representing $\mathscr{C}$, all functors that are liftings of finitary set functor F_0 (i.e., with a distributive law of T over F_0) preserve sifted colimits. Indeed, finitary set functors F_0 preserve all sifted colimits [6, Proposition 6.30]. Since $\mathscr{C}$ is cocomplete and the forgetful functor $U \colon \mathscr{C} \to$ Set preserves and reflects sifted colimits, it follows that every lifting of F_0 preserves sifted colimits, too. The following examples are not liftings of set functors.

(2) The functor $FX = X + X$, where $+$ denotes the coproduct of $\mathscr{C}$ preserves sifted colimits. More generally, every coproduct of sifted colimit preserving functors preserves them too. Similarly, for finite products of sifted colimit preserving functors. Thus, all polynomial functors on $\mathscr{C}$ preserve sifted colimits.

(3) Let $\mathscr{C}$ is an *entropic* variety, i.e. such that the usual tensor product makes it symmetric monoidal closed. (Examples include sets, vector spaces, join-semilattices, or abelian groups.) Then the functor $FX = X \otimes X$ preserves sifted colimits. To see this, it suffices to show that (a) F is finitary and (b) it preserves reflexive coequalizers (see [5]). First note that since $\mathscr{C}$ is symmetric monoidal closed, we know that each functor $X \otimes -$ and $- \otimes X$ is a left adjoint and therefore preserves all colimits.

Ad (a). Suppose that $D : \mathscr{D} \to C$ is a filtered diagram with colimit injections $a_d : Dd \to A$ for $d \in \mathscr{D}$. We need to prove that all $a_d \otimes a_d : Dd \otimes Dd \to A \otimes A$ form a colimit cocone. That is, for every morphism $f : X \to A \otimes A$ with X fp, (i) there exists some $d \in \mathscr{D}$ and $g : X \to Dd \otimes Dd$ with $(a_d \otimes a_d) \cdot g = f$ and (ii) given $g, h : X \to Dd \otimes Dd$ that yield f in this way, there exists a morphism $m : d \to d'$ in $\mathscr{D}$ such that $Dm \otimes Dm$ merges g and h.

To prove (i), we use that $- \otimes A$ is finitary to obtain some $d \in \mathscr{D}$ and $f' : X \to A \otimes Dd$ with $(A \otimes a_d) \cdot f' = f$. Now use that $Dd \otimes -$ is finitary to obtain $d' \in \mathscr{D}$ and $f'' : X \to Dd \otimes Dd'$ with $(Dd \otimes a_{d'}) \cdot f'' = f'$. Since $\mathscr{D}$ is filtered, we can choose morphisms $m : d \to \bar{d}$ and $n : d' \to \bar{d}$ in $\mathscr{D}$. Let $g = (Dm \otimes Dn) \cdot f''$. Then we have

$$(a_{\bar{d}} \otimes a_{\bar{d}}) \cdot g = (a_{\bar{d}} \otimes a_{\bar{d}}) \cdot (Dm \otimes Dn) \cdot f'' = (a_d \otimes a_{d'}) \cdot f''$$
$$= (a_d \otimes A) \cdot (Dd \otimes a_{d'}) \cdot f'' = (a_d \otimes A) \cdot f' = f$$

as desired.

For (ii), use first that $- \otimes A$ is finitary and choose some morphism $o : d \to d'$ such that

$$(Do \otimes A) \cdot ((Dd \otimes a_d) \cdot g) = (Do \otimes A) \cdot ((Dd \otimes a_d) \cdot h).$$

It follows that $(Dd' \otimes a_d)$ merges $(Do \otimes Dd) \cdot g$ and $(Do \otimes Dd) \cdot h$. Now use that $Dd' \otimes -$ is finitary and choose a morphism $p : d \to d''$ in $\mathscr{D}$ such that $(Dd' \otimes Dp)$ also merges those two morphisms. Finally, use that $\mathscr{D}$ is filtered to choose two morphisms $q : d' \to \bar{d}$ and $r : d'' \to \bar{d}$ such that $q \cdot o = r \cdot p$, and let us call this last morphism $m : d \to \bar{d}$. It is then easy to see that $Dm \otimes Dm$ merges g and h:

$$(Dm \otimes Dm) \cdot g = (D(q \cdot o) \otimes D(r \cdot p)) \cdot g = (Dq \otimes Dr) \cdot (Do \otimes Dp) \cdot g$$
$$= (Dq \otimes Dr) \cdot (Dd' \otimes Dp) \cdot (Do \otimes Dd) \cdot g$$
$$= (Dq \otimes Dr) \cdot (Dd' \otimes Dp) \cdot (Do \otimes Dd) \cdot h$$
$$= (Dm \otimes Dm) \cdot h.$$

Ad (b). Let $f, g : A \to B$ be any (not necessarily reflexive) parallel pair of morphisms, and let $c : B \to C$ be their coequalizer. Use that all functors $- \otimes X$ and $X \otimes -$ preserve coequalizers to see that in the following diagram, whose parts commute in the obvious way, all rows and columns are coequalizers:

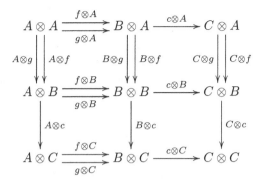

By the '3-by-3 lemma' [16, Lemma 0.17], it follows that the diagonal yields a coequalizer too, i.e., $c \otimes c$ is a coequalizer of the pair $f \otimes f, g \otimes g$ as desired.

(4) Combining the previous argument with induction, we see that sifted colimit preserving functors on an entropic variety $\mathscr{C}$ are stable under finite tensor products. Thus, all tensor-polynomial functors on $\mathscr{C}$ preserve sifted colimits.

Under our assumptions we know that φF is a fixed point of F and we will henceforth denote the inverse of its coalgebra structure by $t \colon F(\varphi F) \to \varphi F$.

Definition 4.3. *By an* ffg-equation *is meant a morphism* $e \colon X \to FX + A$ *where* X *is an ffg object. An* ffg-Elgot algebra *is a triple* $(A, a, \dagger)$ *where* (A, a) *is an* F-algebra and $\dagger$ *is an operation*

$$\frac{e \colon X \to FX + A}{e^{\dagger} \colon X \to A}$$

assigning to every ffg-equation in A a solution and satisfying Weak Functoriality 3.3(1) and Compositionality 3.3(2) with X, Y and Z restricted to ffg objects.

Remark 4.4. Note that in categories where fp objects are ffg, e.g. in the category of sets or vector spaces, (ordinary) Elgot algebras and ffg-Elgot algebras are the same concept. However, in the present setting this may not be the case. Moreover, we do not know whether, for ffg-Elgot algebras, weak functoriality implies functoriality. The proofs of our main results (in particular Proposition 4.8 and Theorem 4.12) do not work when weak functoriality is replaced by functoriality.

Remark 4.5. In the case where $F \colon \mathbf{Set}^T \to \mathbf{Set}^T$ is a lifting of a functor $F_0 \colon \mathbf{Set} \to \mathbf{Set}$ (via a distributive law λ), then an F-algebra is given by a set A equipped with both a T-algebra structure $\alpha \colon TA \to A$ and an F_0-algebra structure $a \colon F_0 A \to A$ such that a is a T-algebra homomorphism, i.e. one has $\alpha \cdot Ta = a \cdot F\alpha \cdot \lambda_A$. Morphisms of F-algebras are those maps that are both T-algebra and F_0-algebra homomorphisms. Now one may think of ffg-equations and their solutions as modelling *effectful iteration*. Indeed, let X_0 be a finite set of variables and consider any map

$$e_0 \colon X_0 \to T(F_0 X_0 + A).$$

Then this may be regarded as a system of recursive equations with variables X_0 and parameters in A, where for any recursive call a side effect in T might happen. If (A, α, a) is an F-algebra, a solution to such a recursive system should assign to each variable in X_0 an element of A, i.e. we have a map $e_0^\dagger \colon X_0 \to A$, such that the square below commutes (here we write $+$ for disjoint union and $\oplus$ for the coproduct in $\mathscr{C}$, which may be different):

$$
\begin{array}{ccc}
X_0 & \xrightarrow{\;e_0^\dagger\;} & A \\
\downarrow{\scriptstyle e_0} & & \uparrow{\scriptstyle \alpha} \\
& & TA \\
& & \uparrow{\scriptstyle T[a,A]} \\
T(F_0 X_0 + A) & \xrightarrow[\;T(F_0 e_0^\dagger + A)\;]{} & T(F_0 A + A)
\end{array}
$$

Indeed, from e_0 we may form the map

$$
\bar{e} = \left(X_0 \xrightarrow{\;e_0\;} T(F_0 X_0 + A) \xrightarrow{\;\cong\;} T F_0 X_0 \oplus T A \xrightarrow{\;\lambda_X \oplus \alpha\;} F T X_0 \oplus A \right).
$$

Then its unique extension $T X_0 \to F T X_0 \oplus A$ to a T-algebra morphism is an ffg-equation, and a solution $T X_0 \to A$ of this in the sense of Definition 4.3 is precisely the same as an extension of a solution for e_0 in the above sense.

Construction 4.6. We aim at proving that φF is an initial ffg-Elgot algebra. For that we first construct a solution $e^\dagger \colon X \to \varphi F$ for every given ffg-equation $e \colon X \to FX + \varphi F$. The colimit cocone of φF is denoted by $c^\sharp \colon C \to \varphi F$ for (C, c) in $\mathsf{Coalg}_{\mathrm{ffg}}\, F$.

Since X is an ffg-object, $\mathscr{C}(X, -)$ preserves the sifted colimit

$$
FX + \varphi F = \mathrm{colim}(FX + C), \qquad (C, c) \text{ in } \mathsf{Coalg}_{\mathrm{ffg}}\, F.
$$

Every ffg-equation $e \colon X \to FX + \varphi F$ thus factorizes through one of the colimit injections $FX + c^\sharp$, i.e. for some $c \colon C \to FC$ in $\mathsf{Coalg}_{\mathrm{ffg}}\, F$ and $w \colon X \to FX + C$ we have the commutative triangle below:

$$
\begin{array}{ccc}
X & \xrightarrow{\;e\;} & FX + \varphi F \\
& \searrow{\scriptstyle w} & \uparrow{\scriptstyle FX + c^\sharp} \\
& & FX + C
\end{array}
\tag{4.1}
$$

We see that w is an ffg-equation. We combine it with the ffg-equation c (having the initial object 0 as parameter) to $w \blacksquare c \colon X + C \to F(X + C)$, which is an object of $\mathsf{Coalg}_{\mathrm{ffg}}\, F$. Finally, we put

$$
e^\dagger = \left(X \xrightarrow{\;\mathrm{inl}\;} X + C \xrightarrow{\;(w \blacksquare c)^\sharp\;} \varphi F \right).
\tag{4.2}
$$

The proofs of the following results can be found in [18].

Lemma 4.7. *The definition of $e^\dagger$ in (4.2) is independent of the choice of the factorization (4.1), and $e^\dagger$ is a solution of e in φF.*

Proposition 4.8. *The algebra $t\colon F(\varphi F) \to \varphi F$ together with the solution operator $\dagger$ from Construction 4.6 is an ffg-Elgot algebra.*

Definition 4.9. *A morphism of ffg-Elgot algebras from $(A, a, \dagger)$ to $(B, b, \ddagger)$ is a morphism $h\colon A \to B$ in $\mathscr{C}$ preserving solutions, i.e. for every ffg-equation $e\colon X \to FX + A$ we have*

$$(h \bullet e)^\ddagger = h \cdot e^\dagger.$$

Identity morphisms are clearly ffg-Elgot algebra morphisms, and ffg-Elgot algebra morphisms compose. Therefore ffg-Elgot algebras form a category, which we denote by

$$\textsf{ffg-Elgot } F.$$

The next lemma shows that the above category is a subcategory of the category **Alg** F of algebras for F.

Lemma 4.10. *Morphisms of ffg-Elgot algebras are F-algebra homomorphisms.*

Note that the converse fails in general. In fact, [1, Example 4.4] exhibits an (ffg-) Elgot algebra for the identity functor on Set and an algebra morphism on it which is not solution-preserving.

Theorem 4.11. *The triple $(\varphi F, t, \dagger)$ is the initial ffg-Elgot algebra for F.*

Proof. (Sketch). Let $(A, a, \ddagger)$ be an ffg-Elgot algebra. We obtain a cocone over the diagram

$$\textsf{Coalg}_{\textsf{ffg}} F \rightarrowtail \textsf{Coalg } F \xrightarrow{U} \mathscr{C}$$

(where U is the forgetful functor) as follows: to every ffg-coalgebra $c\colon C \to FC$ assign the solution

$$(i_A \bullet c)^\ddagger \colon C \to A$$

of $i_A \bullet c\colon C \to FC + A$, where $i_A\colon 0 \to A$ is the unique morphism. Thus there exists a unique morphism $h\colon \varphi F \to A$ in $\mathscr{C}$ such that the triangle below commute for every ffg-coalgebra $c\colon C \to FC$:

One then shows that the morphism h is solution-preserving and is the unique such morphism. $\qquad\square$

The following result is the key to constructing free ffg-Elgot algebras. In the case where $\mathscr{C}_{\text{ffg}} = \mathscr{C}_{\text{fp}}$, this yields a new result about ordinary Elgot algebras.

Theorem 4.12. *Let $a: FA \to A$ be an F-algebra and let Y be a free object of $\mathscr{C}$. For any morphism $h: Y \to A$, there is a bijective correspondence between*

(i) solution operators $\dagger$ such that $(A, a, \dagger)$ is an ffg-Elgot algebra for F, and
(ii) solution operators $\ddagger$ such that $(A, [a, h], \ddagger)$ is an ffg-Elgot algebra for $F(-)+Y$.

Proof. (Sketch). (1) Given an ffg-Elgot algebra $(A, a, \dagger)$ for F, we define a solution operator $\ddagger$ w.r.t. $F(-) + Y$ as follows. For any ffg-equation $e: X \to FX + Y + A$, let

$$e_h \equiv X \xrightarrow{e} FX + Y + A \xrightarrow{FX+[h,A]} FX + A$$

and put

$$e^{\ddagger} := e_h^{\dagger}.$$

Then one can prove that $(A, [a, h], \ddagger)$ is an ffg-Elgot algebra for $F(-) + Y$. (In order to verify weak functoriality, the assumption that Y is free is critical.)

(2) Conversely, given an ffg-Elgot algebra $(A, [a, h], \ddagger)$ for $F(-)+Y$, we define a solution operator $\dagger$ w.r.t. F as follows. For any ffg-equation $e: X \to FX + A$, let

$$\bar{e} \equiv X \xrightarrow{e} FX + A \xrightarrow{\text{inl}+A} FX + Y + A$$

and put

$$e^{\dagger} := \bar{e}^{\ddagger}.$$

Then one can prove that $(A, a, \dagger)$ is an ffg-Elgot algebra.

(3) Finally, one shows that the two passages $\dagger \mapsto \ddagger$ and $\ddagger \mapsto \dagger$ are mutually inverse. $\square$

For the forgetful functor of ffg-Elgot algebras

$$U_F: \text{ffg-Elgot}\, F \to \mathscr{C}$$

recall that the slice category Y/U_F has as objects all morphisms $y: Y \to U_F(A, a, \dagger)$, and morphisms into $y': Y \to U_F(B, b, \ddagger)$ are the solution-preserving morphisms $p: (A, a, \dagger) \to (B, b, \ddagger)$ with $p \cdot y = p'$. Denote by $\pi: Y/U_F \to \mathscr{C}$ the projection functor.

Corollary 4.13. *For every free object Y of $\mathscr{C}$, there is an isomorphism I of categories such that*

$$U_{F(-)+Y} = \left(\text{ffg-Elgot}(F(-) + Y) \xrightarrow{I} Y/U_F \xrightarrow{\pi} \mathscr{C} \right).$$

It is given by $(A, [a, h], \ddagger) \mapsto (h: Y \to U_F(A, a, \dagger))$.

Construction 4.14. For any object Y of $\mathscr{C}$ denote by ΦY the colimit of all ffg-coalgebras for $F(-)+Y$, that is, $\Phi Y = \varphi(F(-)+Y)$. Its coalgebra structure is invertible, and we denote by $t_Y \colon F\Phi Y \to \Phi Y$ and $\eta_Y \colon Y \to \Phi Y$ the components of its inverse.

The F-algebra $(\Phi Y, t_Y)$ is endowed with a canonical solution operation $\dagger$: given an ffg-equation $e\colon X \to FX + \Phi Y$, put $\bar{e} \equiv X \xrightarrow{e} FX + \Phi Y \xrightarrow{FX+\mathrm{inl}} FX + Y + \Phi Y$. This ffg-equation for $F(-)+Y$ has a solution $\bar{e}^{\ddagger}\colon X \to \Phi Y$ in the ffg-Elgot algebra $\Phi Y = \varphi(F(-)+Y)$, and we put $e^{\dagger} := \bar{e}^{\ddagger}$.

The next result shows that all ffg-Elgot algebras form an algebraic category over the given variety $\mathscr{C}$.

Theorem 4.15. *For every free object Y of $\mathscr{C}$, the algebra $(\Phi Y, t_Y)$ with the solution operation $\dagger$ is a free ffg-Elgot algebra for F on Y with η_Y as the universal morphism.*

Proof. (Sketch). ΦY is an ffg-Elgot algebra since it, together with η_Y, corresponds to the initial ffg-Elgot algebra $\varphi(F(-)+Y)$ under the isomorphism of Corollary 4.13. To verify its universal property, let $(A, a, \dagger)$ be an ffg-Elgot algebra for F and $h\colon Y \to A$ a morphism. Corollary 4.13 gives an ffg-Elgot algebra $(A, [a, h], \oplus)$ for $F(-)+Y$ with $e^{\dagger} = \bar{e}^{\oplus}$ for all ffg-equations $e\colon X \to FX + A$. Furthermore, Corollary 4.13 states that a morphism $p\colon \Phi Y \to A$ in $\mathscr{C}$ is solution-preserving w.r.t. $F(-)+Y$ if and only if it is solution-preserving w.r.t. F and satisfies $p \cdot \eta_Y = h$. Therefore the universal property of ΦY w.r.t. F follows from the initiality of ΦY w.r.t. $F(-)+Y$ (see Theorem 4.11). $\qquad\square$

Theorem 4.16. *The forgetful functor $U_F \colon$ ffg-Elgot $F \to \mathscr{C}$ is monadic.*

Proof. (Sketch). (1) First, one readily proves that U_F creates sifted colimits. Moreover, U_F has a left adjoint. Indeed, for every ffg object Y there exists a free ffg-Elgot algebra on Y by Theorem 4.15, which defines the corresponding functor $\Phi\colon \mathscr{C}_{\mathrm{ffg}} \to$ ffg-Elgot F. We can extend it to a left adjoint of U_F as follows. Given an object Y of $\mathscr{C}$ expressed as a sifted colimit $y_i\colon Y_i \to Y$ $(i \in I)$ of ffg objects, then the image of that sifted diagram under Φ has a colimit $\mathrm{colim}_{i \in I}\, \Phi Y_i$ which, since U_F creates sifted colimits, is an ffg-Elgot algebra. It follows easily that this colimit is a free ffg-Elgot algebra on Y.

(2) By Beck's theorem it remains to prove that U_F creates coequalizers of U_F-split pairs of morphisms. Thus let $f, g\colon (A, a, \dagger) \to (B, b, \ddagger)$ be solution-preserving morphisms of ffg-Elgot algebras and suppose that morphisms $c\colon B \to C$, $s\colon C \to B$ and $t\colon B \to A$ in $\mathscr{C}$ are given with $c \cdot f = c \cdot g$, $c \cdot s = \mathrm{id}_C$, $g \cdot t = \mathrm{id}_B$ and $s \cdot c = f \cdot t$.

$$A \mathrel{\substack{\xrightarrow{f} \\ \xleftarrow{t} \\ \xrightarrow[g]{}}} B \mathrel{\substack{\xrightarrow{c} \\ \xleftarrow[s]{}}} C$$

Since the category **Alg** F of F-algebras and their morphisms is monadic over $\mathscr{C}$ [8] we know that there is a unique F-algebra structure $\gamma\colon FC \to C$ such that

C is an F-algebra homomorphisms from (B, b) to (C, γ) and c is, moreover, a coequalizer of f and g in **Alg** F. Define a solution operator $*$ for (C, γ) as follows. Given an ffg-equation $e \colon X \to FX + C$, put $e^* = c \cdot (s \bullet e)^{\ddagger}$. One then proves that $*$ is the unique solution operator making $(C, \gamma, *)$ an ffg-Elgot algebra and c a solution-preserving morphism from $(B, b, \ddagger)$ to $(C, \gamma, *)$. Moreover, c is a coequalizer of f and g in ffg-Elgot F.   □

5   Conclusions and Further Work

For a functor F on a variety preserving sifted colimits, the concept of an Elgot algebra [1] has a natural weakening obtained by working with iterative equations having ffg objects of variables. We call such algebras ffg-Elgot algebras. We have proved that the locally ffg fixed point φF of an endofunctor, constructed by taking the colimit of all F-coalgebras with an ffg carrier, is the initial ffg-Elgot algebra for F. Furthermore, we have proved that all free ffg-Elgot algebras exist, and we have shown that the colimit of all ffg-carried coalgebras for $F(-)+Y$ yield a free ffg-Elgot algebra on Y whenever Y is a free object of $\mathscr{C}$ on some (possibly infinite) set. Finally, we have proved that the forgetful functor ffg-Elgot $H \to \mathscr{C}$ is monadic.

We leave the task of giving a coalgebraic construction of arbitrary free ffg-Elgot algebras for further work. In addition, the study of the properties of the ensuing free ffg-Elgot algebra monad is also left for the future. The monad of ordinary free Elgot algebras (cf. Sect. 3) yields the free Elgot monad on the given endofunctor F; it should be interesting to see whether the above monad of free ffg-Elgot algebras is characterized by a similar universal property.

Finally, in the current setting we have the following picture of categories and forgetful functors: ffg-Elgot $F \hookrightarrow$ Alg $F \to \mathscr{C} \to$ Set. Each of those functors has a left-adjoint and is in fact monadic, and we have shown that the composite of the first two is monadic, too. We leave the question whether the composite of all three of the functors is monadic for further work.

References

1. Adámek, J., Milius, S., Velebil, J.: Elgot algebras. Log. Methods Comput. Sci. **2**(5:4), 1–31 (2006)
2. Adámek, J., Milius, S., Velebil, J.: Iterative algebras at work. Math. Struct. Comput. Sci. **16**(6), 1085–1131 (2006)
3. Adámek, J., Milius, S., Velebil, J.: Semantics of higher-order recursion schemes. Log. Methods Comput. Sci. **7**(11:5), 1–43 (2011)
4. Adámek, J., Rosický, J.: Locally Presentable and Accessible Categories. Cambridge University Press, Cambridge (1994)
5. Adámek, J., Rosický, J., Vitale, E.: What are sifted colimits? Theory Appl. Categ. **23**, 251–260 (2010)
6. Adámek, J., Rosický, J., Vitale, E.: Algebraic Theories. Cambridge University Press, New York (2011)

7. Applegate, H.: Acyclic models and resolvent functors. Ph.D. thesis, Columbia University (1965)
8. Barr, M.: Coequalizers and free triples. Math. Z. **116**, 307–322 (1970)
9. Bonsangue, M.M., Milius, S., Silva, A.: Sound and complete axiomatizations of coalgebraic language equivalence. ACM Trans. Comput. Log. **14**(1:7), 7:1–7:52 (2013)
10. Courcelle, B.: Fundamental properties of infinite trees. Theoret. Comput. Sci. **25**, 95–169 (1983)
11. Ésik, Z., Maletti, A.: Simulation vs. equivalence. In: Proceedings of 6th International Conference on Foundations of Computer Science, pp. 119–122. CSREA Press (2010)
12. Ésik, Z., Maletti, A.: Simulations of weighted tree automata. In: Domaratzki, M., Salomaa, K. (eds.) CIAA 2010. LNCS, vol. 6482, pp. 321–330. Springer, Heidelberg (2011). https://doi.org/10.1007/978-3-642-18098-9_34
13. Fiore, M., Plotkin, G.D., Turi, D.: Abstract syntax and variable binding. In: Proceedings of LICS 1999, pp. 193–202. IEEE Press (1999)
14. Freyd, P.: Rédei's finiteness theorem for commutative semigroups. Proc. Amer. Math. Soc. **19**(4), 1003 (1968)
15. Johnstone, P.T.: Adjoint lifting theorems for categories of algebras. Bull. Lond. Math. Soc. **7**, 294–297 (1975)
16. Johnstone, P.T.: Topos Theory. Academic Press, London (1977)
17. Lambek, J.: A fixpoint theorem for complete categories. Math. Z. **103**, 151–161 (1968)
18. Milius, S., Adámek, J., Urbat, H.: On algebras with effectful iteration. https://www8.cs.fau.de/staff/milius/publications/files/amv_ffg-elgot_2018.pdf
19. Milius, S.: Completely iterative algebras and completely iterative monads. Inform. Comput. **196**, 1–41 (2005)
20. Milius, S.: A sound and complete calculus for finite stream circuits. In: Proceedings of LICS 2010, pp. 449–458. IEEE Computer Society (2010)
21. Milius, S.: Proper functors and fixed points for finite behaviour (2018, submitted). arXiv:1705.09198
22. Milius, S.: Proper functors and their rational fixed point. In: Bonchi, F., König, B. (eds.) Proceedings of 7th Conference on Algebra and Coalgebra in Computer Science (CALCO 2017), LIPIcs, vol. 72, pp. 18:1–18:15. Schloss Dagstuhl (2017)
23. Milius, S., Pattinson, D., Wißmann, T.: A new foundation for finitary corecursion and iterative algebras (2018, submitted). arXiv:1802.08070
24. Milius, S., Pattinson, D., Wißmann, T.: A new foundation for finitary corecursion. In: Jacobs, B., Löding, C. (eds.) FoSSaCS 2016. LNCS, vol. 9634, pp. 107–125. Springer, Heidelberg (2016). https://doi.org/10.1007/978-3-662-49630-5_7
25. Milius, S., Schröder, L., Wißmann, T.: Regular behaviours with names: on rational fixpoints of endofunctors on nominal sets. Appl. Categ. Struct. **24**(5), 663–701 (2016)
26. Milius, S., Wißmann, T.: Finitary corecursion for the infinitary lambda calculus. In: Proceedings of CALCO 2015, LIPIcs, vol. 35, pp. 336–351 (2015)
27. Nelson, E.: Iterative algebras. Theoret. Comput. Sci. **25**, 67–94 (1983)
28. Plotkin, G.D., Turi, D.: Towards a mathematical operational semantics. In: Proceedings of Logic in Computer Science (LICS 1997), pp. 280–291 (1997)
29. Rédei, L.: The Theory of Finitely Generated Commutative Semigroups. Oxford, Pergamon, Edinburgh, New York (1965)

30. Silva, A., Bonchi, F., Bonsangue, M.M., Rutten, J.J.M.M.: Generalizing determinization from automata to coalgebras. Log. Methods Comput. Sci **9**(1:9), 1–27 (2013)
31. Sokolova, A., Woracek, H.: Congruences of convex algebras. J. Pure Appl. Algebra **219**(8), 3110–3148 (2015)
32. Sokolova, A., Woracek, H.: Proper semirings and proper convex functors. In: Baier, C., Dal Lago, U. (eds.) FoSSaCS 2018, LNCS, vol. 10803, pp. 331–347. Springer, Heidelberg (2018). https://doi.org/10.1007/978-3-319-89366-2_18
33. Tiuryn, J.: Unique fixed points vs. least fixed points. Theoret. Comput. Sci. **12**, 229–254 (1980)
34. Urbat, H.: Finite behaviours and finitary corecursion. In: Proceedings of CALCO 2017, LIPIcs, vol. 72, pp. 24:1–24:15 (2017)

Monoidal Computer III: A Coalgebraic View of Computability and Complexity (Extended Abstract)

Dusko Pavlovic$^{(\boxtimes)}$ and Muzamil Yahia

University of Hawaii,Honolulu, USA
{dusko,muzamil}@hawaii.edu

Abstract. Monoidal computer is a categorical model of *intensional* computation, where many different programs correspond to the same input-output behavior. The upshot of yet another model of computation is that a categorical formalism should provide a high-level language for theory of computation, flexible enough to allow abstracting away the low level implementation details when they are irrelevant, or taking them into account when they are genuinely needed. A salient feature of the approach through monoidal categories is the formal graphical language of *string diagrams*, which supports geometric reasoning about programs and computations. In the present paper, we provide a coalgebraic characterization of monoidal computer. It turns out that the availability of interpreters and specializers, that make a monoidal category into a monoidal computer, is equivalent with the existence of a *universal state space*, that carries a weakly final state machine for all types of input and output. Being able to program state machines in monoidal computers allows us to represent Turing machines, and capture the time and space needed for their executions. The coalgebraic view of monoidal computer thus provides a convenient diagrammatic language for studying not only computability, but also complexity.

1 Introduction

In theory of computation, an *extensional* model reduces computations to their set theoretic extensions, *computable functions*, whereas an *intensional* model also takes into account the multiple *programs* that describe each computable function [4, 29, II.3].

In computer science, this semantical gamut got refined on the extensional side by *denotational* models, that take into account not just computable functions but also some computational effects, and on the intensional side by *operational* models, where the meaning of a program is specified up to an operational equivalence [9, 48]. Categorical semantics of computation arose from the realization

---

Partially supported by AFOSR and NSF.

C. Cîrstea (Ed.): CMCS 2018, LNCS 11202, pp. 167–189, 2018.
https://doi.org/10.1007/978-3-030-00389-0_10

that *cartesian closed* categories provide a simple and effective framework for studying the extensional models [24]. Both denotational and operational semantics naturally developed as extensions of this categorical framework [26,45].

The goal of the monoidal computer project is to provide categorical semantics of intensional computation. This turns out to be surprisingly simple technically, but subtle conceptually. In this section, we describe the structure of monoidal computer informally, and try to explain it in the context of categorical semantics. In the rest of the paper, we spell out some of its features formally, in particular the *coalgebraic* part.

1.1   Categorical Computability: Context and Concept

The step from a cartesian closed category $\mathcal{C}$, as an extensional model of computation, to a monoidal computer $\mathbb{C}$, as an intensional model, can be summarized as follows:

$$\frac{\mathcal{C}\,(X,[A,B]) \underset{\lambda_X^{AB}}{\overset{\varepsilon_X^{AB}}{\underset{\cong}{\rightleftarrows}}} \mathcal{C}(X \times A, B)}{\mathbb{C}^{\bullet}(X,\mathbb{P}) \overset{\gamma_X^{AB}}{\twoheadrightarrow} \mathbb{C}(X \otimes A, B)} \tag{1}$$

The first line says that a category $\mathcal{C}$ is cartesian closed when it has the (cartesian) products $X \times A$ and a family of bijections, natural in X and indexed over the types A and B, between the morphisms $X \times A \to B$ and $X \to [A,B]$. If a morphism $X \times A \overset{f}{\to} B$ is thought of as an X-indexed family of computations with the inputs from A and the outputs in B, then the corresponding morphism $X \xrightarrow{\lambda_X^{AB}(f)} [A,B]$ can be thought of as the X-indexed family of programs for these computations. This structure is the categorical version of the simply typed extensional lambda calculus: λ_X^{AB} corresponds to the operation of *abstraction*, whereas ε_X^{AB} corresponds to the *application* [24, Part I]. The equation $\varepsilon_X^{AB} \circ \lambda_X^{AB} = \mathrm{id}$ says that if we *abstract* a computation into a program, and then *apply* that program to some data, then we will get the same result as if we executed the original computation on the data. This is the β-rule of the lambda calculus, the crux of Alonzo Church's representation of program evaluations as function applications of λ-abstractions [10]. The equation $\lambda_X^{AB} \circ \varepsilon_X^{AB} = \mathrm{id}$ says that if we apply a program, and then abstract out of the resulting computation a program, then we will get the same program that we started from. This is the η-rule of the lambda calculus: the extensionality. Dropping the second equation thus corresponds to modeling the *non-extensional* typed lambda calculus, with *weak* exponent types. While this structure was sometimes interpreted as a model of intensional computation, and interesting results were obtained [18], the main result was that every such non-extensional model is *essentially extensional*, in the sense that it contains an extensional model as a retract [15]. In genuinely intensional models, identifying extensionally equivalent programs is not computable.

The structure of a monoidal computer $\mathbb{C}$ is displayed in the second line of (1). There are **three changes** with respect to the cartesian closed structure:

(a) the bijections ε_X^{AB} are relaxed to surjections γ_X^{AB};
(b) the exponents $[A, B]$ are replaced with the type $\mathbb{P}$ of *programs*, the same for all types A and B; and
(c) the product $\times$ is replaced with a tensor $\otimes$, and $\mathbb{C}$ is not a cartesian category, *but* $\mathbb{C}^\bullet$ on the left is its largest cartesian subcategory with $\otimes$ as the product.

We try to clarify these changes in the next three paragraphs.

Change (a) means that we have not only dropped the extensionality equation $\lambda_X^{AB} \circ \varepsilon_X^{AB} = \mathrm{id}$, but eliminated the abstraction operation λ_X^{AB} altogether. All that is left of the bijection between the abstractions and the applications, displayed in the first line of (1), is a surjection from programs to computations, displayed in the second line of (1): for every X-indexed family of computations $X \otimes A \xrightarrow{f} B$ there is an X-indexed family of programs $X \xrightarrow{F} \mathbb{P}$ such that $f = \gamma_X^{AB}(F)$. Could we get away with less? No, because the program evaluation γ_X^{AB} has a left inverse λ_X^{AB} if and only if the model is essentially extensional (i.e., it contains an extensional retract). We will see in Sect. 3.1 that the program evaluation γ_X^{AB} is in fact executed by a universal evaluator $\{\}^{AB} \in \mathbb{C}(\mathbb{P} \otimes A, B)$, and thus takes the form $\gamma_X^{AB}(F) = \{F\}^{AB} = f$.

Change (b) means that all programs are of the same type $\mathbb{P}$. The central feature of intensional computation is that any program can be applied to any data, and in particular to itself. The main constructions of computability theory depend on this, as we shall see in Sect. 3.4. If computations of type $A \rightarrow B$ were encoded by programs of a type depending on A and B, let us write it in the form $\lceil A, B \rceil$, then such programs could not be applied to themselves, but they could only be processed by programs typed in the form $\lceil \lceil A, B \rceil, C \rceil$. That is why all programs must be of the same type $\mathbb{P}$. We will see in Sect. 3.3 that this implies that all types must be retracts of $\mathbb{P}$. This does not imply that the type structure of a monoidal computer can be completely derived from an applicative structure on $\mathbb{P}$, as an essentially untyped model of computation [24, I.15–I.17]. The type structure of monoidal computer, can be derived from internal structure of $\mathbb{P}$ if and only if the model is essentially extensional (i.e., it contains an extensional retract, like before). But where does the monoidal structure come from?

Change (c) makes monoidal computers into monoidal categories, not cartesian. Just like cartesian categories, monoidal computers have the diagonals and the projections for all types, which are necessary for data copying and deleting, as explained in Sect. 2. Unlike in cartesian categories, though, the diagonals and the projections in monoidal computers are *not natural*. The projections are not natural because intensional computations may not terminate: they are not *total* morphisms. The diagonals are not natural when the computations are not deterministic: they are then not *single-valued* as morphisms. While intensional computations can be deterministic, and the diagonals in a monoidal computer can all be natural, if all projections are natural, i.e. if all computations are total, then the model contains an extensional retract. A monoidal computer is thus a cartesian category if and only if it is essentially extensional. That is why a genuinely intensional monoidal computer must be genuinely monoidal. On the other hand, even a computation that is nowhere defined has a program, and programs are always well-defined values. So while the indexed families of intensional computations cannot all be total functions, the corresponding indexed families of programs must all be total functions. That is why the category $\mathbb{C}^\bullet$ on the left in (1) is different from $\mathbb{C}$: it is the largest subcategory of $\mathbb{C}$ for which $\otimes$ is the cartesian product.

In **summary**, dropping or weakening any of the changes described in (a–c) leads to the same outcome: an essentially extensional model. For a genuinely intensional model it is thus *necessary* to have (c) a genuinely monoidal structure, (b) untyped programs, and (a) no computable program abstraction operators. It was shown in [36,41] that this is also *sufficient* for a categorical reconstruction of the basic concepts of computability. Sections 2 and 3 provide a brief overview of this. But our main concern in this paper is *complexity*.

1.2   Categorical Complexity: A Coalgebraic View

To capture complexity, we must capture dynamics, i.e. access the actual process of computation. This, of course, varies from model to model, and different models of computation induce different notions of complexity. Abstract complexity [7] provides, in a sense, a model-independent common denominator, which can be viewed as an abstract notion of complexity; but the categorical view of computations as morphisms at the first sight does not even provide a foothold for abstract complexity. We attempted to mitigate the problem by extending the structure of monoidal computer by *grading* [38], but the approach turned out to be impractical for our goals (indicated in the next section). Now it turns out to also be unnecessary, since dynamics of computation can be captured using the *coalgebraic* tools available in any monoidal computer.

Coalgebra is the categorical toolkit for studying dynamics in general [42,44], and dynamics of computation in particular [23,40,45]. Coalgebras, as morphisms in the form $X \to EX$ for an endofuctor E, provide a categorical view of automata, state machines, and processes with state update [20,39]; the other way around, all coalgebras can be thought of as processes with state update. In the framework on this paper, only a very special class of coalgebras will be considered, as the morphisms in the form $X \times A \to X \times B$, corresponding to what is usually called *Mealy machines* [8,14,17, …]. In the presence of the exponents, such morphisms can be transposed to proper coalgebras in the form $X \to [A, X \times B]$. But coalgebra provides a categorical reconstruction of state machines even without the exponents, since the homomorphisms remain the same, and the category of machines is isomorphic to a category of coalgebras even if the objects are not presented as coalgebras in the strict sense. Our "coalgebras" will thus be in the form $X \times A \to X \times B$, or more generally $X \otimes A \to X \otimes B$.

The crucial step in moving the monoidal computer story into the realm of coalgebra is to replace the X-indexed functions $X \times A \xrightarrow{f} B$ with X-state machines $X \times A \xrightarrow{m} X \times B$. While a function f mapped for each index x an input a to an output b, a machine m now maps at each state x an input a to an output b, *and updates the state to* x'. This state update provides an abstract view of dynamics. Continuous dynamics can be captured in varying the same approach [39,42]. This step from X-indexed functions to X-state machines is displayed in the first row of the following table.

models	static	dynamic
extensional models: cartesian closed	$[A, B] \times A \xrightarrow{\varepsilon} B$ $\exists! \lambda f \times A \nwarrow \quad \nearrow \forall f$ $X \times A$ — abstractions $\xleftrightarrow{\varepsilon}{\lambda}$ applications	$[A^+, B] \times B$ $\xi \nearrow \quad \nwarrow \exists![\![m]\!] \times B$ $[A^+, B] \times A \qquad X \times B$ $\exists![\![m]\!] \times A \nwarrow \quad \nearrow \forall m$ $X \times A$ — behaviors $\xleftarrow{[\![-]\!]}$ machines
intensional models: monoidal computers	$\mathbb{P} \otimes A \xrightarrow{\{\}} B$ $\exists F \times A \nwarrow \quad \nearrow \forall f$ $X \otimes A$ — programs $\twoheadrightarrow$ computations	$\mathbb{P} \otimes B$ $\{\!|\ |\!\} \nearrow \quad \nwarrow \exists M \otimes B$ $\mathbb{P} \otimes A \qquad X \otimes B$ $\exists M \otimes A \nwarrow \quad \nearrow \forall m$ $X \otimes A$ — adaptive programs $\rightarrowtail$ processes

The representation of functions from A to B by the elements of $[A, B]$ lifts to the representation of machines with inputs in A and outputs in B by the induced behaviors in $[A^+, B]$, where A^+ is the inductive type of the nonempty sequences from A. Behaviors are thus construed as *functions extended in time* [20,41,44]. In the presence of list constructors, the representation of functions using the exponents $[A, B]$ induces the representation of machines using the final machines $[A^+, B]$. The other way around, the final machines induce the exponents as soon as the idempotents split.

The rows of the table depict the step from static models to dynamic models. The columns depict the step from the extensional to the intensional. The left-hand column is just a different depiction of (1): the upper triangle unpacks the bijection in the first line of (1), whereas the lower triangle unpacks the surjection in the second line. The right-hand column is the step from the extensional coinduction of final state machines to the intensional coinduction as implemented in the structure of monoidal computer. The bottom row of the table is the step from the monoidal computer structure presented in terms of universal evaluators, the content of Sect. 3, to the monoidal computer structure presented in terms of *universal processes*, the content of Sect. 4. The fact that the two presentations are equivalent is stated in Theorem 9. This coalgebraic view of intensional computation opens an alley towards capturing dynamics of Turing machines in Sect. 5, and a direct internalization of time and space complexity measures in Sect. 6. A general approach through abstract complexity is provided in the full version of the paper. A comment about the role of coalgebra in this effort is in Sect. 7. Some proofs are in the Appendix.

1.3 Background and Related Work

While computability and complexity theorists seldom felt a need to learn about categories, there is a rich tradition of categorical research in computability theory, starting from one of the founders of category theory and his students [13,31], through extensive categorical investigations of realizability [16,19,30], to the recent work on Turing categories [12], and on a monoidal structure of Turing machines [5]. A categorical account of time complexity was proposed in [11], using a special structure called *timed sets*, introduced for the purpose. While our approach in [38] used grading in a similar way, our current approach seems closer in spirit to [2], even if that work is neither coalgebraic nor explicitly categorical. Our effort originated from a need for a framework for reasoning about logical depth of cryptographic protocols and algorithms [34]. The scope of the project vastly exceeded the original cost estimates [37], but also the original benefit expectations. The unexpectedly simple diagrammatic formalism of monoidal computer turned out to be a very convenient teaching tool in several courses.[1]

This extended abstract is shortened to fit the conference proceedings format. The full text is available on arxiv:1704.04882.

2 Preliminaries

A monoidal computer is a *symmetric monoidal category* with some additional structure. As a matter of convenience, and with no loss of generality, we assume that it is a *strict* monoidal category. Monoidal categories are presented in many textbooks, e.g. [25, Sect. VII.1 and Ch. XI].

We call *data service* the structure that allows passing the data around in a monoidal category. In computer programs and in mathematical formulas, the data are usually passed around using variables. They allow copying and propagating the data values where they are needed, or deleting them when they are not needed. The basic features of a variable are thus that it can be freely copied or deleted. The basic data services over a type A in a monoidal category $\mathbb{C}$ are the *copying* operation $A \xrightarrow{\Delta} A \otimes A$, and the *deleting* operation $A \xrightarrow{\top} I$, which together form a *commutative comonoid*, i.e. satisfy the equations

$$\Delta \, ; (\Delta \otimes A) = \Delta \, ; (A \otimes \Delta) \qquad \Delta \, ; (\top \otimes A) = \Delta \, ; (A \otimes \top) = \mathrm{id}_A \qquad \Delta \, ; \sigma \; = \; \Delta$$

---

[1] The course materials are available from http://www.asecolab.org/courses/222/, and the textbook [41] is in preparation.

The correspondence between variables and comonoids was formalized and explained in [32]. The algebraic properties of the binary copying induce unique n-ary copying $A \xrightarrow{\Delta} A^{\otimes n}$, for all $n \geq 0$. The tensor products $\otimes$ in $\mathbb{C}$ are the cartesian products $\times$ if and only if every A in $\mathbb{C}$ carries a canonical comonoid $A \times A \xleftarrow{\Delta} A \xrightarrow{\top} \mathbb{1}$, where $\mathbb{1}$ is the final object of $\mathbb{C}$, and all morphisms of $\mathbb{C}$ are comonoid homomorphisms, or equivalently, the families $A \xrightarrow{\Delta} A \times A$ and $A \xrightarrow{\top} \mathbb{1}$ are natural. Cartesian categories are thus just monoidal categories with natural families of copying and deleting operations.

Definition 1. *A data service of type A in a monoidal category $\mathbb{C}$ is a commutative comonoid structure $A \otimes A \xleftarrow{\Delta} A \xrightarrow{\top} I$, where Δ provides the copying service, and $\top$ provides the deleting service.*

Definition 2. *A morphism $f \in \mathbb{C}(A, B)$ is a map if it is a comonoid homomorphism with respect to the data services on A and B, which means that it satisfies the following equations*

$$f \,;\Delta_B \;=\; \Delta_A \,;(f \otimes f) \qquad f \,;\top_B \;=\; \top_A$$

Given a symmetric monoidal category $\mathbb{C}$ with data services, we denote by $\mathbb{C}^\bullet$ the subcategory spanned by the maps with respect to its data services, i.e. by those $\mathbb{C}$-morphisms that preserve copying and deleting.

Remark. If $\mathbb{C}$ is the category of relations, then the first equation says that f is a single-valued relation, whereas the second equation says that it is total. Hence the name. Note that the morphisms Δ and $\top$ from the data services are maps with respect to the data service that they induce. They are thus contained in $\mathbb{C}^\bullet$, and each of them forms a natural transformation with respect to the maps. This just means that the tensor $\otimes$, restricted to $\mathbb{C}^\bullet$, is the cartesian product.

3 Monoidal Computer

3.1 Evaluation and Evaluators

Notation. When no confusion seems likely, we write AB instead of $A \otimes B$, and $\mathbb{C}(X)$ instead of $\mathbb{C}(I, X)$. We omit the typing superscripts whenever the types are clear from the context.

Definition 3. *A monoidal computer is a (strict) symmetric monoidal category $\mathbb{C}$, with a data service $A \otimes A \xleftarrow{\Delta} A \xrightarrow{\top} I$ on every A, and a distinguished type of programs $\mathbb{P}$, given with, for every pair of types A, B, an X-natural family of surjections $\mathbb{C}^\bullet(X, \mathbb{P}) \xrightarrow{\gamma_X^{AB}} \mathbb{C}(X \otimes A, B)$, representing program evaluations.*

The following proposition says that program evaluations can be construed as a categorical view of Turing's *universal computer* [46], or of Kleene's *acceptable enumerations* [29,43, II.5], or of *interpreters* and *specializers* from programming language theory [21].

Proposition 4. *Let $\mathbb{C}$ be a symmetric monoidal category with data services. Then specifying the program evaluations $\gamma_X^{AB} : \mathbb{C}^\bullet(X, \mathbb{P}) \twoheadrightarrow \mathbb{C}(X \otimes A, B)$ that make $\mathbb{C}$ into a monoidal computer, as defined in Definition 3, is equivalent to specifying for any three types $A, B, C \in |\mathbb{C}|$ the following two morphisms:*

(a) a universal evaluator $\{\}^{AB} \in \mathbb{C}(\mathbb{P}A, B)$ such that for every computation $f \in \mathbb{C}(A, B)$ there is a program $F \in \mathbb{C}^\bullet(\mathbb{P})$ such that $f(a) = \{F\}^{AB} a$

(b) a partial evaluator $[]^{AB} \in \mathbb{C}^\bullet(\mathbb{P}A, \mathbb{P})$ with $\{G\}^{(AB)C}(a, b) = \{[G]^{AB} a\}^{BC} b$

Remark. Note that the partial evaluators [] are maps, i.e. total and single valued morphisms in $\mathbb{C}^\bullet$, whereas the universal evaluators $\{\}$ are ordinary morphisms in $\mathbb{C}$. A recursion theorist will recognize the universal evaluators as Turing's *universal machines* [46], and the partial evaluators as Gödel's primitive recursive *substitution function S*, enshrined in Kleene's S_n^m-theorem [22]. A programmer can think of the universal evaluators as *interpreters*, and of the partial evaluators as *specializers* [21]. In any case, *(a)* can be understood as saying that every computation can be programmed; and then *(b)* says that any program with several inputs can be evaluated on any of its inputs, and reduced to a program that waits for the remaining inputs:

$$h(x, a) \quad = \quad \{H\}(x, a) \quad = \quad \{[H] x\} a$$

(2)

Together, the two conditions thus equivalently say that for every computation $h \in \mathbb{C}(X \otimes A, B)$ there is an X-indexed program $\Xi \in \mathbb{C}^\bullet(X, \mathbb{P})$ such that $h(x, a) = \{\Xi x\} a$, namely $\Xi = [H]$.

Branching. By extending the λ-calculus constructions as in [36], we can extract from $\mathbb{P}$ the convenient types of natural numbers, truth values, etc. E.g., if the truth values $\mathtt{t}$ and $\mathtt{f}$ are defined to be some programs for the two projections, then the role of the $\mathtt{if}$-branching command can be played by the universal evaluator:

$$\mathtt{if}(b, x, y) = \{b\}(x, y) \quad = \quad \begin{cases} x & \text{if } b = \mathtt{t} \\ \\ y & \text{if } b = \mathtt{f} \end{cases}$$

3.2   Examples of Monoidal Computer

Let S be a cartesian category and $T : S \to S$ a commutative monad. Then the Kleisli category S_T of free algebras is monoidal, with the data services induced by the cartesian structure of S. The standard model of monoidal computer C is obtained by taking S to be the category of finite and countable sets, and $TX = \bot + X$ to be the *maybe* monad, adjoining a fresh element to every set. The category $S_\bot$ is the category of partial functions, and the monoidal computer $C \subseteq S_\bot$ is the subcategory of *computable* partial functions:

$$|C| = \{A \subseteq \mathbb{N} \mid \exists e \in \mathbb{N}. \, \{e\}a\downarrow \iff a \in A\} \qquad C(A, B) = \{f : A \rightharpoonup B \mid \exists e. \, \{e\} = f\}$$

The category $C^\bullet$ is then the category of computable *total* functions. Assuming that the programs are encoded as natural numbers, the type of programs is $\mathbb{P} = \mathbb{N}$; but any language containing a Turing complete set of expressions would do, *mutatis mutandis*. The sequence $\{0\}, \{1\}, \{2\}, \ldots$ denotes an acceptable enumeration of computable partial functions [29, II.5]. The universal evaluators can be implemented as partial recursive functions; the partial evaluators are the total recursive functions, constructed in Kleene's S_n^m-theorem [22]. Other commutative monads $T : S \to S$ induce monoidal computers in a similar way, capturing intensional computations together with the corresponding computational effects: exceptions, nondeterminism, randomness [26]. Some of the familiar computational monads need to be restricted to finite support. The distribution monad must be factored modulo computational indistinguishability. A simple quantum monoidal computer can be constructed using a relative monad for finite dimensional vector spaces [1]. However, in the model where the universal evaluators are quantum Turing machines, the program evaluations cannot be surjective in the usual sense, but only in the topologically enriched sense, i.e., they are dense [6]. We do not know how to derive this model from a computational monad, albeit relative. Another interesting feature is that most computational effects induce nonstandard data services, corresponding to complementary bases, which are, of course, used in randomized, quantum, but also in nondeterministic algorithms [33,35]. More examples are in [36], but most work is still ahead.

3.3   Encoding All Types

Proposition 5. *Every type B in a monoidal computer is a retract of the type of programs $\mathbb{P}$. More precisely, for every type $B \in |C|$ there are computations $e^B : B \rightleftarrows \mathbb{P} : d^B$ such that e^B is a map, and $e^B \,; d^B = \mathrm{id}_B$. We often call e^B the encoding of B and $d^B \in C(\mathbb{P}, B)$ is the corresponding* decoding.

Remark. In [36] we only considered the *basic* monoidal computer, where all types are powers of $\mathbb{P}$. In the standard model, programs are encoded as natural numbers, and all data are tuples of natural numbers, which can be recursively encoded as natural numbers. Proposition 5 says that this must be the case in every computer. Note that there is no claim that either e^B or d^B is unique. Indeed, in nondegenerate monoidal computers, each type B has many different encoding pairs e^B, d^B. However, once such a pair is chosen, the fact that e^B is total and single-valued means that it assigns a unique program code to each element of B. The fact that d^B is not total means that some programs in $\mathbb{P}$ may not correspond to elements of B. Since Proposition 5 says that the program evaluations make every type into a retract of $\mathbb{P}$, and Proposition 4 reduced the structure of monoidal computer to the evaluators for all types, it is natural to ask if the evaluators of all types can be reduced to the evaluators over the type $\mathbb{P}$ of programs. Can all of the structure of a monoidal computer be derived from the structure of the type $\mathbb{P}$ of programs? E.g., can the program evaluations be *"uniformized"* by always encoding the input data of all types in $\mathbb{P}$, performing the evaluations to get the outputs in $\mathbb{P}$, and then decoding the outputs back to the originally given types? Can the type structure and the evaluation structure of a monoidal computer be reconstructed by unfolding the structure of $\mathbb{P}$, as it is the case in models of λ-calculus. Is monoidal computer yet another categorical view of a partial applicative structure? The answer to all these question is positive *just* in the degenerate case of an essentially extensional monoidal computer. If the type structure of monoidal computer can be faithfully encoded in $\mathbb{P}$, then there is a retract of $\mathbb{P}$ which supports an extensional model of computation, i.e. allows assigning a unique program to each computation. If all evaluators can be derived by decoding the evaluators with the output type $\mathbb{P}$, and if the decoding preserves the original evaluators on $\mathbb{P}$, then all computation representable in monoidal computer must be provably total and single valued: it degenerates into a cartesian closed category derived from a C-monoid. For details see [24, I.15-I.17], and the references therein.

3.4   The Fundamental Theorem of Computability

In this section we show that every monoidal computer validates the claim of Kleene's "Second Recursion Theorem" [22,27].

Theorem 6. *In every monoidal computer $\mathbb{C}$, every computation $g \in \mathbb{C}(\mathbb{P} \otimes A, B)$ has a Kleene fixed point, i.e. a program $\Gamma \in \mathbb{C}(\mathbb{P})$ such that $g(\Gamma, a) = \{\Gamma\}\, a$.*

Proof. Let G be a program such that
$$g(\,[p]\,p, a) = \{G\}(p, a)$$

A Kleene fixed program Γ can now be constructed by partially evaluating G on itself, i.e. as $\Gamma = [G]\,G$, because

$$g(\Gamma, a) \;=\; g([G]G, a) \;=\; \{G\}(G, a) \;=\; \{[G]G\}a \;=\; \{\Gamma\}\,a$$

This theorem induces convenient representations of integers, arithmetic, primitive recursion, unbounded search and thus shows that monoidal computer is Turing complete [41]. In [36], this was done by using the λ-calculus constructions. Next section provides yet another proof, through Turing machines.

4   Coalgebraic View

So far, we formalized the *programs* $\twoheadrightarrow$ *computations* correspondence from the left hand column of the table in the Introduction. But presenting computations in the form $XA \xrightarrow{\{F\}} B$ only displays their interfaces, and hides the actual process of computation. To capture that, we switch to the right hand column of the table, and study the correspondence *adaptive programs* $\twoheadrightarrow$ *processes*. A *process* is presented as a morphism in the form $X \otimes A \to X \otimes B$. We interpreted the morphisms in the form $X \otimes A \to B$ as X-indexed families of computations with the inputs from A and the outputs in B. The indices of type X can be thought of as the states of the world, determining which of the family of computations should be run. Interpreted along the same lines, a process $X \otimes A \xrightarrow{p} X \otimes B$ does not only provide the output of type B, but it also updates the state in X. This is what state machines also do, and that is why the morphisms $X \times A \xrightarrow{m} X \times B$ in cartesian categories are interpreted as machines. In a sufficiently complete cartesian category, every such machine m induces a machine homomorphism $X \xrightarrow{[\![m]\!]} [A^+, B]$, which assigns to each state $x \in X$ a *behavior* $[\![m]\!]x \in [A^+, B]$, unfolded by the final AB-machine $[A^+, B] \times A \xrightarrow{\xi} [A^+, B] \times B$. The table in the Introduction displayed this. A monoidal computer, though, turns out to provide a much stronger form of representation for its morphisms in the form $X \otimes A \xrightarrow{p} X \otimes B$: each of them induces a machine homomorphism $X \xrightarrow{P} \mathbb{P}$. This P is a *program* for the process p. Note that there may be many programs for each process; but on the other hand, all programs, for all processes of all possible input types A and output types B, are represented in the same type of programs $\mathbb{P}$. This makes a fundamental difference, distinguishing machines m from computational processes p, which include life $[28, 47]^2$. Every family of machines is designed in a suitable

---

[2] Both Turing and von Neumann devoted a lot of attention to studying life as a computational process. Their ideas have been adopted in biology [3], but most computer scientists remain skeptical.

engineering language; but all computational processes can be programmed in any Turing complete language, just like all processes of life are programmed in the language of genes. That is why the morphisms $X \otimes A \xrightarrow{p} X \otimes B$ are *processes*, and not merely machines. Their representations $X \xrightarrow{P} \mathbb{P}$ are not merely X-indexed programs, but they are *adaptive* programs, since they adapt to the state changes, in the sense that we now describe.

Definition 7. *A morphism $XA \xrightarrow{p} XB$ in a monoidal category $\mathbb{C}$ is an AB-process. If $YA \xrightarrow{r} YB$ is another AB-process, then an AB-process homomorphism is a $\mathbb{C}$-morphism $X \xrightarrow{f} Y$ such that $(f \otimes A) ; r = p ; (f \otimes B)$. We denote the category of AB-processes by $\mathbb{C}_{AB}$.*

Definition 8. *A universal process in a monoidal category $\mathbb{C}$ is carried by a universal state space $\mathbb{S} \in |\mathbb{C}|$, which comes with a weakly final AB-process $\mathbb{S}A \xrightarrow{\{\}\}} \mathbb{S}B$ for every pair $A, B \in |\mathbb{C}|$. The weak finality means that for every $p \in \mathbb{C}(X \otimes A, X \otimes B)$ there is an X-adaptive program $P \in \mathbb{C}^{\bullet}(X, \mathbb{S})$ where*

$$\{P(x)\}_{\mathbb{S}}\, a = P(p_X(x, a))$$
$$\{P(x)\}_{B}\, a = p_B(x, a)$$

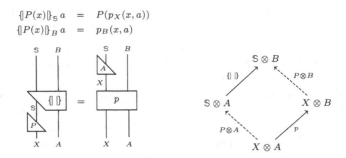

Theorem 9. *Let $\mathbb{C}$ be a symmetric monoidal category with data services. Then $\mathbb{C}$ is a monoidal computer if and only if it has a universal process. The type $\mathbb{P}$ of programs coincides with the universal state space $\mathbb{S}$.*

Proof. Given a weakly final AB-process $\mathbb{S} \otimes A \xrightarrow{\{\}\}} \mathbb{S} \otimes B$, we show that

$$\{\}^{AB} = \left(\mathbb{S} \otimes A \xrightarrow{\{\}\}^{AB}} \mathbb{S} \otimes B \xrightarrow{\top \otimes B} B \right)$$

is a universal evaluator, and thus makes $\mathbb{C}$ into a monoidal computer. Towards proving (2), suppose that we are given a computation $X \otimes A \xrightarrow{h} B$, and consider the process

$$\widehat{h} = \left(X \otimes A \xrightarrow{\Delta \otimes A} X \otimes X \otimes A \xrightarrow{X \otimes h} X \otimes B \right)$$

By Definition 8, there is then an X-adaptive program $\Xi = [\![h]\!] \in \mathbb{C}^{\bullet}(X, \mathbb{S})$ satisfying the rightmost equation in the next diagram.

The middle equation holds because $[\![h]\!]$ is in $\mathbb{C}^{\bullet}$, i.e. a comonoid homomorphism. Deleting the state update from the process yields (2). The other way around, if $\mathbb{C}$ is a monoidal computer, with universal evaluators for all pairs of types, we claim that the weakly final AB-process is

$$\{\!\!\}^{AB} = \left(\mathbb{P} \otimes A \xrightarrow{\{\}^{A(\mathbb{P}B)}} \mathbb{P} \otimes B \right)$$

To prove the claim, take an arbitrary AB-process $X \otimes A \xrightarrow{p} X \otimes B$, and post-compose it with the partial evaluator on X, to get

$$\widehat{p} = \left(\mathbb{P} \otimes X \otimes A \xrightarrow{\mathbb{P} \otimes p} \mathbb{P} \otimes X \otimes B \xrightarrow{[]^{X B \mathbb{P}} \otimes B} \mathbb{P} \otimes B \right)$$

Using the Fundamental Theorem of Computability, Theorem 6, construct a Kleene's fixed point $\widehat{P} \in \mathbb{C}(\mathbb{P})$ of $\widehat{p}$.

The X-adaptive program $P \in \mathbb{C}^{\bullet}(\mathbb{P})$ corresponding to the process $p \in \mathbb{C}(XA, XB)$ is now $P(x) = \left[\widehat{P}, x \right]^{X B \mathbb{P}}$.

This completes the proof that $\{\}^{A(B\mathbb{P})}$ satisfies Definition 8 of weakly final AB-process, and that $\mathbb{P}$ is thus not only a type of programs, but also a universal state space.

5 Computability

In the remaining two sections we show how to run Turing machines in a monoidal computer, and how to measure their complexity. But a coalgebraic treatment of Turing machines as machines, in the sense discussed at the beginning of Sect. 4, would only display their behaviors, i.e. what rewrite and which move of the machine head will happen on which input, and it obliterates the configurations of the tape, where the actual computation happens. In terms of Sect. 4, a Turing machine as a model of actual computation should not be viewed as a machine, but as a process. So we call them *Turing processes* here. While changing well established terminology is seldom a good idea, and we may very well regret this decision, the hope is that it will be a useful reminder that we are doing something unusual: relating Turing machines with adaptive programs, coalgebraically. The presented constructions go through in an arbitrary monoidal computer, but require spelling out a suitable representation of the integers, and some arithmetic. This was done in [36], and can be done more directly; but for the sake of brevity, we work here with the category C of recursively enumerable sets and computable partial functions from Sect. 3.2. The monoidal structure and the data services are induced by the cartesian products of sets, which are, however, not categorical products any more, since the singleton set, providing the tensor unit, is not a terminal object for partial functions. The monoidal category $(\mathbb{C}, \otimes, I)$ will thus henceforth be $(C, \otimes, \mathbb{1})$.

Recall that Turing's definition of his machines can be recast [40, Appendix] to processes in the form $Q_\rho \otimes \Sigma \xrightarrow{\rho} Q_\rho \otimes \Sigma \otimes \Theta$, where

- Q_ρ is the finite set of states, always including the *final* state $\checkmark \in Q_\rho$;
- Σ is a fixed alphabet, always including the blank symbol $\sqcup \in \Sigma$;
- $\Theta = \{\triangleleft, \square, \triangleright\}$ are the directions in which the head can move along the tape.

Let us recall the execution model: how these machines and processes compute. A Mealy machine $Q_\kappa \times I \xrightarrow{\kappa} Q_\kappa \times O$ inputs a string $n \xrightarrow{\iota} I$, where $n = \{0, 1, \ldots, n-1\}$ sequentially, e.g. it reads the inputs ι_0, then ι_1 etc, and it outputs a string $n \xrightarrow{\omega} O$ in the same order, i.e. ω_0, ω_1, etc. In contrast, a Turing process in principle overwrites its inputs, and outputs the results of overwriting when it halts; therefore, in a Turing process, the input alphabet I and its output alphabet O must be the same, say $I = O = \Sigma$. Both the inputs, and the outputs, and the intermediary data of a Turing process are in the form $w : \mathbb{Z} \to \Sigma$, where all but finitely many values $w(z)$ must be $\sqcup$. So each word $w : \mathbb{Z} \to \Sigma$ is still a finite string of symbols, like in the Mealy machine model. The difference is that w is written on the infinite 'tape', here represented by the set of integers $\mathbb{Z}$, which allows the processing 'head' to move in both directions, or to stay stationary (while in a Mealy machine the head moves in the same direction at each step). We represent the position of the head by the integer 0, and the symbol that the head reads on that position is thus denoted by $w(0)$. If the process $Q_\rho \otimes \Sigma \xrightarrow{\rho} Q_\rho \otimes \Sigma \otimes \Theta$, which is a triple of functions $\rho = \langle \rho_Q, \rho_\Sigma, \rho_\Theta \rangle$, is defined on a given state $q \in Q_\rho$ and a given input $\sigma = w(0)$, then it will

- overwrite σ with $\sigma' = \rho_\Sigma(q, \sigma)$,
- transition to the state $q' = \rho_Q(q, \sigma)$, and
- move the head to the next cell in the direction $\theta = \rho_\Theta(q, \sigma)$.

If $q = \checkmark$, then $\rho(\checkmark, \sigma) = \langle \checkmark, \sigma, \square \rangle$, which means that the process must halt at the state $\checkmark$, if it ever reaches it. To capture this execution model formally, we extend Turing processes over the alphabet Σ, first to processes over the set $\widetilde{\Sigma}$ of Σ-words written on a tape, and then to computations with the inputs and the outputs from $\widetilde{\Sigma}$

$$\frac{Q_\rho \otimes \Sigma \xrightarrow{\rho} Q_\rho \otimes \Sigma \otimes \Theta}{Q_\rho \otimes \widetilde{\Sigma} \xrightarrow{\widetilde{\rho}} Q_\rho \otimes \widetilde{\Sigma}}$$
$$\overline{Q_\rho \otimes \widetilde{\Sigma} \xrightarrow{\overline{\rho}} \widetilde{\Sigma}}$$

where $\widetilde{\Sigma} = \left\{ w : \mathbb{Z} \to \Sigma \mid \mathsf{supp}(w) < \infty \right\}$ is the set of Σ-words written on a tape, and $\mathsf{supp}(w) = \{ z \mid w(z) \neq \sqcup \}$. The elements of $\widetilde{\Sigma}$ are often also called the *tape configurations*. Writing the tuples in the form $\widetilde{\rho} = \langle \widetilde{\rho}_Q, \widetilde{\rho}_{\widetilde{\Sigma}} \rangle$, define

$$\widetilde{\rho}_Q(q, w) = \rho_Q(q, w(0))$$

$$\widetilde{\rho}_{\widetilde{\Sigma}}(q, w) = w' \text{ where } w'(z) = \begin{cases} \widetilde{w}(z-1) & \text{if } \rho_\Theta(q, w(0)) = \triangleleft \\ \widetilde{w}(z) & \text{if } \rho_\Theta(q, w(0)) = \square \\ \widetilde{w}(z+1) & \text{if } \rho_\Theta(q, w(0)) = \triangleright \end{cases} \text{ and}$$

$$\widetilde{w}(z) = \begin{cases} \rho_\Sigma(q, w(0)) & \text{if } z = 0 \\ w(z) & \text{otherwise} \end{cases}$$

$$\overline{\rho}(q, w) = \begin{cases} w & \text{if } q = \checkmark \\ \overline{\rho}(\widetilde{\rho}(q, w)) & \text{otherwise} \end{cases}$$

The execution of all Turing processes can now be captured as a single process $Q \otimes \widetilde{\Sigma} \xrightarrow{P} Q \otimes \widetilde{\Sigma}$, where the state space Q is the disjoint union of the state spaces Q_ρ of all Turing processes $\rho \in \mathcal{T}$, i.e. $Q = \coprod_{\rho \in \mathcal{T}} Q_\rho$ where $\mathcal{T} = \{ Q_\rho \otimes \Sigma \xrightarrow{\rho} Q_\rho \otimes \Sigma \otimes \Theta \}$, so that the elements of Q are the pairs $\langle \rho, q \rangle$, where $q \in Q_\rho$, and $Q \otimes \widetilde{\Sigma} \xrightarrow{P} Q \otimes \widetilde{\Sigma}$ is the pair $p = \langle p_Q, p_{\widetilde{\Sigma}} \rangle$ which, when applied to $\langle \rho, q \rangle \in Q$ and $w \in \widetilde{\Sigma}$, gives $p(\langle \rho, q \rangle, w) = \langle \langle \rho, q' \rangle, w' \rangle$ where $q' = \widetilde{\rho}_Q(q, w)$ and $w' = \widetilde{\rho}_{\widetilde{\Sigma}}(q, w)$. By applying Theorem 9 to the process $Q \otimes \widetilde{\Sigma} \xrightarrow{P} Q \otimes \widetilde{\Sigma}$, we get the following

Proposition 10. *There is an adaptive program $\widetilde{P} \in C^\bullet(Q, \mathbb{P})$ such that $\widetilde{P}(\rho, q)$ executes any Turing process ρ starting from the initial state $q \in Q_\rho$. This means that for every tape configuration $w \in \widetilde{\Sigma}$ holds $\{\!|\widetilde{P}(\rho, q)|\!\}_\mathbb{P}\, w = \widetilde{P}(\rho, q')$ and $\{\!|\widetilde{P}(\rho, q)|\!\}_{\widetilde{\Sigma}} w = w'$, where $q' = \rho_Q(q, w(0))$ is the next state of ρ, and $w' = \widetilde{\rho}_{\widetilde{\Sigma}}(q, w)$ is the next tape configuration. (The string diagram is the same as the one in Definition 8.)*

Corollary 1. *The monoidal computer C is Turing complete.*

6 Complexity

6.1 Evaluating Turing Processes

Using the process $Q \otimes \widetilde{\Sigma} \xrightarrow{P} Q \otimes \widetilde{\Sigma}$, which according to Proposition 10 executes the single step transitions of Turing processes, we would now like to define a computation $Q \otimes \widetilde{\Sigma} \xrightarrow{\overline{P}} \widetilde{\Sigma}$ that will evaluate Turing processes all the way; i.e. should execute all transitions that a process executes, and halt and deliver the output if the process halts, or diverge if the process diverges. The idea is to run something like the following pseudocode

$$\overline{p}(\langle \rho, q \rangle, w) = \Big(x := \langle \rho, q \rangle; \; y := w;$$
$$\text{while } \big(p_Q(x, y) \neq \checkmark \big)$$
$$\Big\{ x := p_Q(x, y); \; y := p_{\widetilde{\Sigma}}(x, y) \Big\};$$
$$\text{print } y \Big) \tag{3}$$

We implement this program using the Fundamental Theorem of Computability. The function $\overline{p}$ is derived as a Kleene's fixed program for an intermediary function $\widetilde{p}$, lifting the derivation from Sect. 5, as follows

$$\frac{Q \otimes \widetilde{\Sigma} \xrightarrow{P} Q \otimes \widetilde{\Sigma}}{\frac{P \otimes Q \otimes \widetilde{\Sigma} \xrightarrow{\widetilde{P}} \widetilde{\Sigma}}{Q \otimes \widetilde{\Sigma} \xrightarrow{\overline{P}} \widetilde{\Sigma}}} \quad \text{where } \widetilde{p}(\Upsilon, \langle \rho, q \rangle, w) = \begin{cases} w & \text{if } \rho_Q(q, w(0)) = \checkmark \\ \{\Upsilon\}(\langle \rho, q' \rangle, w') & \text{otherwise} \\ \quad \text{where } q' = \rho_Q(q, w(0)) \\ \quad \text{and } w' = p_{\widetilde{\Sigma}}(\langle \rho, q \rangle, w) \end{cases}$$

Using the `if`-branching from Sect. 3.1, this schema can be expressed in a monoidal computer, as illustrated in the following Figure

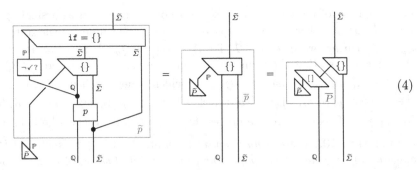

$$\tag{4}$$

The first equation is obtained by setting Υ to be Kleene's fixed program $\widetilde{P}$ of $\widetilde{p}$, and defining $\overline{p} = \{\widetilde{P}\}$. Given $\langle \rho, q \rangle \in Q$ and $w \in \widetilde{\Sigma}$, this $\overline{p}$ thus runs ρ on w, starting from q and halting at $\checkmark$, at which point it outputs the current w. If it does not reach $\checkmark$, then ρ runs forever. The second equation proves the following proposition.

Proposition 11. *There is an adaptive program* $\overline{P} \in C^\bullet(\mathbb{Q}, \mathbb{P})$ *that evaluates any Turing process ρ starting from a given initial state $q \in Q_\rho$. This means that for every tape configuration $w \in \widetilde{\Sigma}$ holds $\{\overline{P}(\rho, q)\}w = \overline{\rho}(q, w)$.*

6.2 Counting Time

To count the steps in the executions of Turing processes, we add a counter $i \in \mathbb{N}$ to the Turing process evaluator $\overline{p}$. The counter gets increased by 1 at each execution step, and thus counts them. We call $\overline{t}$ the computation which outputs the final count. If $\overline{p}$ halts, then $\overline{t}$ outputs the value of the counter i; if $\overline{p}$ does not halt, then $\overline{t}$ diverges as well. The pseudocode for $\overline{t}$ could thus look something like this:

$$\overline{t}(\langle \rho, q \rangle, w) = \Big(x := \langle \rho, q \rangle; \ y := w; \ i := 0; \tag{5}$$
$$\texttt{while } \big(p_\mathbb{Q}(x, y) \neq \checkmark \big)$$
$$\Big\{ x := p_\mathbb{Q}(x, y); \ y := p_{\widetilde{\Sigma}}(x, y); \ i := i + 1 \Big\};$$
$$\texttt{print } i \Big)$$

The implementation of $\overline{t}$ in a monoidal computer is similar to the implementation of $\overline{p}$. It follows a similar derivation pattern:

$$\frac{\dfrac{\mathbb{Q} \otimes \widetilde{\Sigma} \xrightarrow{P} \mathbb{Q} \otimes \widetilde{\Sigma}}{\mathbb{P} \otimes \mathbb{Q} \otimes \widetilde{\Sigma} \otimes \mathbb{N} \xrightarrow{\widetilde{t}} \mathbb{N}}}{\mathbb{Q} \otimes \widetilde{\Sigma} \xrightarrow{\overline{t}} \mathbb{N}} \quad \text{where } \widetilde{t}(\Upsilon, \langle \rho, q \rangle, w, i) = \begin{cases} i & \text{if } \rho_\mathbb{Q}(q, w(0)) = \checkmark \\ \{\Upsilon\}(\langle \rho, q' \rangle, w', i+1) & \text{otherwise} \end{cases}$$

with $q' = \rho_\mathbb{Q}(q, w(0))$ and $w' = p_{\widetilde{\Sigma}}(\langle \rho, q \rangle, w)$. Like before, we set $\overline{t}(\langle \rho, q \rangle, w) = \{\widetilde{T}\}(\langle \rho, q \rangle, w, 0)$, where and $\widetilde{T}$ is a Kleene fixed program of $\widetilde{t}$. It is easy to see, and prove, that $\overline{t}(\langle \rho, q \rangle, w)$ halts if and only if $\overline{p}(q, w)$ halts, and if it does halt, then it outputs the number of steps that ρ made before halting, having started from q and w. The string diagrams that implement $\widetilde{t}$, $\widetilde{T}$, $\overline{t}$ and $\overline{T}$ are similar to those in figure (4): just rename ps to ts and Ps to Ts, and add a string of type $\mathbb{N}$ on the right, with the successor operation on it, to increase the counter at each run. The added string outputs the time complexity $\overline{t}$. Hence

Proposition 12. *There is an adaptive program $\overline{T} \in C^\bullet(\mathbb{Q}, \mathbb{P})$ that outputs the number of steps that a Turing process ρ makes in any run from a given initial state $q \in Q_\rho$ to the halting state $\checkmark$. If the Turing process ρ starting from q diverges, then the computation $\{\overline{T}(\rho, q)\}$ diverges as well. This means that, for every tape configuration $w \in \widetilde{\Sigma}$ holds $\{\overline{T}(\rho, q)\}w = \overline{t}(\langle \rho, q \rangle, w)$.*

6.3   Counting Space

So far, we used the integers $\mathbb{Z}$ as the index set for the tape configurations w : $\mathbb{Z} \to \Sigma$. The position of the head has always been $0 \in \mathbb{Z}$, and whenever the head moves, the tape configuration w gets updated to $w' = \tilde{\rho}_{\tilde{\Sigma}}(q, w)$, where $w'(0)$ is the new position of the head, and the rest of the word w is reindexed accordingly, as described in Sect. 5. At each point of the computation w thus describes the tape content *relative to the current position of the head*; there is no record of the prior positions or contents. To count the tape cells used by Turing processes, we must make the tape itself into a first class citizen. The simplest way to do this seems to be to add a counter $m \in \mathbb{Z}$, which denotes the offset of the current position of the head with respect to the initial position. This allows us to record how far up and down the tape, how far from its original position, does the head ever travel in either direction during the computation. To record these maximal offsets of the head, we need two more counters: let $r \in \mathbb{Z}$ be the highest value that the head offset m ever takes; and let $\ell \in \mathbb{Z}$ be the lowest value that the head offset m ever takes. The number of cells that the head has visited during the computation is then clearly $r - \ell$. To implement this space counting idea, we need to run a program roughly like this:

$$\overline{s}(\langle \rho, q \rangle, w) = \Bigg(x := \langle \rho, q \rangle; \; y := w; \; \ell, m, r := 0;$$

$$\texttt{while } \big(p_Q(x, y) \neq \checkmark\big)$$

$$\begin{cases} x := p_Q(x, y); \; y := p_{\tilde{\Sigma}}(x, y); \end{cases}$$

$$\texttt{if } \big(\rho_\Theta(q, w(0)) = \triangleleft\big)$$

$$\Big\{ \texttt{if } (m = \ell)\{\ell := \ell - 1\}; \; m := m - 1 \Big\}$$

$$\texttt{if } \big(\rho_\Theta(q, w(0)) = \triangleright\big)$$

$$\Big\{ \texttt{if } (m = r)\{r := r + 1\}; \; m := m + 1 \Big\} \Big\}$$

$$\texttt{print } r - \ell \Bigg) \tag{6}$$

The derivation now becomes

$$\frac{\dfrac{Q \otimes \tilde{\Sigma} \xrightarrow{P} Q \otimes \tilde{\Sigma}}{P \otimes Q \otimes \tilde{\Sigma} \otimes \mathbb{Z}^3 \xrightarrow{\tilde{s}} \mathbb{N}}}{Q \otimes \tilde{\Sigma} \xrightarrow{\overline{s}} \mathbb{N}} \quad \text{where } \tilde{s}(\Upsilon, \langle \rho, q \rangle, w, \ell, m, r) = \begin{cases} r - \ell & \text{if } \rho_Q\,(q, w(0)) = \checkmark \\ \{\Upsilon\}(\langle \rho, q' \rangle), \ldots \\ \ldots w', \ell', m', r') & \text{otherwise} \end{cases}$$

and where

$$q' = \rho_Q(q, w(0))$$

$$\ell' = \begin{cases} \ell - 1 & \text{if } m = \ell \text{ and } \rho_\Theta(q, w(0)) = \triangleleft \\ \ell & \text{otherwise} \end{cases}$$

$$w' = p_{\widetilde{\Sigma}}(\langle \rho, q \rangle, w)$$

$$m' = \begin{cases} m - 1 & \text{if } \rho_\Theta(q, w(0)) = \triangleleft \\ m & \text{if } \rho_\Theta(q, w(0)) = \square \\ m + 1 & \text{if } \rho_\Theta(q, w(0)) = \triangleright \end{cases}$$

$$r' = \begin{cases} r + 1 & \text{if } m = r \text{ and } \rho_\Theta(q, w(0)) = \triangleright \\ r & \text{otherwise} \end{cases}$$

Kleene's fixed point $\widetilde{S}$ of $\widetilde{s}$ defines $\overline{s}(\langle \rho, q \rangle, w) = \{\widetilde{S}\}(\langle \rho, q \rangle, w, 0, 0, 0)$. The construction is summarized in the following figure:

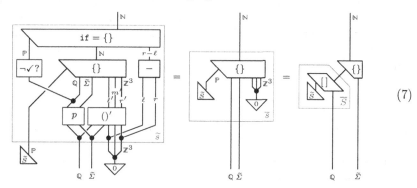

$$(7)$$

The box $()'$, which computes ℓ', m' and r' in Fig. (7), is implemented by composing several branching commands, e.g. as described at the end of Sect. 3.1. Implementing this box is an easy but instructive exercise in programming monoidal computers. Together, these constructions prove the following proposition.

Proposition 13. *There is an adaptive program $\overline{S} \in C^\bullet(\mathbb{Q}, \mathbb{P})$ that outputs the number of cells that a Turing process ρ uses in any run from a given initial state $q \in Q_\rho$ to the halting state $\checkmark$. If the Turing process ρ starting from q diverges, then the computation $\{\overline{S}(\rho, q)\}$ diverges as well. This means that, for every tape configuration $w \in \widetilde{\Sigma}$ holds $\{\overline{S}(\rho, q)\}w = \overline{s}(\langle \rho, q \rangle, w)$.*

Remark. There are many variations of the above definitions in the literature, and several different counting conventions. E.g., an alternative to the above definition of $\overline{s}$ would be something like

$$\overline{s}'(\langle \rho, q \rangle, w) = \{\widetilde{S}\}(\langle \rho, q \rangle, w, w_\ell, 0, w_r) \quad \text{where}$$
$$w_\ell = \min\{i \in \mathbb{Z} \mid w(i) \neq \sqcup\} \qquad w_r = \max\{i \in \mathbb{Z} \mid w(i) \neq \sqcup\}$$

In contrast with $\overline{s}$, where the space counting convention is that a memory cell counts as used if and only if it is ever reached by the head, the space counting convention behind $\overline{s}'$ is that every computation uses at least $|w| = w_r - w_\ell$ cells, on which its initial input is written. If a Turing process halts without reading all of its input w, or even without reading any of it, the space used will still be $|w|$.

Some textbooks adhere to the $\bar{s}$-counting convention, some to the $\bar{s}'$-counting convention, but many do not describe the process in enough detail to discern this difference. This is perhaps justified by the fact that the resulting complexity classes and their hierarchies are the same for all such subtly different counting conventions. E.g., the difference between $\bar{s}$ and $\bar{s}'$ is absorbed by the $\mathcal{O}$-notation, and only arises for computations that do not read their inputs.

7   Final Comments

A bird's eye view of algebra and coalgebra in computer science suggests that algebra provides *denotational* semantics of computation, whereas coalgebra provides *operational* semantics [23,40,45]. Denotational semantics goes beyond the purely extensional view of computations (as maps from inputs to outputs), and models certain computational effects (such as non-termination, exceptions, non-determinism, etc.). Operational semantics goes further, and models computational operations. While computational effects are thus presented using the suitable algebraic operations in denotational semantics, computational behaviors are represented as elements of final coalgebras in operational semantics. But although both the denotational and the operational approaches go beyond the purely *extensional* view, neither has supported a genuinely *intensional* view, envisioned by Turing and von Neumann, where programs are data. Therefore, in spite of the tremendous successes in understanding and systematizing computational structures and behaviors, categorical semantics of computation has remained largely disjoint from theories of computability and complexity.

The claim put forward in this paper is that coalgebra provides a natural categorical framework for a fully intensional categorical theory of computability and complexity. The crucial step that enables this theory leads beyond *final* coalgebras, that assign *unique* descriptions to computational behaviors of *fixed* types, to *universal* coalgebras, that assign *non-unique* descriptions to computations of *arbitrary* types. These descriptions are what we usually call *programs*. Our message is thus that *programmability is a coalgebraic property*, just like *computational behaviors are coalgebraic*. This message is formally expressed through *universal processes*; it can perhaps be expressed more generally through *universal coalgebras*, as families of weakly final coalgebras, all carried by the same *universal state space*. Theorem 9 spells out in the framework of monoidal computer the fact that every Turing complete programming language provides a universal coalgebra for computable functions of all types; and vice versa, every universal coalgebra induces a corresponding notion of program. Just like abstract computational behaviors of a given type are precisely the elements of a final coalgebra of that type, abstract programs are precisely the elements of a universal coalgebra. Just like final coalgebras can be used to define semantics of computational behaviors [40], universal coalgebras can be used to define semantics of programs. From a slightly different angle, the fact that universal coalgebras characterize monoidal computers, proven in Theorem 9, can also be viewed as a coalgebraic characterization of computability. There are, of course, many characterizations of computability. The upshot of this one is, however, in Propositions 12 and 13: the coalgebaic view of computability opens an alley towards complexity. In any universe of computable functions, normal complexity measures [38] can be programmed coalgebraically. Combining this coalgebraic view of complexity with the algebraic view of randomized computation seems to open up a path towards a categorical model of one-way functions, and towards categorical cryptography, which has been the original goal of this project [34].

References

1. Altenkirch, T., Chapman, J., Uustalu, T.: Monads need not be endofunctors. Log. Methods Comput. Sci. **11**(1) (2015)
2. Asperti, A.: The intensional content of Rice's Theorem. In: Proceedings of the 35th Annual ACM SIGPLAN-SIGACT Symposium on Principles of Programming Languages, POPL 2008, pp. 113–119. ACM, New York (2008)
3. Barbieri, M.: Code Biology: A New Science of Life. Springer, Cham (2015). https://doi.org/10.1007/978-3-319-14535-8
4. Barendregt, H.P.: The Lambda Calculus: Its Syntax and Semantics. vol. 103, North Holland (1984)
5. Bartha, M.: The monoidal structure of turing machines. Math. Struct. Comput. Sci. **23**(2), 204–246 (2013)
6. Bernstein, E., Vazirani, U.: Quantum complexity theory. In: Proceedings of the Twenty-fifth Annual ACM Symposium on Theory of Computing, pp. 11–20. ACM (1993)
7. Blum, M.: A machine-independent theory of the complexity of recursive functions. J. ACM **14**(2), 322–336 (1967)
8. Bonsangue, M.M., Rutten, J., Silva, A.: Coalgebraic logic and synthesis of mealy machines. In: Amadio, R. (ed.) FoSSaCS 2008. LNCS, vol. 4962, pp. 231–245. Springer, Heidelberg (2008). https://doi.org/10.1007/978-3-540-78499-9_17
9. Bruni, R., Montanari, U.: Models of Computation. Texts in Theoretical Computer Science. An EATCS Series. Springer, Cham (2017). https://doi.org/10.1007/978-3-319-42900-7
10. Church, A.: An unsolvable problem of elementary number theory. Am. J. Math. **58**, 345–363 (1936)
11. Robin, J., Cockett, B., Díaz-Boïls, J., Gallagher, J., Hrubes, P.: Timed sets, functional complexity, and computability. Electr. Notes Theor. Comput. Sci. **286**, 117–137 (2012)
12. Robin, J., Cockett, B., Hofstra, P.J.W.: Introduction to turing categories. Ann. Pure Appl. Logic **156**(2–3), 183–209 (2008)
13. Eilenberg, S., Elgot, C.C.: Recursiveness. ACM Monograph. Academic Press (1970)
14. Hansen, H.H., Costa, D., Rutten, J.J.M.M.: Synthesis of mealy machines using derivatives. Electr. Notes Theor. Comput. Sci. **164**(1), 27–45 (2006)
15. Hayashi, S.: Adjunction of semifunctors: categorical structures in nonextensional lambda calculus. Theor. Comput. Sci. **41**, 95–104 (1985)
16. Hofstra, P.J.W., Warren, M.A.: Combinatorial realizability models of type theory. Ann. Pure Appl. Logic **164**(10), 957–988 (2013)
17. Holcombe, W.M.L.: Algebraic Automata Theory. Cambridge Studies in Advanced Mathematics. Cambridge University Press, Cambridge (1982)

18. Hoofman, R.: Comparing models of the intensional typed λ-calculus. Theor. Comput. Sci. **166**(1), 83–99 (1996)
19. Hyland, M.: The effective topos. In: Troelstra, A.S., van Dalen, D. (eds.) L. E. J. Brouwer Centenary Symposium, Studies in Logic and the Foundations of Mathematics, vol. 110, pp. 165–216. North-Holland (1982)
20. Jacobs, B.: Introduction to Coalgebra: Towards Mathematics of States and Observation. Cambridge Tracts in Theoretical Computer Science. Cambridge University Press, Cambridge (2016)
21. Jones, N.D.: Computability and Complexity: From a Programming Perspective. Foundations of Computing. The MIT Press, Cambridge (1997)
22. Kleene, S.C.: On notation for ordinal numbers. J. Symb. Log. **3**(4), 150–155 (1938)
23. Klin, B.: Bialgebras for structural operational semantics. Theor. Comput. Sci. **412**(38), 5043–5069 (2011)
24. Lambek, J., Scott, P.: Introduction to Higher Order Categorical Logic. Cambridge Studies in Advanced Mathematics, vol. 7. Cambridge University Press, Cambridge (1986)
25. Mac Lane, S.: Categories for the Working Mathematician. Graduate Texts in Mathematics, vol. 5. Springer, New York (1971)
26. Moggi, E.: Notions of computation and monads. Inf. Comput. **93**, 55–92 (1991)
27. Moschovakis, Y.N.: Kleene's amazing second recursion theorem. Bull. Symb. Log. **16**(2), 189–239 (2010)
28. von Neumann, J., Burks, A.W.: Theory of Self-Reproducing Automata. University of Illinois Press, Champaign (1966)
29. Odifreddi, P.: Classical Recursion Theory : The Theory of Functions and Sets of Natural Numbers. Studies in Logic and the Foundations of Mathematics, vol. 125. North-Holland, Amsterdam, New-York, Oxford, Tokyo (1989)
30. van Oosten, J.: Realizability: An Introduction to its Categorical Side. Studies in Logic and the Foundations of Mathematics. vol. 152. Elsevier Science (2008)
31. Di Paola, R.A., Heller, A.: Dominical categories: recursion theory without elements. J. Symb. Log. **52**(3), 594–635 (1987)
32. Pavlovic, D.: Categorical logic of names and abstraction in action calculus. Math. Struct. Comput. Sci. **7**, 619–637 (1997)
33. Pavlovic, D.: Quantum and classical structures in nondeterministic computation. In: Bruza, P., Sofge, D., van Rijsbergen, K. (eds.) Proceedings of Quantum Interaction 2009. Lecture Notes in Artificial Intelligence, vol. 5494, pp. 143–158. Springer Verlag (2009). arxiv.org:0812.2266
34. Pavlovic, D.: Gaming security by obscurity. In: Gates, C., Hearley, C. (eds.) Proceedings of NSPW 2011, pp. 125–140. ACM, New York (2011). arxiv:1109.5542
35. Pavlovic, D.: Geometry of abstraction in quantum computation. Proc. Symp. Appl. Math. **71**, 233–267 (2012). arxiv.org:1006.1010
36. Pavlovic, D.: Monoidal computer I: basic computability by string diagrams. Inf. Comput. **226**, 94–116 (2013). arxiv:1208.5205
37. Pavlovic, D.: Chasing diagrams in cryptography. In: Casadio, C., Coecke, B., Moortgat, M., Scott, P. (eds.) Categories and Types in Logic, Language, and Physics. LNCS, vol. 8222, pp. 353–367. Springer, Heidelberg (2014). https://doi.org/10.1007/978-3-642-54789-8_19
38. Pavlovic, D.: Monoidal computer II: Normal complexity by string diagrams. Technical report, ASECOLab (2014). arxiv:1402.5687
39. Pavlovic, D., Fauser, B.: Smooth coalgebra: testing vector analysis. Math. Struct. Comput. Sci. **26**, 1–41, 2 (2016). arxiv:1402.4414

40. Pavlovic, D., Mislove, M., Worrell, J.B.: Testing semantics: connecting processes and process logics. In: Johnson, M., Vene, V. (eds.) AMAST 2006. LNCS, vol. 4019, pp. 308–322. Springer, Heidelberg (2006). https://doi.org/10.1007/11784180_24

41. Pavlovic, D., Yahia, M.: Basic Concepts of Computer Science (with Pictures), December 2018. Draft textbook; chapters available as lecture notes at http://www.asecolab.org/courses/ics-222/

42. Pavlović, D., Escardó, M.: Calculus in coinductive form. In: Pratt, V. (eds.), Proceedings. Thirteenth Annual IEEE Symposium on Logic in Computer Science, pp. 408–417. IEEE Computer Society (1998)

43. Rogers Jr., H.: Theory of Recursive Functions and Effective Computability. MIT Press, Cambridge (1987)

44. Rutten, J.J.M.M.: Universal coalgebra: a theory of systems. Theor. Comput. Sci. **249**(1), 3–80 (2000)

45. Turi, D., Plotkin, G.D.: Towards a mathematical operational semantics. In: Proceedings of the Twelfth Annual IEEE Symposium on Logic in Computer Science (LICS 1997), pp. 280–291. IEEE Computer Society Press, June 1997

46. Turing, A.M.: On computable numbers, with an application to the Entscheidungsproblem. Proc. Lond. Math. Soc. Second Ser. **42**, 230–265 (1936)

47. Turing, A.M.: The chemical basis of morphogenesis. Philos. Trans. R. Soc. Lond. Seri. B, Biol. Sci. B **237**(641), 37–72 (1952)

48. Winskel, G.: The Formal Semantics of Programming Languages: An Introduction. Foundations of Computing. Zone Books, U.S. (1993)

Fibrational Bisimulations
and Quantitative Reasoning

David Sprunger[1]([✉]), Shin-ya Katsumata[1], Jérémy Dubut[1,2], and Ichiro Hasuo[1]

[1] National Institute of Informatics, Tokyo, Japan
{sprunger,s-katsumata,dubut,hasuo}@nii.ac.jp
[2] Japanese-French Laboratory for Informatics, CNRS, Tokyo, Japan

Abstract. *Bisimulation* and *bisimilarity* are fundamental notions in comparing state-based systems. Their extensions to a variety of systems have been actively pursued in recent years, a notable direction being *quantitative* extensions. In this paper we present an abstract categorical framework for such extended (bi)simulation notions. We use coalgebras as system models and fibrations for organizing predicates—following the seminal work by Hermida and Jacobs—but our focus is on the *structural* aspect of fibrational frameworks. Specifically we use morphisms of fibrations as well as canonical liftings of functors via Kan extensions. We apply this categorical framework by deriving some known properties of the Hausdorff pseudometric and approximate bisimulation in control theory.

1 Introduction

In the study of transition systems, *bisimulation relations* are a fundamental concept, and their categorical study revealed the importance of *coalgebras*. One approach to characterise bisimilarity is via *liftings* of the coalgebra functor along *fibrations* [12], which are a well-established framework to attach relational structures on categories for modelling transition systems and programming languages [14].

Recently, there is emerging interest in quantitative analysis of transition systems. *Behavioural metrics* were introduced in [5,7] to refine bisimilarity for probabilistic transition systems. Metrics give a real number for each pair of states in a transition system, while a relation can only provide a bit for each pair (whether the pair is in the relation or not). Therefore a metric can indicate a degree to which the behaviour of two states differ, whereas a bisimilarity relation can only indicate whether or not those behaviours differ. From this observation, a common desideratum for behavioral metrics associated with coalgebras is that two states should have distance 0 if and only if they are bisimilar.

Bisimilarity and behavioural metrics are also analogous on a categorical level. Behavioural metrics were recently shown to be constructible from liftings of the coalgebra functor to categories of (pseudo)metrics [2,3], similar to how Hermida-Jacobs bisimulations are constructed from liftings of a functor to the category of

C. Cîrstea (Ed.): CMCS 2018, LNCS 11202, pp. 190–213, 2018.
https://doi.org/10.1007/978-3-030-00389-0_11

relations. This type of construction is known generally as a coalgebraic predicate and can be performed when a lifting of the coalgebra functor is known.

These developments present two natural issues. The first is an open-ended quest for liftings of functors in general fibrations. These liftings are the rare ingredient in forming coalgebraic predicates, so having a variety of liftings in a variety of fibrations allows us to express more coalgebraic predicates. The second issue is more recent and concerns the desired relationship between behavioural metrics and bisimilarity mentioned above. Given some liftings in different fibrations, is there a relationship between the liftings we can use to verify a relationship between the coalgebraic predicates they define on a given coalgebra?

The main contributions of this paper pertain to these two issues:

- We propose two methods to lift functors along fibrations, both of which generalise existing constructions. The first is the *codensity lifting of endofunctors*, generalising Baldan et al.'s *Kantorovich lifting* [3] to arbitrary fibrations. This lifting also represents a further development of the codensity lifting of *monads* [15]. The second is the construction of an *enriched left Kan extension* using the *canonical symmetric monoidal closed structure* [17] on the total category in fibrations. This generalises Balan et al.'s construction [2] of enriched left Kan extension for quantale-enriched small categories.

  Apart from these lifting methods, we derive several methods to combine existing liftings. Using these methods, we construct the *Hausdorff metric* as the pushforward of the lifting of the list functor along a particular transformation.
- We propose the use of *predicate morphisms* to translate between these liftings. We use these translations to provide facilities for establishing relationships between the coalgebraic predicates provided by these liftings on coalgebras. We illustrate the utility of this approach with two examples. First, we demonstrate the translation of *approximate functions* to ϵ-*approximate relations*, which is the key technical tool used in control theory. Second, we translate metrics to relations to show the kernels of many behavioural metrics are bisimilarity relations.

Outline. In Sect. 2, we recall the important technical background for this work, particularly focusing on a class of fibrations where each fibre category is a lattice. In Sect. 3, we recall the construction of Hermida-Jacobs bisimulations and general coalgebraic predicates. As mentioned above, these require a lifting of a functor. Existence of such liftings is not guaranteed, and in Sect. 4 we present a few generalizations of extant techniques for producing liftings in particular fibrations to our more general class of fibrations. Finally, in Sect. 5, we use so-called predicate morphisms to establish relationships between coalgebraic predicates, focusing on deriving approximate functions from ϵ-approximate relations and deriving bisimilarity as the kernel of behavioural metrics.

2   Background

In this paper, we are interested in finding data about a wide variety of state-based transition systems. This data comes in a variety of types: relations, unary predicates, and pseudometrics are frequently found in the literature. Data of a particular type can also satisfy a variety of properties. For example, we are interested in both a relation consisting of the states with exactly the same behaviours, and a relation where the behaviours of the first member in the pair is a subset of the behaviours of the second member of the pair.

We capture these degrees of flexibility with three largely orthogonal categorical abstractions. First, we use coalgebras as a means of modeling many kinds of transition systems. Second, fibrations represent the types of data we are interested in deriving about the states of a coalgebra. Finally, functor liftings together with a property of fibrations allow us to model the different ways the same type of data may be created. We review each of these concepts separately here.

We assume familiarity with basic category theory, but not necessarily with the theory of fibrations.

2.1 Coalgebras

Coalgebras are our tool of choice for modeling state-based transition systems. Given a **Set** endofunctor F, an F-*coalgebra* is a pair (I, f) consisting of a set I and a function $f : I \to FI$. The set is often called the *carrier* of the coalgebra, while the function provides the *transition structure* of the coalgebra.

This pair is usually interpreted as a transition system under the following scheme. The (object part of the) functor F is thought of as an operation which sends a set of states to the set of all possible transition structures on that set. The set I is the set of states of a transition system. Under this interpretation, FI is then set of all the possible transition structures available using the set of states I, so the transition structure map $f : I \to FI$ assigns one of these possible transition structures to every state in I.

A *coalgebra morphism* $\phi : (I, f) \to (J, g)$ is a function on the underlying state sets $\phi : I \to J$ which respects the transitions in the source coalgebra, meaning $g \circ \phi = F\phi \circ f$. F-coalgebras together with their morphisms form a category we denote by $\mathbf{Coalg}(F)$.

By varying the functor F, we can capture a wide variety of transition system types, including deterministic and nondeterministic finite automata, Mealy and Moore machines, probabilistic transition systems, Markov decision processes, Segala systems and many more. For more background on the theory of coalgebra, we recommend consulting [20].

2.2 Fibrations

A fibration over a category $\mathbb{B}$ is a functor $\pi : \mathbb{E} \to \mathbb{B}$ with a *cartesian lifting property*. We will describe this property later, but intuitively, it allows us to take the inverse image of objects in $\mathbb{E}$ along morphisms in $\mathbb{B}$. The source category of the fibration, $\mathbb{E}$, is referred to as the "total category" and the target is the "base category".

Often the total category of a fibration is depicted vertically above the base category and language referencing this physical configuration is common.

An object or morphism $\dot{a}$ in $\mathbb{E}$ is *above* or *over* an object or morphism a in $\mathbb{B}$ means $\pi\dot{a} = a$. The collection of objects and morphisms above an object I and the id_I morphism is called the *fibre over* I. Each of these fibres is itself a subcategory of $\mathbb{E}$, denoted by $\mathbb{E}_I$.

Now we discuss the cartesian lifting property. In a fibration[1] over $\mathbb{B}$, for every morphism $f : I \to J$ in $\mathbb{B}$ and every object Y in $\mathbb{E}$ above J, there is a morphism $\dot{f} : f^*Y \to Y$ such that $\dot{f}$ is above f (hence f^*Y is above I). This morphism is called a *cartesian lifting of* f *with* Y and is further required to satisfy the following universal property: for all morphisms $g : K \to I$ in $\mathbb{B}$ and $\dot{h} : Z \to Y$ in $\mathbb{E}$ above $f \circ g$, there is a unique morphism $\dot{g} : Z \to f^*Y$ above g such that $\dot{h} = \dot{f} \circ \dot{g}$.

The operation sending Y to f^*Y is often called *pullback*.[2]

Pullback also sends morphisms in $\mathbb{E}_J$ to morphisms in $\mathbb{E}_I$ by the universal property. Straightforward checks show that the assignment $Y \mapsto f^*Y$ extends to a functor $f^* : \mathbb{E}_J \to \mathbb{E}_I$. When $g^*f^* = (f \circ g)^*$ and $\mathrm{id}_I^* = \mathrm{Id}_{\mathbb{E}_I}$ holds, we say that the fibration is *split*.

A functor π is a *cofibration* if $\pi^{op} : \mathbb{E}^{op} \to \mathbb{B}^{op}$ is a fibration, and *bifibration* if π and π^{op} are fibrations. Pullback in π^{op} is denoted by f_*, and called *pushforward*. In a bifibration, the pullback f^* is a right adjoint to pushforward f_* [14, Lemma 9.1.2].

A common scenario encountered in the study of fibrations is that each fibre $\mathbb{E}_I$ has a categorical structure, say X, and pullback functors preserve these fibrewise structures. When this is the case, we say that the fibration *has fibred* X. For instance, a fibration $\pi : \mathbb{E} \to \mathbb{B}$ has *fibred final objects* if (1) each fibre $\mathbb{E}_I$ has a final object, and (2) for any morphism $f : I \to J$, the pullback functor $f^* : \mathbb{E}_J \to \mathbb{E}_I$ preserves final objects. The fibrewise structure and the structure on the total category often have a close relationship. We state it next for the case $X = $ "limit".

Theorem 1 ([14]). *Let* $\pi : \mathbb{E} \to \mathbb{B}$ *be a fibration and* $\mathbb{D}$ *be a category. If* $\mathbb{B}$ *has limits of shape* $\mathbb{D}$, *and* π *has fibred limits of shape* $\mathbb{D}$, *then* $\mathbb{E}$ *also has limits of shape* $\mathbb{D}$.

The dual version of this theorem also holds, replacing fibration with cofibration, limit with colimit and pullback with pushforward.

We also mention the preservation of fibrations by functor-category construction:

Theorem 2. *For any fibration* $\pi : \mathbb{E} \to \mathbb{B}$ *and category* $\mathbb{C}$, $\pi \circ - : [\mathbb{C}, \mathbb{E}] \to [\mathbb{C}, \mathbb{B}]$ *is also a fibration.*

In this work, we are interested in state-based transition systems. Hence, the fibrations we are most interested in have $\mathbb{B} = \mathbf{Set}$. Indeed, most of the total

---

[1] In this work we always assume that a cleavage is given to a fibration.

[2] In this paper we shall use the word *pullback* in this fibrational sense. This usage generalizes the word's common meaning as a limit of a cospan in a category. Specifically, the latter gives a (fibrational) pullback in a codomain fibration. See [14].

categories we are interested in are sets equipped with some extra structure, such as sets with relations or sets with a metric. In these cases, the forgetful functor is usually a fibration.

Example 3. The forgetful functors from the following categories to **Set** are fibrations:

- **Pre** is the category of preorders and monotone functions between them.
- **ERel** is the category of endorelations. An object is a pair (I, R) of a set I and a relation $R \subseteq I \times I$. Morphisms are functions which preserve the relation, meaning $f : (I, R) \to (J, S)$ is a function $f : I \to J$ such that $i \; R \; i'$ implies $f(i) \; S \; f(i')$.
- **RERel** is the category of $\mathbb{R}^+$-indexed endorelations.[3] That is, an object is a pair (I, R) of $I \in \mathbf{Set}$ and a $\mathbb{R}^+$-indexed family $\{R_\epsilon\}_{\epsilon \in \mathbb{R}^+}$ of endorelations on I monotone in the index, so $\delta \leq \epsilon$ implies $R_\delta \subseteq R_\epsilon$. Morphisms are required to preserve the relation at each value ϵ, meaning $i \; R_\epsilon \; i'$ implies $f(i) \; S_\epsilon \; f(i')$ for all ϵ.
- **BVal** is the category of all $\mathbb{R}^+$-valued binary endorelations. Objects in this category are pairs (I, r) of a set I together with a function $r : I \times I \to \mathbb{R}^+$, with no constraints. Morphisms in this category are required to be non-expansive, so $f : (I, r) \to (J, s)$ satisfies $s(f(i), f(i')) \leq r(i, i')$ for all $i, i' \in I$.
- **PMet**$_b$ is the full subcategory of **BVal** consisting of b-bounded pseudometric spaces, for a fixed bound $b \in (0, \infty]$.[4] An *extended pseudometric* is an ∞-bounded pseudometric, and the category of extended pseudometrics is called **EPMet**.
- **Top**, **Meas** are the categories of all topological/measurable spaces and continuous/measurable functions between them, respectively.
- $\mathbb{V}$-**Cat**, with a commutative quantale $\mathbb{V}$, is the category of small $\mathbb{V}$-categories and $\mathbb{V}$-functors between them. The forgetful functor extracts the object part of small $\mathbb{V}$-categories. This category is used in [2] as a generalisation of metric spaces.

Technically, a fibration is a functor, particularly the forgetful functor in the examples above. In these examples, however, the functor is relatively unremarkable, so we will abuse terminology slightly and refer to the fibration by the name of the total category.

Cartesian morphisms in **ERel** preserve *and reflect* their source relation, in **RERel** they preserve and reflect the relation at each index, and in **BVal** and **PMet**$_b$ they are isometries, replacing the inequality in the condition for non-expansiveness with equality.

---

[3] Throughout this paper, we write $\mathbb{R}^+ = [0, \infty]$.

[4] A *b-bounded pseudometric* on a set I is a function $r : I \times I \to [0, b]$ which satisfies the axioms of a pseudometric: (1) $r(i, i) = 0$, (2) $r(i, i') = r(i', i)$, and (3) $r(i, i'') \leq r(i, i') + r(i', i'')$ for all $i, i', i'' \in I$. A pseudometric drops only the definiteness condition of a metric, so $r(i, i') = 0$ does not imply $i = i'$ when r is a pseudometric. This is crucial for our intended application to coalgebras where distinct states may have identical behaviours and we wish the distance between two states to reflect the difference in their behaviours only.

2.3 CLat$_\wedge$-Fibrations Over Set

In this paper, we focus on the fibrations over **Set** such that (1) each fibre category is a complete lattice and (2) pullbacks preserve all meets in fibres. Such fibrations bijectively correspond to functors of type **Set**$^{op} \to$ **CLat**$_\wedge$ via the Grothendieck construction, where the codomain is the category of complete lattices and meet-preserving functions between them. Following [1, Sect. 4.3], we call such fibrations **CLat**$_\wedge$-*fibrations* over **Set**, or simply **CLat**$_\wedge$-fibrations in this paper. This is a restricted class of *topological functors* to **Set** [13], where each fibre category is a poset.

There are indeed many examples of **CLat**$_\wedge$-fibrations over **Set**, covering a wide range of mathematical objects, including preorders, predicates, relations, pseudometrics, topologies, σ-fields and so on. In particular, every fibration listed in Example 3 is a **CLat**$_\wedge$-fibration.

We introduce a notation: for objects $X, Y \in \mathbb{E}$ in a **CLat**$_\wedge$-fibration $\pi : \mathbb{E} \to$ **Set** and a function $f : \pi X \to \pi Y$, by $f : X \dot\to Y$ we mean the sentence: "there exists a (necessarily unique) $\dot f : X \to Y$ such that $\pi \dot f = f$". For instance, in the **CLat**$_\wedge$-fibration $\pi :$ **Top** $\to$ **Set**, $f : X \dot\to Y$ is equivalent to the sentence "a function $f : \pi X \to \pi Y$ is a continuous function from X to Y.

Despite their simple definition, **CLat**$_\wedge$-fibrations have many useful properties. Let $\pi : \mathbb{E} \to$ **Set** be a **CLat**$_\wedge$-fibration. The following properties are well-known:

- π is a split bifibration. (Each fibre is a poset and each pullback functor $f^* : \mathbb{E}_J \to \mathbb{E}_I$ has a left adjoint $f_* : \mathbb{E}_I \to \mathbb{E}_J$ by the adjoint functor theorem.)
- π is faithful and has left and right adjoints, mapping $I \in$ **Set** to the least and greatest elements in $\mathbb{E}_I$, respectively. We name the left adjoint $\Delta :$ **Set** $\to \mathbb{E}$. Intuitively, it constructs discrete spaces of given sets.
- $\mathbb{E}$ has small limits and colimits by Theorem 1.
- π uniquely lifts arbitrary limits and colimits that exist in **Set**, including large ones. We describe this for the case of colimits. For any diagram $F : \mathbb{D} \to \mathbb{E}$ and a colimiting cocone $\{\iota_D : \pi FD \to C\}_{D \in \mathbb{D}}$ of πF in **Set**, there exists a unique colimiting cocone $\{i_D : FD \to \dot C\}_{D \in \mathbb{D}}$ of F in $\mathbb{E}$ such that $\pi i_D = \iota_D$. The colimit $\dot C$ is given as $\bigvee_{D \in |\mathbb{D}|} (\iota_D)_* (FD)$. The same statement holds for coends instead of colimits.
- The change-of-base of a **CLat**$_\wedge$-fibration $\pi : \mathbb{E} \to$ **Set** along *any* $F :$ **Set** $\to$ **Set** is again a **CLat**$_\wedge$-fibration.

Another less known, but important fact is that the total category $\mathbb{E}$ of any **CLat**$_\wedge$-fibration over **Set** carries a canonical (affine) *symmetric monoidal closed* (*SMC* for short) structure. The one on **Top** is described in [4,21]. The following construction of the SMC structure is a reformulation of the one given in [17] using fibred category theory.

The tensor unit is a chosen terminal object 1.

The tensor product of $X, Y \in \mathbb{E}$ is constructed as follows. Let us define $\pi X \cdot Y$ to be the coproduct of πX-many copies of Y. We explicitly construct it above $\pi X \times \pi Y$ by

$$\pi X \cdot Y = \bigvee_{x \in \pi X} (x, -)_* Y,$$

where $(x, -) : \pi Y \to \pi X \times \pi Y$ is the function that pairs an argument with a specified $x \in \pi X$. We similarly define $X \cdot \pi Y$ to be the coproduct of πY-many copies of X, again constructed above $\pi X \times \pi Y$. We then define the tensor product of X and Y to be the join of these two in the fibre over $\pi X \times \pi Y$:

$$X \otimes Y = (\pi X \cdot Y) \vee (X \cdot \pi Y).$$

This tensor product classifies bi-$\mathbb{E}$-morphisms in the following sense: a function f satisfies $f : X \otimes Y \dashrightarrow Z$ if and only if $f(x, -) : Y \dashrightarrow Z$ and $f(-, y) : X \dashrightarrow Z$ holds for any $x \in \pi X$ and $y \in \pi Y$.

The closed structure of $X, Y \in \mathbb{E}$ is constructed as follows. We first construct the product $\pi X \pitchfork Y$ of πX-many copies of Y above $\mathbf{Set}(\pi X, \pi Y)$ by

$$\pi X \pitchfork Y = \bigwedge_{x \in \pi X} (-(x))^* Y,$$

where $-(x) : \mathbf{Set}(\pi X, \pi Y) \to \pi Y$ is the function that evaluates an argument function with a specified $x \in \pi X$. We then define the closed structure $X \multimap Y$ to be the pullback of $\pi X \pitchfork Y$ along the morphism part $\pi_{X,Y} : \mathbb{E}(X, Y) \to \mathbf{Set}(\pi X, \pi Y)$ of π:

$$X \multimap Y = \pi_{X,Y}^*(\pi X \pitchfork Y).$$

We note that both $\pi : \mathbb{E} \to \mathbf{Set}$ and its left adjoint $\Delta : \mathbf{Set} \to \mathbb{E}$ are strict symmetric monoidal (for $\mathbf{Set}$ we take the cartesian monoidal structure).

Example 4. We illustrate the bifibrational structure of **BVal**. Let us recall the order relation in the fibre categories. The following are equivalent: (1) in the fibre $\mathbf{BVal}_I$, $(I, r) \leq (I, s)$ holds, (2) id_I is a nonexpansive function from (I, r) to (I, s), and (3) $s(x, y) \leq r(x, y)$ holds for all $x, y \in I$. Note the apparent disparity between (1) and (3): though $r \leq s$ in the fibre order, s has smaller values than r pointwise.

Next, let $(I, r) \in \mathbf{BVal}$ and $H \xrightarrow{f} I \xrightarrow{g} J$ be functions. The pullback (H, f^*r) and the pushforward (J, g_*r) are given by

$$f^*r(x, y) = r(f(x), f(y)), \quad g_*r(x, y) = \inf_{\substack{g(p)=x \\ g(q)=y}} r(p, q).$$

The fibrational construction of the canonical SMC structure on **BVal** yields the following tensor product and closed structure:

$$(I, r) \otimes (J, s) = \left(I \times J, \ \lambda((x, y), (x', y')) \cdot \begin{cases} \infty & x \neq x' \wedge y \neq y' \\ s(y, y') & x = x' \wedge y \neq y' \\ r(x, x') & x \neq x' \wedge y = y' \\ \min(r(x, x'), s(y, y')) & x = x' \wedge y = y' \end{cases} \right)$$

$$(I, r) \multimap (J, s) = \left(\mathbf{BVal}((I, r), (J, s)), \ \lambda(f, f') \cdot \sup_{x \in I} s(\pi f(x), \pi f'(x)) \right)$$

2.4 Liftings

Another major object of study in this work are liftings of a functor. Given a **Set** endofunctor F and two functors $\pi : \mathbb{E} \to \mathbf{Set}$ and $\rho : \mathbb{F} \to \mathbf{Set}$, a *lifting* of F is a functor $\dot{F} : \mathbb{E} \to \mathbb{F}$ such that $\rho \circ \dot{F} = F \circ \pi$. In many of the cases we consider $\pi = \rho$, so $\dot{F}$ is an endofunctor on the domain of π. To emphasize this particular situation we will call such an $\dot{F}$ an *endolifting*. In [11], endoliftings were called modalities.

Obviously, we will also usually be considering a situation where π and ρ are **CLat**$_\wedge$-fibrations. In such a case, restricting a lifting to a particular fibre yields a functor between fibre categories: $\dot{F}|_I : \mathbb{E}_I \to \mathbb{F}_{FI}$. Some liftings also specially respect the cartesian morphisms of the fibrations they operate between. A lifting is called a *fibration morphism* if it sends cartesian morphisms in $\mathbb{E}$ to cartesian morphisms in $\mathbb{F}$.

Notation. We pause here to set out some notational conventions, some of which have already been used. **Set** is the category of sets and functions. Typical objects of **Set** are denoted I, J, and K and typical morphisms are denoted f, g, and h. Generally, F is a **Set** endofunctor, C_I is the constant-to-I **Set** endofunctor, $-^*$ is the list functor, and P_{fin} is the finite (covariant) powerset functor.

Abstract categories are denoted $\mathbb{D}$, $\mathbb{E}$, and $\mathbb{F}$ and are often the total category for a **CLat**$_\wedge$-fibrations over **Set** with functors π or ρ. In such a case, applying a dot or two over a **Set**-related entity denotes an entity in the total category above the named **Set**-related entity. For example, $\dot{I}$ is an object in the total category, $\dot{f}$ is a morphism in the total category, $\dot{\times}$ is the binary product in the total category, and $\dot{F}$ is a lifting of F to the total category. We will also generally use X, Y, and Z as objects in a total category.

Two fibrations—**BVal** and **ERel**—are important enough to merit their own notations. Generally, r and s will denote the function in a **BVal** object, while R and S are the relation in an **ERel** object. Generally, writing a $\hat{\ }$ or $\tilde{\ }$ over a **Set**-related entity has the same meaning as a dot over that entity, but particularly for the total categories **BVal** and **ERel**, respectively.

The length of a list is denoted len and subscripts shall be used to select an element from a list at the indicated (zero-indexed) list position.

3 Endoliftings and Invariants

In this section, we describe how **CLat**$_\wedge$-fibrations and liftings of a functor F to that fibration can be used to define data about every F-coalgebra. Perhaps the best-known instance of this construction creates Hermida-Jacobs bisimulations from the canonical relation lifting of a functor along **ERel** $\to$ **Set**. We describe this example first, particularly for polynomial functors F.

3.1 Relation Liftings Define Coalgebraic Relations

Recall from the previous section the fibration **ERel** has objects consisting of pairs (I, R) where I is a set and $R \subseteq I \times I$ is a relation on that set. The fibre category $\mathbf{ERel}_I$ is (isomorphic to) the lattice of relations on I with a vertical morphism from (I, R) to (I, S) if and only if $R \subseteq S$.

A consequence of the equivalence between inclusion of relations in a fibre and the existence of a vertical morphism between them is that any functor between fibre categories in $\mathbf{ERel}$ is necessarily a monotone function on relations with respect to the usual inclusion ordering.

Two important cases where we can apply this fact are (1) $\mathbf{ERel}$ liftings of functors restricted to a fibre and (2) pullbacks along $\mathbf{Set}$ functions, since these are both functors between fibre categories. If $\widetilde{F}$ is an $\mathbf{ERel}$ lifting of F, then $\widetilde{F}|_{\mathbf{ERel}_I}$ is a monotone function $\mathbf{ERel}_I \to \mathbf{ERel}_{FI}$. Similarly, if $f : I \to FI$ is an F-coalgebra structure on I, pullback along f is a monotone function $f^* : \mathbf{ERel}_{FI} \to \mathbf{ERel}_I$.

Composing the above functions yields a monotone function $f^* \circ \widetilde{F}|_{\mathbf{ERel}_{FI}}$ on $\mathbf{ERel}_I$. Since $\mathbf{ERel}_I$ is a complete lattice, this composite monotone function has a greatest fixed point, which we denote by $(I, \nu\widetilde{F}_{(I,f)})$. The relation $\nu\widetilde{F}_{(I,f)}$ picked out in this greatest fixed point has historically turned out examples of great interest.

Perhaps foremost among these examples is the so-called canonical relation lifting, which yields bisimilarity as its greatest fixed point. We recall the description of this lifting for polynomial endofunctors.[5] The *polynomial* $\mathbf{Set}$ *endofunctors* are precisely those generated by the following grammar:

$$P ::= \mathrm{Id} \mid C_A \mid \coprod_i P_i \mid P_1 \times P_2$$

We can create an $\mathbf{ERel}$ lifting for any polynomial P with constructions for each of the inductive cases.

Definition 5 (canonical relation lifting). *Let $\widetilde{\coprod}$ and $\widetilde{\times}$ be the coproduct and binary product operations in $\mathbf{ERel}$, respectively. (These exist by Theorem 1.) The canonical relation lifting of a polynomial $\mathbf{Set}$ functor P is:*

$$\mathrm{Rel}(P) = \begin{cases} \widetilde{\mathrm{Id}} = \mathrm{Id}_{\mathbf{ERel}} & if P = \mathrm{Id}_{\mathbf{Set}}, \\ \widetilde{C}_A : (I, R) \mapsto (A, \Delta_A) & if P = C_A, \\ \widetilde{\coprod}_i \mathrm{Rel}(P_i) & if P = \coprod_i P_i, and \\ \mathrm{Rel}(P_1) \widetilde{\times} \mathrm{Rel}(P_2) & if P = P_1 \times P_2. \end{cases}$$

Given a polynomial $\mathbf{Set}$ functor P and a P-coalgebra (I, f), we can use the canonical relation lifting $\mathrm{Rel}(P)$ to form the function $f^* \circ \mathrm{Rel}(P)|_{\mathbf{ERel}_I}$. Postfixed points of this function in $\mathbf{ERel}_I$ give a useful general definition of bisimulation on (I, f) [12]. The greatest postfixed point $\nu\mathrm{Rel}(P)_{(I,f)}$ is bisimilarity on this coalgebra.

---

[5] The canonical relation lifting can in fact be defined for all functors using image factorization in $\mathbf{Set}$. We use the less-general inductive version as we will need it in Sect. 5.

Example 6. As two examples of the canonical relation lifting, we present bisimulation on coalgebras of the list functor and on coalgebras of the finite powerset functor. These examples will be referenced later when we construct behavioural metrics on the same coalgebra types similarly to how we do it here.

The list functor is defined as $(-)^* = \coprod_{n \in \omega} \prod_{i \in n} \mathrm{Id}$. Following the inductive definition above, each summand in the coproduct, $\prod_n \mathrm{Id}$, sends a relation R to the n-fold repetition of R: two lists k, ℓ of length n are related by $\mathrm{Rel}(\prod_n \mathrm{Id})(R)$ if $k_i R \ell_i$ holds for $0 \leq i < n$. The canonical relation lifting for $\coprod_{n \in \omega} \prod_n \mathrm{Id}$ then relates two lists k, ℓ of arbitrary length if and only if $\mathrm{len}(k) = \mathrm{len}(\ell)$ (they come from the same index in the coproduct), and $k_i R \ell_i$ holds for $0 \leq i < \mathrm{len}(k)$. In other words, $(k, \ell) \in \mathrm{Rel}(\prod_n \mathrm{Id})(R)$.

The finite powerset functor is the quotient of the list functor by the transformation $set_I : I^* \to P_{\mathrm{fin}} I$ given by $set_I : (i_1, \ldots, i_n) \mapsto \{i_1, \ldots, i_n\}$. The pushforward of the lifting for the list functor along this natural transformation is the usual definition of bisimulation for the finite powerset functor. Explicitly,

$$\mathrm{Rel}(P_{\mathrm{fin}})(R) = \{(J, K) \in P_{\mathrm{fin}} I \times P_{\mathrm{fin}} I : \forall j \in J, \exists k \in K.jRk \wedge \forall k \in K, \exists j \in J.jRk\}$$

3.2   Generalizing Hermida-Jacobs Bisimulation

The necessary components to define Hermida-Jacobs bisimulation conveniently can be found in any $\mathbf{CLat}_\wedge$-fibration with any endolifting of any functor. Thus, we can define the abstract counterpart of a bisimulation. This terminology is intended to echo [11].

Definition 7. *Let $\dot{F}$ be a endolifting for F. An $\dot{F}$-invariant [on an F-coalgebra (I, f)] is an $\dot{F}$-coalgebra (X, α) [such that $\pi X = I$ and $\pi \alpha = f$].*
An $\dot{F}$-invariant morphism is an $\dot{F}$-coalgebra morphism.

$\dot{F}$-invariants and $\dot{F}$-invariant morphisms together form a category, in fact exactly the category $\mathbf{Coalg}(\dot{F})$. $\dot{F}$-invariants also evidently sit over F-coalgebras according to π, so we name the functor sending $\mathbf{Coalg}(\dot{F})$ to $\mathbf{Coalg}(F)$.

Definition 8. *Given a endolifting $\dot{F}$ on a functor F, the underlying coalgebra functor $\mathbf{Coalg}(\pi) : \mathbf{Coalg}(\dot{F}) \to \mathbf{Coalg}(F)$ is defined as*

$$\mathbf{Coalg}(\pi)(X, \alpha) = (\pi X, \pi \alpha), \quad \mathbf{Coalg}(\pi)h = h.$$

Since π is faithful in a $\mathbf{CLat}_\wedge$-fibration, the coalgebra structure α of an $\dot{F}$-invariant (X, α) on (I, f) is unique. Therefore, an alternative definition of an $\dot{F}$-invariant on (I, f) is an object X above I such that there exists a (necessarily unique) morphism $\alpha : X \to \dot{F} X$ above f.

Yet another definition of an $\dot{F}$-invariant can be derived from the lattice structure of $\mathbb{E}_I$. For each coalgebra (I, f), there is a monotone function $f^* \circ \dot{F}|_{\mathbb{E}_I} : \mathbb{E}_I \to \mathbb{E}_{FI} \to \mathbb{E}_I$ as described above. An $\dot{F}$-invariant on (I, f) is then precisely a postfixed point for this function.

A useful consequence of this last characterization is the observation that since each fibre $\mathbb{E}_I$ is a complete lattice, Knaster-Tarski ensures the $\dot{F}$-invariants on (I, f) form a complete sublattice. In particular

Definition 9. *The greatest $\dot{F}$-invariant on an F-coalgebra (I, f) always exists and is called the $\dot{F}$-coinductive invariant. We denote the $\dot{F}$-coinductive invariant on (I, f) by $\nu\dot{F}_{(I,f)}$.*

We can alternatively reach $\nu\dot{F}_{(I,f)}$ by the final sequence argument inside the fibre $\mathbb{E}_I$; this is the approach taken in [3]. In [11], coinducive invariants were called coinductive predicates.

$\dot{F}$-similarities give final objects within each fibre category, but there is no assurance of a final object in the total category, nor that final objects are preserved by coalgebra morphisms. The next result, which reorganizes results presented in [11, Sect. 4], sets out some conditions entailing these desiderata.

Theorem 10. *Let $\dot{F}$ be a endolifting for F. If it preserves cartesian morphisms,*

1. *[11, Proposition 4.1]. The underlying coalgebra functor $\mathbf{Coalg}(\pi) : \mathbf{Coalg}(\dot{F}) \to \mathbf{Coalg}(F)$ is a fibration where pullbacks are the same as in the fibration π.*
2. *Each pullback functor of $\mathbf{Coalg}(\pi)$ preserves final objects (hence $\mathbf{Coalg}(\dot{F})$ has fibered final objects).*
3. *If additionally $\mathbf{Coalg}(F)$ has a final object νF, then $\mathbf{Coalg}(\dot{F})$ has a final object.*

For the item 2 and 3 of the above theorem, see also [11, Corollary 4.3].

This theorem is a fibred counterpart of some results in Sect. 6 of [3]. To see this, we instantiate Theorem 10 with the following data: the $\mathbf{CLat}_{\wedge}$-fibration $\pi : \mathbf{PMet}_b \to \mathbf{Set}$ (Sect. 2), a functor $F : \mathbf{Set} \to \mathbf{Set}$ having a final F-coalgebra νF and a lifting $\dot{F}$ of F along π that preserves cartesian $\mathbf{PMet}_b$ morphisms (isometries). Then

- Theorem 6.1 in [3] is equivalent to the conclusion of (this instance of) item 3 of Theorem 10.
- Let $I = (I, f)$ be an F-coalgebra, and $!_I : I \to \nu F$ be the unique F-coalgebra morphism. The *behavioural distance* of I in [3] corresponds to the pullback $!_I^*(\nu\dot{F}_{\nu F})$ in our fibrational language.
- Theorem 6.2 in [3] corresponds to $\nu F_I = !_I^*(\nu\dot{F}_{\nu F})$, which follows from (this instance of) item 2 of Theorem 10.

4   Constructions of Liftings Along CLat$_{\wedge}$-Fibrations

There are many examples of liftings of functors in well-known fibrations, such as the fibration of relations or pseudometrics. Some of these liftings even form classes which cover all functors, such as the canonical relation lifting or the generalized Kantorovich liftings of [3], which ensure *every* functor has a lifting in **ERel** and **PMet**$_b$ respectively. In this work we are considering a variety of fibrations, so a natural concern is whether liftings of **Set** functors exist in all of these $\mathbf{CLat}_{\wedge}$-fibrations.

In this section, we generalize a variety of constructions known in particular fibrations to arbitrary $\mathbf{CLat}_\wedge$-fibrations. In Sects. 4.1 and 4.2, we give two constructions, the first using enriched left Kan extensions and the second using codensity liftings. Then in Sect. 4.3 we mention how to use the categorical structure of the $\mathbf{CLat}_\wedge$-fibration to create new liftings from old.

Hence, in this section we find ourselves with the ingredients F and π:

$$\mathbf{Set} \xrightarrow{\ F\ } \mathbf{Set} \xleftarrow{\ \pi\ } \mathbb{E} \qquad \pi : \mathbf{CLat}_\wedge\text{-fibration} \qquad (1)$$

and seek to create an endolifting of F in a variety of ways.

4.1 Lifting by Enriched Left Kan Extensions

The canonical SMC structure on $\mathbb{E}$ (Sect. 2) allows us to discuss *enriched liftings* of F to $\mathbb{E}^e$, the self-enriched category of $\mathbb{E}$ with its SMC structure. To discuss this, we introduce some $\mathbb{E}$-categories and $\mathbb{E}$-functors.

- By $\mathbb{E}^e$ we mean the self-enriched $\mathbb{E}$-category of $\mathbb{E}$ (that is, $\mathbb{E}^e(X, Y) = X \multimap Y$).
- Since the left adjoint $\Delta : \mathbf{Set} \to \mathbb{E}$ of π (see Sect. 2.3) is strict monoidal, it yields the *change-of-base* 2-functor $\Delta_* : \mathbf{CAT} \to \mathbb{E}\text{-}\mathbf{CAT}$. It takes a locally small category $\mathbb{C}$ and returns the $\mathbb{E}$-category $\Delta_*\mathbb{C}$ defined by $\mathbf{Obj}(\Delta_*\mathbb{C}) = \mathbf{Obj}(\mathbb{C})$ and $(\Delta_*\mathbb{C})(I, J) = \Delta(\mathbb{C}(I, J))$.
- For any functor $G : \mathbb{C} \to \mathbb{E}$, we define the $\mathbb{E}$-functor $\underline{G} : \Delta_*\mathbb{C} \to \mathbb{E}^e$ by: $\underline{G}I = GI$, and $\underline{G}_{I,J} : (\Delta_*\mathbb{C})(I, J) \to \mathbb{E}^e(GI, GJ)$ is the mate of $G_{I,J} : \mathbb{C}(I, J) \to \mathbb{E}(GI, GJ)$ with the adjunction $\Delta \dashv \pi$; recall that $\pi(\mathbb{E}^e(X, Y)) = \pi(X \multimap Y) = \mathbb{E}(X, Y)$ by construction.

The following is a generalisation of [2, Theorem 3.3].

Theorem 11. *Consider the situation* (1). *Let $C : \mathbf{Set} \to \mathbb{E}$ be a functor such that $\pi C = F$. Then there is an enriched left Kan extension $\dot{F}$ of $\underline{C} : \Delta_*\mathbf{Set} \to \mathbb{E}^e$ along $\underline{\Delta} : \Delta_*\mathbf{Set} \to \mathbb{E}^e$ such that its underlying functor $\dot{F}_0 : \mathbb{E} \to \mathbb{E}$ (see [16]) is a lifting of F along π.*

Proof. Since the codomain $\mathbb{E}^e$ of $\underline{C}$ has $\mathbb{E}$-tensors, the enriched left Kan extension can be computed by the enriched coend:

$$\mathbf{Lan}_{\underline{\Delta}}\underline{C}X = \int^{I \in \Delta_*\mathbf{Set}} \mathbb{E}^e(\underline{\Delta}I, X) \otimes \underline{C}I;$$

see [16, (4.25)]. We define the body of this coend by $B(I, J) = \mathbb{E}^e(\underline{\Delta}I, X) \otimes \underline{C}J$. It is an $\mathbb{E}$-functor of type $(\Delta_*\mathbf{Set})^{op} \otimes \Delta_*\mathbf{Set} \to \mathbb{E}^e$. Similarly, we define an ordinary functor $\overline{B} : \mathbf{Set}^{op} \times \mathbf{Set} \to \mathbb{E}$ by $\overline{B}(I, J) = B(I, J)$ on objects and $\overline{B}(f, g) = \pi B_{(I,J),(I',J')}(f, g)$ on morphisms. A calculation shows that $\overline{B}$ is equal to the ordinary functor $\lambda(I, J) . (LI \multimap X) \otimes CJ$.

Because the codomain of B is $\mathbb{E}^e$, the enriched coend can be computed as an ordinary colimit of the following large diagram in $\mathbb{E}$ [16, Sect. 2.1]:

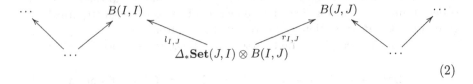

$$\tag{2}$$

where I, J ranges over all objects in **Set**, and $l_{I,J}$ and $r_{I,J}$ are the uncurrying of $B(I, -)_{J,I}$ and $B(-, J)_{I,J}$, respectively.

In $\mathbb{E}$, $\Delta I \otimes X$ is a *tensor* of X with $I \in$ **Set** because

$$\mathbb{E}(\Delta I \otimes X, Y) \simeq \mathbb{E}(\Delta I, X \multimap Y) \simeq \mathbf{Set}(I, \pi(X \multimap Y)) = \mathbf{Set}(I, \mathbb{E}(X, Y)).$$

We name the passage from right to left ϕ. The bottom objects of diagram (2) are thus tensors of $B(I, J)$ with $\mathbf{Set}(J, I)$ for each $I, J \in$ **Set**, and moreover, by easy calculation, we have $l_{I,J} = \phi(\overline{B}(I, -))$ and $r_{I,J} = \phi(\overline{B}(-, J))$. Therefore a colimit of the diagram (2) can be computed as an ordinary *coend* of $\overline{B}$: $\mathbf{Set}^{op} \times \mathbf{Set} \to \mathbb{E}$.

To compute this (large) coend of $\overline{B}$, it suffices to show that the coend of $\pi\overline{B}$ exists in **Set**, because π uniquely lifts coends. We have a natural isomorphism $\iota_{I,J} : \pi\overline{B}(I, J) \to \mathbf{Set}(I, \pi X) \times FJ$, and the right hand side has a coend $\{i_I : \mathbf{Set}(I, \pi X) \times FI \to F\pi X\}_{I \in \mathbf{Set}}$ defined by $i_I(f, x) = Ffx$. Therefore since π uniquely lifts colimits (Sect. 2.3), we obtain a coend of $\overline{B}$. To summarise, the enriched left Kan extension is computed as

$$\mathbf{Lan}_{\underline{\Delta}}\underline{C}X = \bigvee_{I \in \mathbf{Set}} (i_I \circ \iota_{I,I})_*((LI \multimap X) \otimes CI).$$

Example 12. Let $\pi : \mathbf{Pre} \to \mathbf{Set}$ be the $\mathbf{CLat}_\wedge$-fibration from the category **Pre** of preorders and $F : \mathbf{Set} \to \mathbf{Set}$ be a functor. We compute the enriched left Kan extension $\mathbf{Lan}_{\underline{\Delta}}\underline{\Delta}F$. For $(X, \leq) \in \mathbf{Pre}$, the enriched left Kan extension $\mathbf{Lan}_{\underline{\Delta}}\underline{\Delta}F(X, \leq_X)$ is the preorder on FX generated from the following binary relation:

$$\{(Ffa, Fga) \mid I \in \mathbf{Set}, a \in FI, f, g \in \mathbf{Set}(I, X), \forall i \in I \,.\, fi \leq_X gi\}$$
$$= \{(Fp_1a, Fp_2a) \mid a \in F(\leq_X)\}$$

where $p_i : (\leq_X) \to X$ is the composite of the inclusion $(\leq_X) \hookrightarrow X \times X$ of the preorder relation and the projection function $\pi_i : X \times X \to X$.

When F is the powerset functor P, the enriched left Kan extension $\mathbf{Lan}_{\underline{\Delta}}\underline{\Delta}P(X, \leq_X)$ gives the *Egli-Milner* preorder $\sqsubseteq_X$ on PX, as computed in [2, Example 3.8]:

$$V \sqsubseteq_X W \iff (\forall v \in V \,.\, \exists w \in W \,.\, v \leq_X w) \wedge (\forall w \in W \,.\, \exists v \in V \,.\, v \leq_X w).$$

4.2   Codensity Lifting of Endofunctors

As an analog to the codensity lifting of *monads* along $\mathbf{CLat}_\wedge$-fibrations [15, Proposition10], we give a method to lift $\mathbf{Set}$-*endofunctors* along $\mathbf{CLat}_\wedge$-fibrations. We retain the name and call it the *codensity lifting* (of $\mathbf{Set}$-endofunctors). We demonstrate in Example 15 that it subsumes the Kantorovich lifting in [3].

Consider the situation (1). We take the category $\mathbf{Set}^F$ of F-algebras and the associated forgetful functor $U : \mathbf{Set}^F \to \mathbf{Set}$. It comes with a natural transformation $\alpha : FU \to U$, whose components are defined by the F-algebra structure: $\alpha_{(A,a)} = a : FA \to A$.

The codensity lifting of F is defined with respect to a *lifting parameter* for F, which is a pair (R, S) of functors from a discrete category $\mathbb{A}$ such that $\pi S = UR$:

$$
\begin{array}{ccc}
\mathbb{A} & \xrightarrow{\ S\ } & \mathbb{E} \\
{\scriptstyle R}\big\downarrow & & \big\downarrow{\scriptstyle \pi} \\
\mathbf{Set}^F & \xrightarrow[\ U\ ]{} & \mathbf{Set}
\end{array}
\qquad
\begin{array}{l}
\mathbf{Set}^F : \text{category of } F\text{-algebras} \\
U : \text{forgetful functor} \\
\pi S = UR
\end{array}
\tag{3}
$$

Then the codensity lifting $F^{[R,S]}$ of F with respect to the above lifting parameter (R, S) is defined by the following fibred meet:

$$
F^{[R,S]}X = \bigwedge_{A \in \mathbb{A},\, f \in \mathbb{E}(X, SA)} (\alpha_{RA} \circ F\pi f)^*(SA).
$$

The codensity lifting can characterise as a vertex of a pullback when the *codensity monad* $\mathbf{Ran}_S S$ exists. Suppose that $\mathbf{Ran}_S S$ exists. Since the $\mathbf{CLat}_\wedge$-fibration $\pi : \mathbb{E} \to \mathbf{Set}$ preserves all limits, $\pi \mathbf{Ran}_S S$ is a right Kan extension of πS along S. We then take the mate of the natural transformation $\alpha R : F\pi S \to \pi S$ with the right Kan extension of πS along S, and obtain $\overline{\alpha R} : F\pi \to \pi \mathbf{Ran}_S S$.

Theorem 13. *Suppose that* $\mathbf{Ran}_S S$ *exists. Then* $F^{[R,S]}$ *is the vertex of the following pullback in the fibration* $[\mathbb{E}, p] : [\mathbb{E}, \mathbb{E}] \to [\mathbb{E}, \mathbf{Set}]$:

$$
\begin{array}{ccccc}
F^{[R,S]} & \cdots\cdots\!\!\!\!\!\longrightarrow & \mathbf{Ran}_S S & \qquad & [\mathbb{E}, \mathbb{E}] \\
& & \big\downarrow{\scriptstyle po-} & & \big\downarrow{\scriptstyle po-} \\
F\pi & \xrightarrow[\ \overline{\alpha R}\ ]{} & \pi\mathbf{Ran}_S S & \qquad & [\mathbb{E}, \mathbf{Set}]
\end{array}
$$

The codensity lifting enjoys the following universal property. First, we introduce a partial order on the liftings of F by: $\dot{F} \le \ddot{F}$ if and only if $\dot{F}X \le \ddot{F}X$ holds for all $X \in \mathbb{E}$. Moreover, we say that a lifting $\dot{F}$ of F along π *makes S an algebra above* R if, $\alpha_{RA} : \dot{F}SA \dot{\to} SA$ holds for all $A \in \mathbb{A}$.

Theorem 14. *Consider the situation (1) and a lifting parameter given as (3). The codensity lifting* $F^{[R,S]}$ *of* F *is the largest lifting of* F *that makes* S *an algebra above* R.

Example 15. Fix a bound $b \in (0, \infty]$ for metrics. We show that the Kantorovich lifting in [3] is a codensity lifting along the $\mathbf{CLat}_\wedge$-fibration $\pi : \mathbf{PMet}_b \to \mathbf{Set}$. Let $\alpha : F[0, b] \to [0, b]$ be an F-algebra; in [3] it is called an evaluation function. We then form the following lifting parameter: $\mathbb{A} = 1$, $R = ([0, b], \alpha)$, and $S = ([0, b], d_e)$, where d_e is the standard Euclidean distance $d_e(x, y) = |x - y|$ on $[0, b]$. Then the codensity lifting with this parameter yields the following construction of pseudometric:

$$F^{[R,S]}(I, r) = (FI, r')$$
$$r'(x, y) = \sup \{|\alpha((F\pi f)(x)) - \alpha((F\pi f)(y))| \mid f \in \mathbf{PMet}_b((I, r), S)\}.$$

This is exactly the Kantorovich lifting in [3, Definition 3.1].

4.3   Combining Liftings

We have seen two methods to lift endofunctors. In this section, we discuss building new liftings from existing ones. Below we set up a suitable category in which these operations are characterised as categorical constructions.

Let $\pi : \mathbb{E} \to \mathbf{Set}$ be a $\mathbf{CLat}_\wedge$-fibration. Then $\pi \circ - : [\mathbb{E}, \mathbb{E}] \to [\mathbb{E}, \mathbf{Set}]$ is a partial order bifibration with fibred meets of arbitrary size. We take the following change-of-base of this fibration along $- \circ \pi$:

$$
\begin{array}{ccc}
\mathbf{Lift}(\pi) & \longrightarrow & [\mathbb{E}, \mathbb{E}] \\
q \downarrow & & \downarrow \pi \circ - \\
[\mathbf{Set}, \mathbf{Set}] & \xrightarrow{\ -\circ\pi\ } & [\mathbb{E}, \mathbf{Set}]
\end{array}
$$

The vertex of this change-of-base is the category $\mathbf{Lift}(\pi)$ of *liftings* along π. An object is a pair $(F, \dot{F})$ of an endofunctor $F : \mathbf{Set} \to \mathbf{Set}$ and its lifting $\dot{F} : \mathbb{E} \to \mathbb{E}$ along π. A morphism from $(F, \dot{F})$ to $(G, \dot{G})$ is a pair $(\alpha, \dot{\alpha})$ of natural transformations $\alpha : F \to G$ and $\dot{\alpha} : \dot{F} \to \dot{G}$ such that $\pi\dot{\alpha} = \alpha\pi$.

The derived vertical leg $q : \mathbf{Lift}(\pi) \to [\mathbf{Set}, \mathbf{Set}]$ is also a partial order bifibration with fibred meets (of arbitrary size). Since $[\mathbf{Set}, \mathbf{Set}]$ has small limits and colimits, by Theorem 1, $\mathbf{Lift}(\pi)$ has small limits and colimits, hence small products and coproducts.

The bifibredness of q, together with these products and coproducts give us a recipe to combine liftings.

Identity and Constant. The lifting of $\mathrm{Id}_{\mathbf{Set}}$ is $\mathrm{Id}_{\mathbb{E}}$, while the lifting of the constant functor $C_I(J) \equiv I$ is $\dot{C}_I(X) \equiv \Delta I$.

Product and Coproduct. Let $(F_i, \dot{F}_i)$ be an I-indexed family of liftings along π. Then their product and coproduct are computed pointwise.

Pullbacks and Pushforwards. For a lifting $(F, \dot{F})$ along π and natural transformations $H \xrightarrow{\alpha} F \xrightarrow{\beta} G$, the pullback lifting $\alpha^* F$ above H and pushforward lifting $\beta_* F$ above G are computed pointwise in the fibration $\pi : \mathbb{E} \to \mathbf{Set}$:

$$(\alpha^* \dot{F})(X) = (\alpha_{\pi X})^* (\dot{F} X), \qquad (\beta_* \dot{F})(X) = (\beta_{\pi X})_* (\dot{F} X).$$

In particular, these constructions ensure all polynomial and finitary functors in **Set** have at least one lifting in every **CLat**$_\wedge$-fibration.

4.4   The Hausdorff Pseudometric

As an example of how liftings can be constructed with these basic operations, we demonstrate the construction of the Hausdorff pseudometric on finite sets in a **BVal** lifting of the finite powerset functor. Our version of the Hausdorff distance will take a **BVal** object (I, d) and create a **BVal** object $(P_{\text{fin}} I, \mathcal{H}d)$.

Recall that the finite powerset can be realized as a quotient of a polynomial functor with the following construction. First, recall the list functor: $(-)^* = \coprod_{n \in \omega} \prod_{i \in n} \text{Id}$. This is patently a polynomial functor. Then the finite powerset functor is the quotient of the list functor by the natural transformation $set_I : I^* \to P_{\text{fin}} I$ from Example 6.

We can build up a **BVal** lifting of the finite powerset functor in parallel with this construction. First, using the product and coproduct in **BVal** we derive a **BVal** lifting for the list functor. Given a **BVal** object (I, d) the lifted distance on lists $k, \ell \in I^*$ is:

$$d^*(k, \ell) = \begin{cases} \max\limits_{0 \le i < \text{len}(k)} d(k_i, \ell_i) & \text{if } \text{len}(k) = \text{len}(\ell) \\ \infty & \text{if } \text{len}(k) \ne \text{len}(\ell) \end{cases}$$

Then a **BVal** lifting for the finite powerset functor arises as the pushforward of the list lifting along the transformation set. In Example 4, we found pushforward in **BVal** explicitly so, $\mathcal{H}d(J, K) = \inf\limits_{\substack{k \in I^*: \; set(k) = J \\ \ell \in I^*: \; set(\ell) = K}} d^*(k, \ell)$. We have denoted this distance $\mathcal{H}d$ since it turns out to be equal to the usual Hausdorff distance. However, this is not the usual formulation for the Hausdorff distance, so we briefly discuss why this is equivalent.

The usual definition of Hausdorff distance for a metric space is

$$\mathcal{H}d(J, K) = \max \left(\sup_{y \in J} \inf_{z \in K} d(y, z), \sup_{z \in K} \inf_{y \in J} d(y, z) \right)$$

where $J, K \subseteq I$. Typically the Hausdorff distance is also restricted to nonempty compact subsets of the metric space so that $\mathcal{H}d$ is truly a metric. (Otherwise $\mathcal{H}d(J, K) = 0$ does not imply $J = K$, for example.) Since we are interested in pseudometrics anyway, we do not place any such restriction on the domain of $\mathcal{H}d$.

In the finite case, the Hausdorff distance has a game theoretic interpretation as the result of a two-turn game played between a lazy walker (Gerry) and an antagonist (Tony). In the first round, Tony picks a starting point from either J or K for Gerry. In the second round, Gerry walks from Tony's starting point in

$J(K)$ to any point in $K(J)$. The result of the game is the distance Gerry walks. Gerry's goal is to minimize this distance; Tony's goal is to maximize it.

Gerry's optimal strategy is straightforward. Given a starting point, Gerry finds the distances to the (finitely many) points in the other set and picks the least one. Since Gerry's optimal strategy is clear, Tony can make a list consisting of all the points in $J \cup K$ and the distance Gerry will have to walk if that point is used as the starting point. Then Tony's optimal strategy is to pick the starting point corresponding to the greatest distance on this list.

This analysis indicates we can interchange the order of the players and obtain a game with the same result: Gerry can first announce where he will walk given every possible choice of starting point, then Tony picks one of the choices offered by Gerry. If Tony is given two lists k and ℓ by Gerry, he will be to force the result of this modified game to be

$$d^*(k, \ell)$$

where d^* is the list distance defined above. Gerry's best strategy is to pick k and ℓ with the closest corresponding distances possible, making the final result of this modified game

$$\inf_{\substack{k \in I^*: \ set(k)=J \\ \ell \in I^*: \ set(\ell)=K}} d^*(k, \ell)$$

where the constraints $set(k) = J$ and $set(\ell) = K$ express the fact that Gerry must make a choice for every single starting point. Since these games have the same result, we know

$$\max\left(\sup_{y \in J} \inf_{z \in K} d(y, z), \sup_{z \in K} \inf_{y \in J} d(z, y)\right) = \inf_{\substack{k \in I^*: \ set(k)=J \\ \ell \in I^*: \ set(\ell)=K}} d^*(k, \ell)$$

Therefore, our formulation of the Hausdorff distance is equal to the usual formulation of the Hausdorff distance, modulo the consideration that we are satisfied with a pseudometric and so do not confine our definition to nonempty compact sets.[6]

5   The Category of Endoliftings

In Sect. 3, we defined endoliftings and their instantiations to $\dot{F}$-invariants on particular coalgebras. We showed that with certain constraints on the ambient

---

[6] A more technical proof of the same result proceeds by first showing the Hausdorff distance on the left is a lower bound for $d^*(k, \ell)$ given the constraints on k and ℓ. The fact that J and K are finite is crucial so that the value of the left-hand side must be witnessed at a particular entry in one of the lists. Then it can be shown that this lower bound is sharp by a particular choice of k and ℓ witnessing Gerry's optimal strategy, so indeed the left-hand side is the greatest lower bound for the collection of values on the right-hand side.

categories, the $\dot{F}$-coinductive invariant exists and is preserved by coalgebra morphisms. In Sect. 4, we showed that endoliftings exist in many $\mathbf{CLat}_\wedge$-fibrations, and gave several constructions and combinators for producing endoliftings in these general conditions. In this section, we we observe that endoliftings can be collected into a category using the following definition.

Definition 16. *A* endolifting morphism *from one endolifting* $(\pi : \mathbb{E} \to \mathbf{Set}, \dot{F})$ *to another* $(\rho : \mathbb{F} \to \mathbf{Set}, \ddot{F})$ *is a lifting* $H : \mathbb{E} \to \mathbb{F}$ *of* $\mathrm{Id}_{\mathbf{Set}}$ *(i.e.* $\pi = \rho \circ H$*) such that* $H \circ \dot{F} = \ddot{F} \circ H$.

Endolifting morphisms do not appear in the story of Hermida-Jacobs bisimulations or the coalgebraic predicates defined analogously, but we will observe this category is a useful abstraction for comparing coalgebraic invariants of various endoliftings.

A concrete goal in this section is to establish some general conditions under which a $\mathbf{BVal}$ coinductive invariant has an $\mathbf{ERel}$ coinductive invariant at its kernel. Results of this type are pursued, for example, in [6].

5.1   Quantitative and Qualitative Liftings

We begin by focusing on three $\mathbf{CLat}_\wedge$-fibrations introduced in Sect. 2.3, namely $\mathbf{ERel}$, $\mathbf{RERel}$, and $\mathbf{BVal}$. These total categories consist of sets together with endorelations, real-indexed families of endorelations, and "distance" functions (which satisfy no metric axioms other than having codomain nonnegative reals), respectively.

These fibrations have many functors between them:

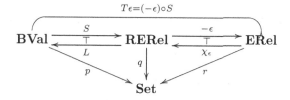

where (the object parts of) each of these functors are given by

$$L(I, R) = (I, \lambda(x, y) . \inf\{\delta \mid (x, y) \in R_\delta\}) \quad S(I, r) = (I, \lambda\epsilon . \{(x, y) \in I \mid r(x, y) \le \epsilon\})$$

$$\chi_\epsilon(I, R) = \left(I, \lambda\delta . \begin{cases} (I, \varnothing) & \text{if } \delta < \epsilon \\ (I, R) & \text{if } \delta \ge \epsilon \end{cases}\right) \quad (I, R)\epsilon = (I, R\epsilon)$$

Note that the two functors between $\mathbf{ERel}$ and $\mathbf{RERel}$ are actually a real-indexed family of functors, where $\epsilon \in [0, \infty)$. It may help to think of S as Stratifying a distance function into a family of relations and L as finding the Least index where the relation holds. As usual, the empty infimum in the definition of L is the maximum element, namely ∞.

These functors patently do not change the index set I associated with each of the objects in the total category. Each of these functors is also defined to be the identity on morphisms.[7] Therefore, these are liftings of the identity on **Set**.

We define the composite functor $T\epsilon = (-\epsilon) \circ S$. This functor sends $(I, r) \in$ **BVal** to $(I, \{(x, y) \mid r(x, y) \leq \epsilon\}) \in$ **ERel**, truncating the distance function r at ϵ. The fact that $T\epsilon$ is a right adjoint, as depicted in the diagram above, will be an important fact later on.

The functor $T0$ gives the *kernel* of a distance function, namely the relation consisting of pairs which are at distance 0. A common desideratum of pseudometric liftings or more generally **BVal** liftings is that the kernel of the $\hat{F}$-coinductive invariant function in **BVal** is bisimilarity in **ERel** (i.e. the $\mathrm{Rel}(F)$-coinductive invariant where $\mathrm{Rel}(F)$ is the canonical relation lifting of F, defined in Sect. 3.1). We show how to establish this result for the Hausdorff metric in a highly reusable manner.

5.2 $T\epsilon$ Is a Endolifting Morphism Between Kripke Polynomial Functors

Next, we show that $T\epsilon$ is a endolifting morphism from every polynomial functor in **BVal** to the polynomial functor of parallel shape in **ERel**. This result is the backbone of our proof that $T\epsilon$ is a endolifting morphism from the Hausdorff lifting of the finite powerset functor to the canonical relation lifting of the finite powerset functor.

Proposition 17. *For all $\epsilon \in [0, \infty)$, $T\epsilon$ is a endolifting morphism:*

1. *from $\mathrm{Id}_{\mathbf{BVal}}$ to $\mathrm{Id}_{\mathbf{ERel}}$,*
2. *from $\hat{C}_A$ to $\widetilde{C}_A$ where C_A is the constant-to-A functor, and*
3. *from $\hat{F}_1 \hat{\times} \hat{F}_2$ to $\mathrm{Rel}(F_1) \widetilde{\times} \mathrm{Rel}(F_2)$, given that it is a morphism from $\hat{F}_i$ to $\mathrm{Rel}(F_i)$, and*
4. *from $\coprod_i \hat{F}_i$ to $\widetilde{\coprod}_i \mathrm{Rel}(F_i)$, given that it is a morphism from $\hat{F}_i$ to $\mathrm{Rel}(F_i)$*

*Therefore, $T\epsilon$ is a endolifting morphism from any polynomial functor in **ERel** to the polynomial functor of the same shape in **BVal**.*

This establishes $T\epsilon$ as a endolifting morphism between polynomial functors, but we also want it to be a endolifting morphism from the Hausdorff lifting of P_{fin} to the canonical relation lifting in **BVal**. That is, we want to show $T\epsilon : \mathcal{H} \to \mathrm{Rel}(P_{\mathrm{fin}})$ is a endolifting morphism. A reasonable strategy, given the proof we just completed, would be to hope that if $T\epsilon$ is a endolifting morphism between two liftings of a functor, then it is a morphism between the pushforward of those

---

[7] Morphisms in each of these categories are **Set**-functions which satisfy conditions regarding the extra data in the total category. (That is, functions which preserve the source relation, shrink the source distance, etc.). It is straightforward to show $f : X \to Y$ is a morphism in $\mathbb{A}$ implies $f : FX \to FY$ is a morphism in $\mathbb{B}$ by writing down the extra conditions on f imposed by $\mathbb{A}$ and $\mathbb{B}$ where $F : \mathbb{A} \to \mathbb{B}$ is any of the four functors defined above.

functors along a natural transformation in the base category. In general this is not true, but liftings satisfying a simple side condition do have this property.

Proposition 18. *Suppose* $\tau : P \to F$ *is a natural transformation in* **Set**, $T\epsilon :$ $\hat{P} \to \widetilde{P}$ *(Sect. 5.1) is a endolifting morphism from a* **BVal** *lift of* P *to an* **ERel** *lift of* P, *and* $\hat{F}$ *and* $\widetilde{F}$ *are the pushforwards of* $\hat{P}$ *and* $\widetilde{P}$ *along* τ. *Further suppose for every set* I, *every* $f, f' \in FI$ *and* $r : I \times I \to \mathbb{R}^+$, *the lower bound for* $\{\hat{Pr}(p, p') : \tau p = f \text{ and } \tau p' = f'\}$ *is achieved in this set. Then* $T\epsilon$ *is a endolifting morphism from* $\hat{F}$ *to* $\widetilde{F}$.

We can now apply this proposition to obtain the following corollary.

Corollary 19. $T\epsilon$ *is a endolifting morphism from* $\mathcal{H}$ *to* $\text{Rel}(P_{\text{fin}})$.

Proof. Proposition 17 shows $T\epsilon$ is a endolifting morphism from the standard **BVal** lifting for the list functor to the standard **ERel** lifting for the list functor. We know $\mathcal{H}$ and $\text{Rel}(P_{\text{fin}})$ are the pushforwards of these list functors along set_X in their respective total categories.[8] Hence to apply Proposition 18 we only need to show for all finite sets $J, K \subseteq I$, and all distances $r : I \times I \to \mathbb{R}^+$, there exist lists $k^\dagger$ and $\ell^\dagger$ such that $set(k^\dagger) = J$, $set(\ell^\dagger) = K$ and $r^*(k^\dagger, \ell^\dagger) = \inf\limits_{\substack{set(k)=J \\ set(\ell)=K}} r^*(k, \ell)$. We noted this in the Hausdorff distance section, where these dagger lists represent Gerry's optimal strategy.

5.3   Approximate Bisimulations: An Example from Control Theory

Here we present an example from a rather different context: *approximate bisimulation* by Girard and Pappas [8]. Defined as a binary relation on a metric space that is subject to the "mimicking" condition, the notion is widely used in control theory as a quantitative relaxation of usual (Milner-Park) bisimulation that allows bounded errors. Its principal use is in bounding errors caused by some abstraction of dynamical systems: given the original dynamics $\mathcal{S}$, one derives its abstraction $\mathcal{A}$; by exhibiting an ϵ-*approximate bisimulation* between $\mathcal{S}$ and $\mathcal{A}$, one then shows that the difference between the trajectory of $\mathcal{A}$ and that of $\mathcal{S}$ is bounded by ϵ. Such abstraction methods include: state space discretization (e.g. in [10]) and ignoring switching delays [18]. See [9] for an overview.

In the above scenario, an ϵ-approximate bisimulation between $\mathcal{S}$ and $\mathcal{A}$ is synthesized through analysis of the continuous dynamics of $\mathcal{S}$: for example the *incremental stability* of $\mathcal{S}$ yields an approximate bisimulation via its Lyapunov-type witness. Another common strategy for finding an approximate bisimulation is via a *bisimulation function*. Our goal here is to describe the latter strategy in the current coalgebraic and fibrational framework.

We fix the set O of output values together with a distance function $d :$ $O \times O \to \mathbb{R}^+$, and a U-labelled finitely branching LTS $(Q, \delta : Q \to U \pitchfork P_{fin} Q)$ with an output function $o : Q \to O$, where $\pitchfork$ denotes the power operation

---

[8] Recall set_X from Example 6.

[19, p. 70]. An ϵ-*approximate bisimulation relation* is a binary relation $R \subseteq Q \times Q$ such that

$$\forall (q, q') \in R . d(o(q), o(q')) \leq \epsilon \wedge \forall l \in U .$$
$$(\forall r \in Q . r \in \delta(l, q) \implies \exists r' \in Q . r' \in \delta(l, q') \wedge (r, r') \in R) \wedge$$
$$(\forall r' \in Q . r' \in \delta(l, q') \implies \exists r \in Q . r \in \delta(l, q) \wedge (r, r') \in R). \quad (4)$$

The difference from the usual Milner-Park bisimulation is that R is additionally required to witness the ϵ-proximity of outputs of related states q and q'.

A bisimulation function is a quantitative (real-valued) witness for an approximate bisimulation. In many settings in control theory where dynamics are smooth and described by ordinary differential equations, such real-valued functions are easier to come up with than an approximate bisimulation itself. For the above LTS, a function $v : Q \times Q \to \mathbb{R}^+$ is a *bisimulation function* if it satisfies, for each $q, q' \in Q$,

$$\max \left(d(o(q), o(q')), \sup_{l \in U} \mathcal{H}v(\delta(l, q), \delta(l, q')) \right) \leq v(q, q') \quad (5)$$

A crucial fact is that a bisimulation function v gives rise to an ϵ-approximate bisimulation $\{(q, q') \mid v(q, q') \leq \epsilon\}$. See e.g. [9].

We move on to give a categorical account of this construction. We use the following functor as a coalgebra signature:

$$F : \mathbf{Set} \to \mathbf{Set}, \quad FX = O \times (U \pitchfork P_{fin}X),$$

We can then package a U-labelled finitely branching LTS and an output function into a single F-coalgebra $Q = (Q, \langle o, \delta \rangle : Q \to FQ)$.

Firstly, the endolifting that captures ϵ-approximate bisimulations consists of

$$r : \mathbf{ERel} \to \mathbf{Set}, \quad \tilde{F}_\epsilon X = T\epsilon(O, d) \times (U \pitchfork Rel(P_{fin})(X)).$$

Secondly, the endolifting that captures bisimulation functions consists of

$$p : \mathbf{BVal} \to \mathbf{Set}, \quad \hat{F}X = (O, d) \times (U \pitchfork \mathcal{H}X).$$

Indeed, by unfolding the definitions the following can be observed: $\tilde{F}_\epsilon$-invariants on Q are nothing but ϵ-approximate bisimulations; and $\hat{F}$-invariants on Q are bisimulation functions. Thanks to Proposition 17 and Corollary 19, the functor $T\epsilon$—that sends a function $v : Q \times Q \to \mathbb{R}^+$ to the relation $\{(q, q') \mid v(q, q') \leq \epsilon\}$—is a endolifting morphism from $\hat{F}$ to $\tilde{F}$. Therefore $T\epsilon$ transfers a $\hat{F}$-invariant v on Q to a $\tilde{F}_\epsilon$-invariant $T\epsilon v$ on Q, that is, a bisimulation function to an ϵ-approximate bisimulation.

5.4   Endolifting Morphisms Preserve Final Coalgebras

We next state a result which we can use to ensure that the coinductive invariant in the source of a endolifting morphism is sent to the coinductive invariant in the target of that morphism.

Lemma 20. *Suppose H is a endolifting morphism from $(\pi : \mathbb{E} \to \mathbf{Set}, \dot{F})$ to $(\rho : \mathbb{F} \to \mathbf{Set}, \ddot{F})$ which is also a fibration map. Suppose additionally that H preserves fibred meets. Then $H(I, \nu\dot{F}_{(I,f)}) = (I, \nu\ddot{F}_{(I,f)})$.*

Proof. Preservation of top elements ensures $H\top_{\mathbb{E}_I} = \top_{\mathbb{F}_I}$. Since H is a fibration map and a endolifting morphism, we get $H(f^*\dot{F}(A_I)) = f^*\ddot{F}(H(A_I))$ for all $A_I \in \mathbb{E}_I$. Combining this with the above observation ensures H sends the final sequence in the fiber $\mathbb{E}_I$ to the final sequence in the fiber $\mathbb{F}_I$. Finally, H preserving meets ensures H will send the $\dot{F}$-coinductive invariant for (I, f) to the $\ddot{F}$-coinductive invariant for (I, f).

Note $T\epsilon$ satisfies most of the conditions in this lemma. Since we are interested in concluding something about the kernel of a behavioural metric, we specialize to the case where $\epsilon = 0$ where these conditions are all satisfied.

Corollary 21. *If $T0$ is a endolifting morphism from $(\hat{F}, \mathbf{BVal})$ to $(\dot{F}, \mathbf{ERel})$, then the behavioural metric induced by $\hat{F}$ has the coalgebraic predicate induced by $\dot{F}$ at its kernel.*

Proof. $T\epsilon$ is a fibration map and a right adjoint, and thus preserves all fibred meets.

All our work from the previous section establishing that $T\epsilon$ is a endolifting morphism from $\mathcal{H}$ to $\mathrm{Rel}(P_{\mathrm{fin}})$ now pays off.

Corollary 22. *The Hausdorff behavioural metric on $P_{\mathrm{fin}}X$ has P_{fin}-bisimilarity at its kernel.*

Above, we also showed $T0$ is a endolifting morphism between many other **BVal** and **ERel** liftings (Proposition 17). Therefore, we could also derive an analog of Corollary 21 for these pairs and conclude behavioural metrics of the **BVal** lift have bisimilarity (the coalgebraic relation of the corresponding **ERel** lift) at their kernels.

6   Conclusions and Future Work

We presented a fibrational framework for various extensions of (bi)simulation notions. On the categorical side our focus has been on structural aspects of fibrations such as fibration morphisms and lifting by Kan extensions; on the application side we took examples from quantitative reasoning about systems. This has allowed us to capture known constructions in more abstract and general terms, such as the Hausdorff pseudometric and approximate bisimulation in control theory.

As future work, we shall investigate conditions under which the two liftings in Sect. 4—one by left Kan extension and the other involving right Kan extension—coincide. We would then compare this coincidence and the Kantorovich-Wasserstein duality, which is the coincidence of the metric on probability distributions computed by sup and inf. We mentioned that **Top** and **Meas** are examples of $\mathbf{CLat}_\wedge$-fibrations; their use in reasoning about systems will also be explored.

Acknowledgement. This research was supported by ERATO HASUO Metamathematics for Systems Design Project (No. JPMJER1603), JST. The authors are grateful to anonymous reviewers, whose constructive comments helped to improve the paper.

References

1. Azevedo de Amorim, A., Gaboardi, M., Hsu, J., Katsumata, S.Y., Cherigui, I.: A semantic account of metric preservation. In: Proceedings of the 44th ACM SIGPLAN Symposium on Principles of Programming Languages (POPL 2017), pp. 545–556 (2017). https://doi.org/10.1145/3009837.3009890
2. Balan, A., Kurz, A., Velebil, J.: Extensions of functors from set to V-cat. In: Proceedings of the 6th Conference on Algebra and Coalgebra in Computer Science (CALCO 2015), pp. 17–34 (2015). https://doi.org/10.4230/LIPIcs.CALCO.2015.17
3. Baldan, P., Bonchi, F., Kerstan, H., König, B.: Behavioral metrics via functor lifting. In: Proceedings of the 34th International Conference on Foundation of Software Technology and Theoretical Computer Science (FSTTCS 2014), pp. 403–415 (2014). https://doi.org/10.4230/LIPIcs.FSTTCS.2014.403
4. Borceux, F.: Handbook of Categorical Algebra 2. In: Encyclopedia of Mathematics and Its Applications, vol. 51. Cambridge University Press, Cambridge (1994)
5. van Breugel, F., Worrell, J.: A behavioural pseudometric for probabilistic transition systems. Theor. Comput. Sci. **331**(1), 115–142 (2005)
6. Chatzikokolakis, K., Gebler, D., Palamidessi, C., Xu, L.: Generalized bisimulation metrics. In: Baldan, P., Gorla, D. (eds.) CONCUR 2014. LNCS, vol. 8704, pp. 32–46. Springer, Heidelberg (2014). https://doi.org/10.1007/978-3-662-44584-6_4
7. Desharnais, J., Jagadeesan, R., Gupta, V., Panangaden, P.: The metric analogue of weak bisimulation for probabilistic processes. In: Proceedings of the 17th IEEE Symposium on Logic in Computer Science (LICS 2002), pp. 413–422 (2002)
8. Girard, A., Pappas, G.J.: Approximation metrics for discrete and continuous systems. IEEE Trans. Autom. Control **52**(5), 782–798 (2007). https://doi.org/10.1109/TAC.2007.895849
9. Girard, A., Pappas, G.J.: Approximate bisimulation: a bridge between computer science and control theory. Eur. J. Control **17**(5–6), 568–578 (2011). https://doi.org/10.3166/ejc.17.568-578
10. Girard, A., Pola, G., Tabuada, P.: Approximately bisimilar symbolic models for incrementally stable switched systems. IEEE Trans. Autom. Control **55**(1), 116–126 (2010)
11. Hasuo, I., Cho, K., Kataoka, T., Jacobs, B.: Coinductive predicates and final sequences in a fibration. Electron. Notes Theor. Comput. Sci. **298**, 197–214 (2013). https://doi.org/10.1016/j.entcs.2013.09.014
12. Hermida, C., Jacobs, B.: Structural induction and coinduction in a fibrational setting. Inf. Comput. **145**(2), 107–152 (1998). https://doi.org/10.1006/inco.1998.2725
13. Herrlich, H.: Topological functors. Gen. Topol. Appl. **4**(2), 125–142 (1974). https://doi.org/10.1016/0016-660X(74)90016-6. http://www.sciencedirect.com/science/article/pii/0016660X74900166
14. Jacobs, B.: Categorical Logic and Type Theory. Studies in Logic and the Foundations of Mathematics, vol. 141. Elsevier Science, Amsterdam (1999)
15. Katsumata, S., Sato, T.: Codensity liftings of monads. In: Proceedings of the 6th Conference on Algebra and Coalgebra in Computer Science (CALCO 2015), pp. 156–170 (2015). https://doi.org/10.4230/LIPIcs.CALCO.2015.156

16. Kelly, G.M.: Basic Concepts of Enriched Category Theory. Lecture Notes in Mathematics, vol. 64. Cambridge University Press, Cambridge (1982)
17. Kelly, G.M., Rossi, F.: Topological categories with many symmetric monoidal closed structures. Bull. Aust. Math. Soc. **31**(1), 41–59 (1985). https://doi.org/10.1017/S0004972700002264
18. Kido, K., Sedwards, S., Hasuo, I.: Bounding Errors Due to Switching Delays in Incrementally Stable Switched Systems (Extended Version) (2017). arXiv:1712.06311
19. Lane, S.M.: Categories for the Working Mathematician. Graduate Texts in Mathematics, vol. 5. Springer, New York (1978). https://doi.org/10.1007/978-1-4757-4721-8
20. Rutten, J.: Universal coalgebra: a theory of systems. Theor. Comput. Sci. **249**(1), 3–80 (2000). https://doi.org/10.1016/S0304-3975(00)00056-6
21. Činčura, J.: Tensor products in the category of topological spaces. Comment. Math. Univ. Carol. **20**(3), 431–446 (1979)

Categorical Büchi and Parity Conditions via Alternating Fixed Points of Functors

Natsuki Urabe[1]([✉]) and Ichiro Hasuo[2]

[1] The University of Tokyo, Tokyo, Japan
urabenatsuki@is.s.u-tokyo.ac.jp
[2] National Institute of Informatics, Tokyo, Japan

Abstract. Categorical studies of recursive data structures and their associated reasoning principles have mostly focused on two extremes: initial algebras and induction, and final coalgebras and coinduction. In this paper we study their in-betweens. We formalize notions of alternating fixed points of functors using constructions that are similar to that of free monads. We find their use in categorical modeling of accepting run trees under the Büchi and parity acceptance condition. This modeling abstracts away from states of an automaton; it can thus be thought of as the "behaviors" of systems with the Büchi or parity conditions, in a way that follows the tradition of coalgebraic modeling of system behaviors.

1 Introduction

Büchi Automata. The *Büchi condition* is a common acceptance condition for automata for infinite words. Let $x_i \in X$ be a state of an automaton $\mathcal{A}$ and $a_i \in \mathsf{A}$ be a character, for each $i \in \omega$. An infinite run $x_0 \xrightarrow{a_0} x_1 \xrightarrow{a_1} \cdots$ satisfies the Büchi condition if x_i is an accepting state (usually denoted by ⓞ) for infinitely many i. An example of a Büchi automaton is shown on the right. The word $(ba)^\omega$ is accepted, while ba^ω is not. A function that assigns each $x \in X$ the set of accepted words from x is called the *trace semantics* of the Büchi automaton.

Categorical Modeling. The main goal of this paper is to give a categorical characterization of such runs under the Büchi condition. This is in the line of the established field of categorical studies of finite and infinite datatypes: it is well-known that finite trees form an initial algebra, and infinite trees form a final coalgebra; and finite/infinite words constitute a special case. These categorical characterizations offer powerful reasoning principles of *(co)induction* for both definition and proof. While the principles are categorically simple ones corresponding to universality of initial/final objects, they have proved powerful and useful in many different branches of computer science, such as functional programming and process theory. See the diagram on

$$
\begin{array}{ccc}
FY & \to & FZ \\
d\uparrow & & \zeta\uparrow\cong \\
Y & \dashrightarrow & Z
\end{array}
$$

C. Cîrstea (Ed.): CMCS 2018, LNCS 11202, pp. 214–234, 2018.
https://doi.org/10.1007/978-3-030-00389-0_12

the right above illustrating coinduction: given a functor F, its final coalgebra $\zeta \colon Z \xrightarrow{\cong} FZ$ has a unique homomorphism to it from an arbitrary F-coalgebra $d \colon Y \to FY$. In many examples, a final coalgebra is described as a set of "infinite F-trees."

Extension of such (co)algebraic characterizations of data structures to the Büchi condition is not straightforward, however. A major reason is the non-local character of the Büchi condition: its satisfaction cannot be reduced to a local, one-step property of the run. For example, one possible attempt of capturing the Büchi condition is as a suitable subobject of the set $\mathrm{Run}(X) = (\mathsf{A} \times X)^{\omega}$ of all runs (including nonaccepting ones). The latter set admits clean categorical characterization as a final coalgebra $\mathrm{Run}(X) \xrightarrow{\cong} F\big(\mathrm{Run}(X)\big)$ for the functor $F = (\mathsf{A} \times X) \times \_$. Specifying its subset according to the Büchi condition seems hard if we insist on the coalgebraic language which is centered around the local notion of transition represented by a coalgebra structure morphism $c \colon X \to FX$.

There have been some research efforts in this direction, namely the categorical characterization of the Büchi condition. In [5] the authors insisted on finality and characterize languages of *Muller automata* (a generalization of Büchi automata) by a final coalgebra in **Sets**^{2}. Their characterization however relies on the lasso characterization of the Büchi condition that works only in the setting of finite state spaces. In [21] we presented an alternative characterization that covers infinite state spaces and automata with probabilistic branching. The key idea was the departure from coinduction, that is, reasoning that relies on the universal property of greatest fixed points. Note that a final coalgebra $\zeta \colon Z \xrightarrow{\cong} FZ$ is a "categorical greatest fixed point" for a functor F.

Our framework in [21] was built on top of the so-called *Kleisli approach* to trace semantics of coalgebras [10–12,16]. There a system is a coalgebra in a Kleisli category $\mathcal{K}\ell(T)$, where T represents the kind of branching the system exhibits (nondeterminism, probability, etc.). A crucial fact in this approach is that homsets of the category $\mathcal{K}\ell(T)$ come with a natural order structure. Specifically, in [21], we characterized trace semantics under the Büchi condition as in the diagrams (1) below[1], where (i) X_1 (resp. X_2) is the set of nonaccepting (resp. accepting) states of the Büchi automaton (i.e. $X = X_1 + X_2$), and (ii) the two diagrams form a *hierarchical equation system* (HES), that is roughly a planar representation of nested and alternating fixed points. In the HES, we first calculate the least fixed point for the left diagram, and then calculate the greatest fixed point for the right diagram with u_1 replaced by the obtained least fixed point. Note that the order of calculating fixed points matters.

$$
\begin{array}{cc}
\begin{array}{ccc}
 & \overline{F}[u_1, u_2] & \\
FX & \dashrightarrow & FZ \\
{\scriptstyle c_1}\big\uparrow & {\scriptstyle =_\mu}\ \ {\scriptstyle J\zeta}\big\uparrow{\scriptstyle\cong} \\
X_1 & \xdashrightarrow{\ u_1\ } & Z
\end{array}
&
\begin{array}{ccc}
 & \overline{F}[u_1, u_2] & \\
FX & \dashrightarrow & FZ \\
{\scriptstyle c_2}\big\uparrow & {\scriptstyle =_\nu}\ \ {\scriptstyle J\zeta}\big\uparrow{\scriptstyle\cong} \\
X_2 & \xdashrightarrow{\ u_2\ } & Z
\end{array}
\end{array}
\qquad (1)
$$

---

[1] We write $f : X \nrightarrow Y$ for a Kleisli arrow $f \in \mathcal{K}\ell(T)(X, Y)$ and $\overline{F} : \mathcal{K}\ell(T) \to \mathcal{K}\ell(T)$ for a lifting of the functor F over $\mathcal{K}\ell(T)$, for distinction.

Contributions: Decorated Trace Semantics by Categorical Datatypes.
In this paper we introduce an alternative categorical characterization to the one
in [21] for the Büchi conditions, where we do not need alternating fixed points
in homsets. This is made possible by suitably refining the value domain, from a
final coalgebra to a novel categorical datatypes $F^{+\oplus}0$ and $F^+(F^{+\oplus}0)$ that have
the Büchi condition built in them. Diagrammatically the characterization looks
as in (2) below. Note that we ask for the greatest fixed point in both squares.

$$
\begin{array}{ccc}
FX \xrightarrow{\overline{F}(v_1 + v_2)} F(F^+(F^{+\oplus}0) + F^{+\oplus}0) & \qquad & FX \xrightarrow{\overline{F}(v_1 + v_2)} F(F^+(F^{+\oplus}0) + F^{+\oplus}0) \quad (2)\\
c_1 \uparrow \quad =_\nu \quad J(\beta_1)_0 \uparrow \cong & & c_2 \uparrow \quad =_\nu \quad J(\beta_2)_0 \uparrow \cong \\
X_1 \xrightarrow{v_1} F^+(F^{+\oplus}0) & & X_2 \xrightarrow{v_2} F^{+\oplus}0
\end{array}
$$

The functors F^+ and $F^{+\oplus}$ used in the datatypes are obtained by applying
two operations $(\_)^+$ and $(\_)^\oplus$ to a functor F. For an endofunctor G on a
category $\mathbb{C}$ with enough initial algebras, G^+X is given by the carrier object of
a (choice of) an initial $G(\_ + X)$-algebra for each $X \in \mathbb{C}$. The universality of
initial algebras allows one to define $G^+f : G^+X \to G^+Y$ for each $f : X \to Y$
and extend G^+ to a functor $G^+ : \mathbb{C} \to \mathbb{C}$. This definition is much similar to that
of a *free monad* G^*, where G^*X is the carrier object of an initial $G(\_) + X$-
algebra for $X \in \mathbb{C}$. The operation $(\_)^\oplus$ is defined similarly: for $G : \mathbb{C} \to \mathbb{C}$ and
$X \in \mathbb{C}$, $G^\oplus X$ is given by the carrier object of a final $G(\_ + X)$-coalgebra. This
construction resembles to that of *free completely iterative algebras* [14].

The constructions of $F^+(F^{+\oplus}0)$ and $F^{+\oplus}0$ has a clear intuitive meaning.
For the specific example of A-labeled nondeterministic Büchi automata, $T = \mathcal{P}$,
$F = \mathsf{A} \times (\_)$, $F^+(F^{+\oplus}0) \cong F^{+\oplus}0 \cong (\mathsf{A}^+)^\omega$. Hence an element in $F^+(F^{+\oplus}0)$
or $F^{+\oplus}0$ is identified with an infinite sequence of finite words. We understand
it as an infinite word "decorated" with information about how accepting states
are visited, by considering that an accepting state is visited at each splitting
between finite words. For example, we regard $(a_0a_1)(a_2a_3a_4)(a_5a_6)(a_7)\ldots \in$
$(\mathsf{A}^+)^\omega \cong F^{+\oplus}0$ as an infinite word decorated as follows.

$$(3)$$

An element in $F^+(F^{+\oplus}0)$ is similarly understood, except that the initial state
is regarded as a nonaccepting state. We note that by its definition, the resulting
"decorated" word always satisfies the Büchi condition.

Thus the arrows $v_1 : X_1 \twoheadrightarrow F^+(F^{+\oplus}0)$ and $v_2 : X_2 \twoheadrightarrow F^{+\oplus}0$ in (2) are
regarded as a kind of trace semantics that assigns each state $x \in X$ the set of
infinite words accepted from x "decorated" with information about the corresponding accepting run. Hence we shall call v_1 and v_2 a *decorated trace semantics*
for the coalgebra c. The generality of the category theory allows us to define decorated trace semantics for systems with other transition or branching types, e.g.
Büchi tree automata or *probabilistic Büchi automata*.

In this paper, we also show the relationship between decorated trace semantics and (ordinary) trace semantics for Büchi automata. For the concrete case of Büchi automata sketched above, there exists a canonical function $(A^+)^\omega \to A^\omega$ that flattens a sequence and hence removes the "decorations". It is easy to see that if we thus remove decorations of a decorated trace semantics then we obtain an ordinary trace semantics. We shall prove its categorical counterpart.

In fact, the framework in [21] also covered the *parity* condition, which generalizes the Büchi condition. A *parity automaton* is equipped with a function $\Omega : X \to [1, 2n]$ that assigns a natural number called a *priority* to each state $x \in X$. Our new framework developed in the current paper also covers parity automata. In order to obtain the value domain for parity automata, we repeatedly apply $(\_)^+$ and $(\_)^\oplus$ to F like $F^{+\oplus\cdots+\oplus}0$.

Compared to the existing characterization shown in (1), one of the characteristics of our new characterization as shown in (2) is that information about accepting states is more explicitly captured in decorated trace semantics, as in (3). This characteristics would be useful in categorically characterizing notions about Büchi or parity automata. For example, we could use it for categorically characterizing (bi)simulation notions for Büchi automata, e.g. *delayed simulation* [8], a simulation notion known to be appropriate for state space reduction.

To summarize, our contributions in this paper are as follows:

- We introduce a new categorical data type $F^{+\oplus}0$, an alternating fixed point of a functor, for characterizing the Büchi acceptance condition.
- Using the data type, we introduce a categorical decorated trace semantics, simply as a greatest fixed point.
- We show the categorical relationship with ordinary trace semantics in [21].
- We instantiate the framework to several types of concrete systems.
- We extend the framework to the parity condition (in the appendix).

Related Work. As we have mentioned, a categorical characterization of Büchi and parity conditions is also found in [5], but adaptation to infinite-state or probabilistic systems seems to be difficult in their framework. There also exist notions which are fairly captured by their characterization but seem difficult to capture in the frameworks in [21] and this paper, such as bisimilarity.

The notion of alternating fixed point of functors is also used in [2,9]. In [9] the authors characterize the set of continuous functions from A^ω to B^ω as an alternating fixed point $\nu X. \mu Y. (B \times X) + Y^A$ of a functor. Although the data type and the one used in the current paper are different and incomparable, the intuition behind them is very similar, because the former comes with a Büchi-like flavor: if $f(a_0 a_1 \ldots) = b_0 b_1 \ldots$ then each b_i should be determined by a *finite* prefix of $a_0 a_1 \ldots$, and therefore f is regarded as an *infinite* sequence of such assignments. In [2, Sect. 7] a sufficient condition for the existence of such an alternating fixed point is discussed.

Organization. Section 2 gives preliminaries. In Sect. 3 we introduce a categorical data type for decorated trace semantics as an alternating fixed point of functors. In Sect. 4 we define a categorical decorated trace semantics, and show a relationship with ordinary categorical trace semantics in [21]. In Sect. 5 we apply the framework to nondeterministic Büchi tree automata. In Sect. 6, we briefly discuss systems with other branching types. In Sect. 7, we conclude and give future work.

All the discussions in this paper also apply to the parity condition. However, for the sake of simplicity and limited space, we mainly focus on the Büchi condition throughout the paper, and defer discussions about the parity condition to the appendix, that is found in the extended version [20] of this paper. We omit a proof if an analogous statement is proved for the parity condition in the appendix. Some other proofs and discussions are also deferred to the appendix.

2 Preliminaries

2.1 Notations

For $m, n \in \mathbb{N}$, $[m, n]$ denotes the set $\{i \in \mathbb{N} \mid m \leq i \leq n\}$. We write $\pi_i : \prod_i X_i \to X_i$ and $\kappa_i : X_i \to \coprod_i X_i$ for the canonical projection and injection respectively. For a set A, A^* (resp. A^ω) denotes the set of finite (resp. infinite) sequences over A, A^∞ denotes $A^* \cup A^\omega$, and A^+ denotes $A^* \setminus \{\langle\rangle\}$. We write $\langle\rangle$ for the empty sequence. For a monotone function $f : (X, \sqsubseteq) \to (X, \sqsubseteq)$, μf (resp. νf) denotes its least (resp. greatest) fixed point (if it exists). We write **Sets** for the category of sets and functions, and **Meas** for the category of measurable sets and measurable functions. For $f : X \to Y$ and $A \subseteq Y$, $f^{-1}(A)$ denotes $\{x \in X \mid f(x) \in A\}$.

2.2 Fixed Point and Hierarchical Equation System

In this section we review the notion of *hierarchical equation system* (HES) [3,6]. It is a kind of a representation of an alternating fixed point.

Definition 2.1 (HES). A *hierarchical equation system* (HES for short) is a system of equations of the following form.

$$
E = \begin{cases}
u_1 =_{\eta_1} f_1(u_1, \ldots, u_m) & \in (L_1, \sqsubseteq_1) \\
u_2 =_{\eta_2} f_2(u_1, \ldots, u_m) & \in (L_2, \sqsubseteq_2) \\
\quad\vdots \\
u_m =_{\eta_m} f_m(u_1, \ldots, u_m) & \in (L_m, \sqsubseteq_m)
\end{cases}
$$

Here for each $i \in [1, m]$, $(L_i, \leq_i)$ is a complete lattice, u_i is a variable that ranges over L_i, $\eta_i \in \{\mu, \nu\}$ and $f_i : L_1 \times \cdots \times L_m \to L_i$ is a monotone function.

Definition 2.2 (Solution). Let E be an HES as in Definition 2.1. For each $i \in [1, m]$ and $j \in [1, i]$ we inductively define $f_i^{\ddagger} : L_i \times \cdots \times L_m \to L_i$ and $l_j^{(i)} : L_{i+1} \times \cdots \times L_m \to L_j$ as follows (no need to distinguish the base case from the step case):

- $f_i^{\ddagger}(u_i, \ldots, u_m) := f_i(l_1^{(i-1)}(u_i, \ldots, u_m), \ldots, l_{i-1}^{(i-1)}(u_i, \ldots, u_m), u_i, \ldots, u_m);$
  and
- $l_i^{(i)}(u_{i+1}, \ldots, u_m) := \eta f_i^{\ddagger}(\_, u_{i+1}, \ldots, u_m)$ where $\eta = \mu$ if i is odd and $\eta = \nu$ if i is even. For $j < i$, $l_j^{(i)}(u_{i+1}, \ldots, u_m) := l_j^{(i-1)}(l_i^{(i)}(u_{i+1}, \ldots, u_m), u_{i+1}, \ldots, u_m).$ If such a least or greatest fixed point does not exist, then it is undefined.

We call $(l_1^{(i)}, \ldots, l_i^{(i)})$ the *i-th intermediate solution*. The *solution* of the HES E is a family $(u_1^{\text{sol}}, \ldots, u_m^{\text{sol}}) \in L_1 \times \cdots \times L_m$ defined by $u_i^{\text{sol}} := l_i^{(m)}(*)$ for each i.

2.3   Categorical Finite and Infinitary Trace Semantics

We review [11,12,16,18] and see how finite and infinitary traces of transition systems are characterized categorically. We assume that the readers are familiar with basic theories of categories and coalgebras. See e.g. [4,13] for details.

We model a system as a (T, F)-*system*, a coalgebra $c : X \to TFX$ where T is a monad representing the branching type and F is an endofunctor representing the transition type of the system. Here are some examples of T and F:

Definition 2.3 ($\mathcal{P}$, $\mathcal{D}$, $\mathcal{L}$ and $\mathcal{G}$). The *powerset monad* is a monad $\mathcal{P} = (\mathcal{P}, \eta^{\mathcal{P}}, \mu^{\mathcal{P}})$ on **Sets** where $\mathcal{P}X := \{A \subseteq X\}$, $\mathcal{P}f(A) := \{f(x) \mid x \in A\}$, $\eta_X^{\mathcal{P}}(x) := \{x\}$ and $\mu_X^{\mathcal{P}}(\Gamma) := \bigcup_{A \in \Gamma} A$. The *subdistribution monad* is a monad $\mathcal{D} = (\mathcal{D}, \eta^{\mathcal{D}}, \mu^{\mathcal{D}})$ on **Sets** where $\mathcal{D}X := \{\delta : X \to [0,1] \mid |\{x \mid \delta(x) > 0\}|$ is countable, and $\sum_x \delta(x) \leq 1\}$, $\mathcal{D}f(\delta)(y) := \sum_{x \in f^{-1}(\{y\})} \delta(x)$, $\eta_X^{\mathcal{D}}(x)(x')$ is 1 if $x = x'$ and 0 otherwise, and $\mu_X^{\mathcal{D}}(\Phi)(x) := \sum_{\delta \in \mathcal{D}X} \Phi(\delta) \cdot \delta(x)$. The *lift monad* is a monad $\mathcal{L} = (\mathcal{L}, \eta^{\mathcal{L}}, \mu^{\mathcal{L}})$ on **Sets** where $\mathcal{L}X := \{\bot\} + X$, $\mathcal{L}f(a)$ is $f(a)$ if $a \in X$ and $\bot$ if $a = \bot$, $\eta_X^{\mathcal{L}}(x) := x$ and $\mu_X^{\mathcal{L}}(a) := a$ if $a \in X$ and $\bot$ if $a = \bot$. The *sub-Giry monad* is a monad $\mathcal{G} = (\mathcal{G}, \eta^{\mathcal{G}}, \mu^{\mathcal{G}})$ on **Meas** where $\mathcal{G}(X, \mathfrak{F}_X)$ is carried by the set of probability measures over $(X, \mathfrak{F}_X)$, $\mathcal{G}f(\varphi)(A) := \varphi(f^{-1}(A))$, $\eta_X^{\mathcal{G}}(x)(A)$ is 1 if $x \in A$ and 0 otherwise, and $\mu_X^{\mathcal{G}}(\Xi)(A) := \int_{\delta \in \mathcal{G}X} \delta(A) d\Xi$.

Definition 2.4 (Polynomial Functors). A *polynomial functor* F on **Sets** is defined by the following BNF notation: $F ::= \text{id} \mid A \mid F \times F \mid \coprod_{i \in I} F$ where $A \in$ **Sets** and I is countable. A *(standard Borel) polynomial functor* F on **Meas** is defined by the following BNF notation: $F ::= \text{id} \mid A \mid F \times F \mid \coprod_{i \in I} F$ where $A \in$ **Meas**, I is countable, and the σ-algebras over products and coproducts are given in the standard manner (see e.g. [18, Definition 2.2]).

A carrier of an *initial F-algebra* models a domain of finite traces [11] while that of a *final F-coalgebra* models a domain of infinitary traces [12]. For example, as we have seen in Sect. 1, for $F = \{\checkmark\} + \mathsf{A} \times (\_)$ on **Sets**, the carrier set of the final F-coalgebra is A^{∞} while that of the initial F-algebra is A^*. The situation is similar for a polynomial functor $F = (\{\checkmark\}, \mathcal{P}\{\checkmark\}) + (A, \mathcal{P}A) \times (\_)$ on **Meas**. The carrier of an initial algebra is $(A^*, \mathcal{P}A^*)$, and that of a final coalgebra is $(A^\infty, \mathfrak{F}_{A^\infty})$ where $\mathfrak{F}_{A^\omega}$ is the standard σ-algebra generated by the cylinder set.

In general, for a certain class of functors, an initial algebra and a final coalgebra are obtained by the following well-known construction.

Theorem 2.5 ([1])

1. Let $(A, (\pi_i : F^i 0 \to A)_{i\in\omega})$ be a colimit of an ω-chain $0 \xrightarrow{\mathsf{i}} F0 \xrightarrow{F\mathsf{i}} F^2 0 \xrightarrow{F^2\mathsf{i}}$ $\dots$. If F preserves the colimit, then the unique mediating arrow $\iota : FA \to A$ from the colimit $(FA, (F\pi_i : F^{i+1}0 \to FA)_{i\in\omega})$ to a cocone $(A, (\pi'_i : F^i 0 \to A)_{i\in\omega})$ where $\pi'_i = \pi_{i+1}$ is an initial F-algebra.

2. Let $(Z, (\pi_i : A \to F^i 1)_{i\in\omega})$ be a limit of an ω^{op}-chain $1 \xleftarrow{!} F1 \xleftarrow{F!} F^2 1 \xleftarrow{F^2!}$ $\dots$. If F preserves the limit, then the unique mediating arrow $\zeta : Z \to FZ$ from a cone $(Z, (\pi'_i : A \to F^i 1)_{i\in\omega})$ where $\pi'_i = \pi_{i+1}$ to the limit $(FZ, (F\pi_i : FZ \to F^{i+1}1)_{i\in\omega})$ is a final F-coalgebra. $\square$

We next quickly review notions about the Kleisli category $\mathcal{K}\ell(T)$.

Definition 2.6 ($\mathcal{K}\ell(T)$, J, U and $\overline{F}$). Let $T = (T, \eta, \mu)$ be a monad on $\mathbb{C}$. The *Kleisli category* $\mathcal{K}\ell(T)$ is given by $|\mathcal{K}\ell(T)| = |\mathbb{C}|$ and $\mathcal{K}\ell(T)(X, Y) = \mathbb{C}(X, TY)$ for $X, Y \in |\mathcal{K}\ell(T)|$. An arrow $f \in \mathcal{K}\ell(T)(X, Y)$ is called a *Kleisli arrow*, and we write $f : X \nrightarrow Y$ for distinction. Composition of arrows $f : X \nrightarrow Y$ and $g : Y \nrightarrow Z$ is defined by $\mu_Z \circ Tg \circ f$, and denoted by $g \odot f$ for distinction. The *lifting functor* $J : \mathbb{C} \to \mathcal{K}\ell(T)$ is defined by: $JX := X$ and $J(f) := \eta_Y \circ f$ for $f : X \to Y$. The *forgetful functor* $U : \mathcal{K}\ell(T) \to \mathbb{C}$ is defined by: $UX := TX$ and $U(g) := \mu_Y \circ Tg$ for $g : X \nrightarrow Y$. A functor $\overline{F} : \mathcal{K}\ell(T) \to \mathcal{K}\ell(T)$ is called a *lifting* of $F : \mathbb{C} \to \mathbb{C}$ if $\overline{F}J = JF$.

Example 2.7. Let $T = \mathcal{P}$ and $F = \sum_{n=0}^{\omega} \Sigma_n \times (\_)^n : \mathbf{Sets} \to \mathbf{Sets}$. A lifting $\overline{F}$ over $\mathcal{K}\ell(T)$ is given by $\overline{F}X = FX$ for $X \in \mathbf{Sets}$ and $\overline{F}f(\sigma, x_0, \dots, x_{n-1}) = \{(\sigma, y_0, \dots, y_{n-1}) \mid \forall i. y_i \in f(x_i)\}$ for $f : X \nrightarrow Y$, $\sigma \in \Sigma_n$ and $x_0, \dots, x_{n-1} \in X$. (see e.g. [11]).

It is well-known that there is a bijective correspondence between a lifting $\overline{F}$ and a *distributive law*, a natural transformation $\lambda : FT \Rightarrow TF$ satisfying some axioms [15]. See Sect. D of the extended version [20] for the details.

In the rest of this section, let F be an endofunctor and T be a monad on a category $\mathbb{C}$, and assume that a lifting $\overline{F} : \mathcal{K}\ell(T) \to \mathcal{K}\ell(T)$ is given.

In [11], a finite trace semantics of a transition system was characterized as the unique homomorphism to the final $\overline{F}$-coalgebra in $\mathcal{K}\ell(T)$, which is obtained by reversing and lifting the initial F-algebra in $\mathbb{C}$.

Definition 2.8 (tr(c)). We say F and T constitute a *finite trace situation* with respect to $\overline{F}$ if the following conditions are satisfied:

- An initial F-algebra $\iota^F : FA \to A$ exists.
- $J(\iota^F)^{-1} : A \twoheadrightarrow \overline{F}A$ is a final $\overline{F}$-coalgebra.

For $c : X \twoheadrightarrow \overline{F}X$, the unique homomorphism from c to $J(\iota^F)^{-1}$ is called the *(coalgebraic) finite trace semantics* of c and denoted by tr(c) $: X \twoheadrightarrow A$.

In [11], a sufficient condition for constituting a finite trace situation is given.

Theorem 2.9 ([11]). *Assume each homset of $\mathcal{Kl}(T)$ carries a partial order $\sqsubseteq$. If the following conditions are satisfied, F and T constitute a finite trace situation.*

- *The functor F preserves ω-colimits in $\mathbb{C}$.*
- *Each homset of $\mathcal{Kl}(T)$ constitutes an ω-cpo with a bottom element $\bot$.*
- *Kleisli composition $\odot$ is monotone, and the lifting $\overline{F}$ is locally monotone, i.e. $f \sqsubseteq g$ implies $\overline{F}f \sqsubseteq \overline{F}g$.*
- *Kleisli composition $\odot$ preserves ω-suprema and the bottom element $\bot$.*

Here by Theorem 2.5, the first condition above implies existence of an initial algebra.

In [11] it was shown that $T \in \{\mathcal{P}, \mathcal{D}, \mathcal{L}\}$ and a polynomial functor F satisfy the conditions in Theorem 2.9 with respect to some orderings and liftings, and hence constitute finite trace situations. We can see the result for $T = \mathcal{D}$ implies that $T = \mathcal{G}$ and a standard Borel polynomial functor F also satisfy the conditions.

An infinitary trace semantics was characterized in [12] as the greatest homomorphism to a weakly final coalgebra obtained by lifting a final coalgebra.

Definition 2.10 (Infinitary Trace Situation). We assume that each homset of $\mathcal{Kl}(T)$ carries a partial order $\sqsubseteq$. We say that F and T constitute an *infinitary trace situation* with respect to $\overline{F}$ and $\sqsubseteq$ if the following conditions are satisfied:

- A final F-coalgebra $\zeta^F : Z \to FZ$ exists.
- $J\zeta^F : Z \twoheadrightarrow \overline{F}Z$ is a weakly final $\overline{F}$-coalgebra that admits the greatest homomorphism, i.e. for an $\overline{F}$-coalgebra $c : X \twoheadrightarrow \overline{F}X$, there exists the greatest homomorphism from c to $J\zeta^F$ with respect to $\sqsubseteq$.

The greatest homomorphism from c to $J\zeta^F$ is called the *(coalgebraic) infinitary trace semantics* of c and denoted by tr$^\infty(c) : X \twoheadrightarrow Z$.

It is known that $T \in \{\mathcal{P}, \mathcal{D}, \mathcal{L}, \mathcal{G}\}$ and a polynomial functor F constitute infinitary trace situations with respect to some orderings and liftings [18]. Differently from finite trace situation, sufficient conditions for infinitary trace situation are not unified. In [18], two sufficient conditions are given. One is applicable for $T = \mathcal{P}$, and the other is for $T \in \{\mathcal{L}, \mathcal{G}\}$. No condition is known for $T = \mathcal{D}$.

Example 2.11. Let $T = \mathcal{P}$ and $F = \{\checkmark\} + \mathsf{A} \times (\_)$. Then a TF-coalgebra $c : X \rightarrow \mathcal{P}(\{\checkmark\} + \mathsf{A} \times X)$ is identified with an A-labeled nondeterministic automaton whose accepting states are given by $\{x \mid \checkmark \in c(x)\}$. The arrow $\mathsf{tr}(c)$ has a type $X \rightarrow \mathsf{A}^*$ and assigns the set of accepted finite words to each state [11]:

$$
\mathsf{tr}(c)(x) = \left\{ a_1 a_2 \ldots a_n \in \mathsf{A}^* \;\middle|\; \begin{array}{l} \exists x_0, \ldots, x_n \in X. \, \forall i \in [1, n-1]. \\ \quad (a_{i+1}, x_{i+1}) \in c(x_i) \text{ and } \checkmark \in c(x_n) \end{array} \right\} .
$$

In contrast, $\mathsf{tr}^\infty(c) : X \rightarrow \mathsf{A}^\infty$ is given as follows [12]:

$$
\mathsf{tr}^\infty(c)(x) = \mathsf{tr}(c)(x)
$$
$$
\cup \{ a_1 a_2 \ldots \in \mathsf{A}^\omega \mid \exists x_0, x_1, \ldots \in X. \, x = x_0, \forall i \in \omega. \, (a_{i+1}, x_{i+1}) \in c(x_i) \} .
$$

2.4   Büchi (T, F)-systems and Its Coalgebraic Trace Semantics

The results in Sect. 2.3 was extended for systems with the parity acceptance condition in [21]. We hereby review the results for the Büchi acceptance condition.

Definition 2.12 (Büchi (T, F)-system). Let $n \in \mathbb{N}$. A *Büchi (T, F)-system* is a pair $(c, (X_1, X_2))$ of a $\overline{F}$-coalgebra $c : X \rightarrow \overline{F}X$ in $\mathcal{K}\ell(T)$ and a partition (X_1, X_2) of X (i.e. $X \cong X_1 + X_2$). For $i \in \{1, 2\}$, we write c_i for $c \circ \kappa_i : X_i \rightarrow \overline{F}X$.

Their coalgebraic trace semantics is given by a solution of an HES.

Definition 2.13 ($\mathsf{tr}_i^\mathsf{B}(c)$). Assume that each homset of $\mathcal{K}\ell(T)$ carries a partial order $\sqsubseteq$. We say that F and T constitute a *Büchi trace situation* with respect to $\overline{F}$ and $\sqsubseteq$ if they satisfy the following conditions:

– A final F-coalgebra $\zeta : Z \rightarrow FZ$ exists.
– For an arbitrary Büchi (T, F)-system $\mathcal{X} = (c, (X_1, X_2))$, the following HES has a solution.

$$
E_c = \begin{cases} u_1 =_\mu \ J\zeta^{-1} \odot \overline{F}[u_1, u_2] \odot c_1 \in (\mathcal{K}\ell(T)(X_1, Z), \sqsubseteq_{X_1, Z}) \\ u_2 =_\nu \ J\zeta^{-1} \odot \overline{F}[u_1, u_2] \odot c_2 \in (\mathcal{K}\ell(T)(X_2, Z), \sqsubseteq_{X_2, Z}) \end{cases}
$$

The solution $(u_1^{\mathsf{sol}} : X_1 \rightarrow Z, u_2^{\mathsf{sol}} : X_2 \rightarrow Z)$ of E_c is called the *(coalgebraic) Büchi trace semantics* of $\mathcal{X}$. We write $\mathsf{tr}_i^\mathsf{B}(c)$ for u_i^{sol} for each i (see also Eq. (1)).

Example 2.14. Let $T = \mathcal{P}$ and $F = \mathsf{A} \times (\_)$. Then a Büchi (T, F)-system $(c : X \rightarrow FX, (X_1, X_2))$ is identified with an A-labeled Büchi automaton. Following Definition 2.2 we shall sketch how the solution of the HES E_c in Definition 2.13 is calculated. Note that $Z \cong \mathsf{A}^\omega$.

– We first calculate an intermediate solution $l_1^{(1)}(u_2) : X_1 \rightarrow \mathsf{A}^\omega$ as the least fixed point of $u_1 \mapsto J\zeta^{-1} \odot \overline{F}[u_1, u_2] \odot c_1$.
– We next define $f_2^\ddagger : \mathcal{K}\ell(T)(X_2, Z) \rightarrow \mathcal{K}\ell(T)(X_2, Z)$ by $f_2^\ddagger(u_2) := J\zeta^{-1} \odot \overline{F}[l_1^{(1)}(u_2), u_2] \odot c_2$.

- We calculate $l_2^{(2)}(*) : X_2 \twoheadrightarrow \mathsf{A}^\omega$ as the greatest fixed point of $f_2^\ddagger$.
- We let $l_1^{(2)}(*) := l_1^{(1)}(l_2^{(2)}) : X_1 \twoheadrightarrow \mathsf{A}^\omega$.

Then for each i, the solution $\mathrm{tr}_i^{\mathsf{B}}(c) = l_i^{(2)}(*)$ is given as follows [21]:

$$\mathrm{tr}_i^{\mathsf{B}}(c)(x) := \left\{ a_1 a_2 \ldots \in \mathsf{A}^\omega \;\middle|\; \begin{array}{l} \exists x_0, x_1, \ldots \in X. \; \forall i \in \omega. \; (a_{i+1}, x_{i+1}) \in c(x_i) \text{ and} \\ x_i \in X_2 \text{ for infinitely many } i \end{array} \right\}.$$

3 Alternating Fixed Points of Functors

3.1 Categorical Datatypes for Büchi Systems

We first introduce the categorical datatypes F^+X and $F^\oplus X$, which are understood as least and greatest fixed points of a functor F.

Definition 3.1 $(F^+, F^\oplus)$. For $F : \mathbb{C} \to \mathbb{C}$, we define functors $F^+, F^\oplus : \mathbb{C} \to \mathbb{C}$ as follows. Given $X \in \mathbb{C}$, the object F^+X is the carrier of (a choice of) an initial algebra $\iota_X^F : F(F^+X + X) \xrightarrow{\cong} F^+X$ for the functor $F(\_ + X)$. Similarly, the object $F^\oplus X$ is the carrier of a final coalgebra $\zeta_X^F : F^\oplus X \xrightarrow{\cong} F(F^\oplus X + X)$. For $f : X \to Y$, $F^+f : F^+X \to F^+Y$ is given as the unique homomorphism from ι_X^F to $\iota_Y^F \circ F(\mathrm{id}_{F^+Y} + f)$. We define $F^\oplus f : F^\oplus X \to F^\oplus Y$ similarly.

$$
\begin{array}{ccc}
F(F^+X + X) \dashrightarrow F(F^+Y + X) & \quad & F(F^\oplus X + Y) \dashrightarrow F(F^\oplus Y + Y) \\
\hspace{1em} {\scriptstyle F(F^+f + \mathrm{id})} \quad \downarrow {\scriptstyle F(\mathrm{id}+f)} & & {\scriptstyle F(\mathrm{id}+f)}\uparrow \quad {\scriptstyle F(F^\oplus f + \mathrm{id})} \\
\iota_X^F \; \Big\downarrow {\cong} \quad = \quad F(F^+Y + Y) & & F(F^\oplus X + X) \quad = \quad \zeta_Y^F \Big\uparrow {\cong} \\
\hspace{1em} \quad \iota_Y^F \hspace{-0.5em}\searrow {\cong} & & \zeta_X^F \nwarrow {\cong} \\
F^+X \dashrightarrow_{F^+f} F^+Y & & F^\oplus X \dashrightarrow_{F^\oplus f} F^\oplus Y
\end{array}
$$

Remark 3.2. The construction F^+ resembles the *free monad* F^* over F. The latter is defined as follows: given $X \in \mathbb{C}$, the object F^*X is the carrier of an initial algebra $F(F^*X) + X \xrightarrow{\cong} F^*X$ for the functor $F(\_) + X$. The notations generalize the usual distinction between $*$ and $+$. Indeed, for $\mathbb{C} = \mathbf{Sets}$ and $F = \Sigma_0 \times \_$ (where Σ_0 is an alphabet), we have $F^+1 = \Sigma_0^+$ (the set of finite words of length ≥ 1) and $F^*1 = \Sigma_0^*$ (the set of all finite words). Similarly, $F^\oplus$ resembles the *free completely iterative monad* [14].

Example 3.3. For $F = \mathsf{A} \times (\_)$, by the construction in Theorem 2.5, $F^+X \cong \mathsf{A}^+X$, $F^\oplus X \cong \mathsf{A}^+X + \mathsf{A}^\omega$ and $F^{+\oplus}X \cong (\mathsf{A}^+)^+X + (\mathsf{A}^+)^\omega$. Especially, if we let $X = 0$ then $F^{+\oplus}0 \cong (\mathsf{A}^+)^\omega$. We identify $(a_{00}a_{01} \ldots a_{0n_0})(a_{10}a_{11} \ldots a_{1n_1}) \ldots \in F^{+\oplus}0 \cong (\mathsf{A}^+)^\omega$ with the following "decorated" sequence:

$$(a_{00}, \textcircled{\circ})(a_{01}, \bigcirc) \ldots (a_{0n_0}, \bigcirc)(a_{10}, \textcircled{\circ})(a_{11}, \bigcirc) \ldots (a_{1n_1}, \bigcirc) \ldots \in (\mathsf{A} \times \{\bigcirc, \textcircled{\circ}\})^\omega.$$

The second component of each element (i.e. *decoration*) represents a break of a word: it is ⊚ if and only if it is the beginning of a word. It is remarkable that in the sequence above, ⊚ always appears infinitely many times. Hence $w \in (\mathsf{A}^+)^\omega$ is understood as an infinite word decorated so that the Büchi condition is satisfied.

We next define Kleisli arrows $\beta_{1\,X}$ and $\beta_{2\,X}$ that are used to define decorated trace semantics (see the diagrams in (2)).

Definition 3.4. We define natural transformations $\beta_1 : F^+(F^{+\oplus} + \mathrm{id}) \Rightarrow F(F^+F^{+\oplus} + F^{+\oplus} + \mathrm{id})$ and $\beta_2 : F^{+\oplus} \Rightarrow F(F^+F^{+\oplus} + F^{+\oplus} + \mathrm{id})$ as follows.

$$\beta_{1\,X} := \left(\quad F^+(F^{+\oplus}X + X) \xrightarrow{(\iota^F_{F+\oplus X})^{-1}} F(F^+F^{+\oplus}X + F^{+\oplus}X + X) \right)$$

$$\beta_{2\,X} := \left(F^{+\oplus}X \xrightarrow{\zeta^{F^+}_X} F^+(F^{+\oplus}X + X) \xrightarrow{(\iota^F_{F+\oplus X})^{-1}} F(F^+F^{+\oplus}X + F^{+\oplus}X + X) \right)$$

Remark 3.5. As a final coalgebra $\zeta^{F^+}_X$ is an isomorphism, we can see from Definition 3.4 that $F^+(F^{+\oplus}X + X) \cong F^{+\oplus}X$. For $F = \mathsf{A} \times (\_)$, if we regard $F^{+\oplus}X$ as $(\mathsf{A}^+)^\omega$ as in Example 3.3, $F^+(F^{+\oplus}X + X)$ would be understood as $\mathsf{A}^+(\mathsf{A}^+)^\omega$, which is indeed isomorphic to $(\mathsf{A}^+)^\omega$. However, in this paper, mainly for the sake of simplicity of notations, we explicitly distinguish them and later write types of a decorated trace semantics of a Büchi (T, F)-system as $\mathsf{dtr}_1(c) : X_1 \nrightarrow F^+(F^{+\oplus}0)$ and $\mathsf{dtr}_2(c) : X_2 \nrightarrow F^{+\oplus}0$. Because of this choice, while an element in $F^{+\oplus}0 \cong (\mathsf{A}^+)^\omega$ is regarded as a decorated word whose first letter is decorated by $\bigcirc$ (Example 3.3), an element $a_0 \ldots a_n \big((a_{00}a_{01} \ldots a_{0n_0})(a_{10}a_{11} \ldots a_{1n_1}) \ldots \big) \in F^+(F^{+\oplus}0) \cong \mathsf{A}^+(\mathsf{A}^+)^\omega$ is understood as the following decorated sequence:

$$(a_0, \bigcirc) \ldots (a_n, \bigcirc)(a_{00}, \circledcirc)(a_{01}, \bigcirc) \ldots (a_{0n_0}, \bigcirc)(a_{10}, \circledcirc)(a_{11}, \bigcirc) \ldots (a_{1n_1}, \bigcirc) \ldots.$$

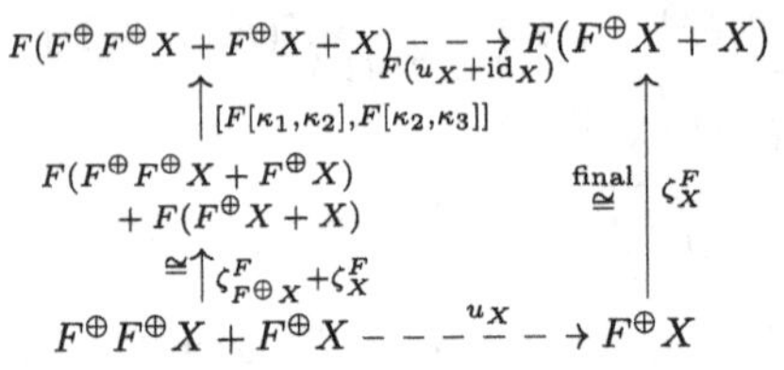

Fig. 1. The unique arrow u_X

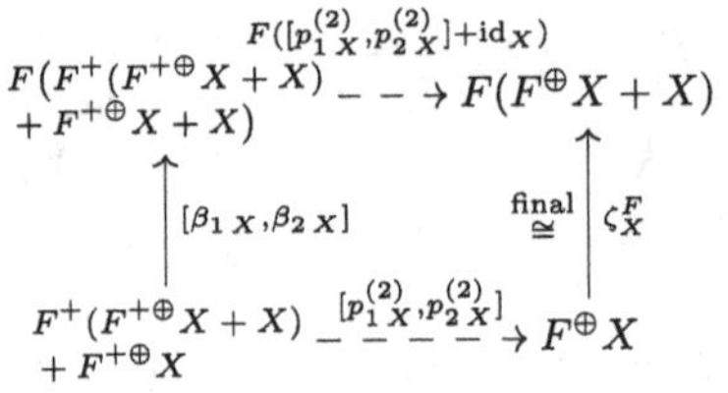

Fig. 2. The unique arrow $[p^{(2)}_{1\,X}, p^{(2)}_{2\,X}]$

3.2 Natural Transformations Regarding to F^+ and $F^\oplus$

We introduce two natural transformations for later use. As mentioned in Remark 3.2, F^+ resembles the free monad F^* while $F^\oplus$ is similar to the free completely iterative monad. The first natural transformation we introduce is analogous to the multiplication of those free monads.

Definition 3.6 ($\mu^{F^\oplus}$). We define a natural transformation $\mu^{F^\oplus} : F^\oplus F^\oplus \Rightarrow F^\oplus$ by $\mu^{F^\oplus} := (u_X \circ \kappa_1)_{X \in \mathbb{C}}$, where u_X is the unique homomorphism from $[F[\kappa_1, \kappa_2] \circ \zeta^F_{F^\oplus X}, F[\kappa_2, \kappa_3] \circ \zeta^F_X]$ to ζ^F_X (see Fig. 1).

Example 3.7. Let $F = \mathsf{A} \times (\_)$. According to the characterizations in Example 3.3 and Remark 3.5, $p^{(1)}_{1\,X}$ has a type $(\mathsf{A}^+)^+(\mathsf{A}^+)^+X + (\mathsf{A}^+)^+(\mathsf{A}^+)^\omega + (\mathsf{A}^+)^\omega \to (\mathsf{A}^+)^+X + (\mathsf{A}^+)^\omega$, and is given by the concatenating function that preserves each finite word.

The second natural transformation is for "removing" decorations.

Definition 3.8 $(p_j^{(i)})$. We define a natural transformation $p_1^{(1)} : F^+ \Rightarrow F^\oplus$ so that $p_{1X}^{(1)} : F^+X \to F^\oplus X$ is the unique homomorphism from $J(\iota_X^F)^{-1}$ to $J\zeta_X^F$. Similarly, we define natural transformations $p_1^{(2)} : F^+(F^{+\oplus} + \mathrm{id}) \Rightarrow F^\oplus$ and $p_2^{(2)} : F^{+\oplus} \Rightarrow F^\oplus$ so that $[p_{1X}, p_{2X}] : F^+(F^{+\oplus}X + X) + F^{+\oplus}X \to F^\oplus X$ is the unique homomorphism from $[\beta_{1X}, \beta_{2X}]$ to ζ_X^F (see Fig. 2).

Example 3.9. Let $F = \mathsf{A} \times (\_)$. According to the characterizations in Example 3.3 and Remark 3.5, $p_{1X}^{(1)}$ has a type $\mathsf{A}^+X \to \mathsf{A}^+X + \mathsf{A}^\omega$ and is given by the natural inclusion. In contrast, $p_{10}^{(2)}$ and $p_{20}^{(2)}$ have types $\mathsf{A}^+(\mathsf{A}^+)^\omega \to \mathsf{A}^\omega$ and $(\mathsf{A}^+)^\omega \to \mathsf{A}^\omega$ respectively, and they are given by the flattening functions. See also Proposition 5.10.

3.3   Liftings $\overline{F^+}$ and $\overline{F^\oplus}$ over $\mathcal{Kl}(T)$

Let $\overline{F} : \mathcal{Kl}(T) \to \mathcal{Kl}(T)$ be a lifting of of a functor F. We show that under certain conditions, it induces liftings $\overline{F^+} : \mathcal{Kl}(T) \to \mathcal{Kl}(T)$ of F^+ and $\overline{F^\oplus} : \mathcal{Kl}(T) \to \mathcal{Kl}(T)$ of $F^\oplus$. Note that a lifting $\overline{F}$ induces a lifting $\overline{F(\_ + A)} : \mathcal{Kl}(T) \to \mathcal{Kl}(T)$ of $F(\_+A)$ which is defined by $\overline{F(\_ + A)}f := \overline{F}(f+\mathrm{id}_A) = \overline{F}([T\kappa_1, T\kappa_2] \circ (f+\eta_A))$ using the coproduct in $\mathcal{Kl}(T)$.

Definition 3.10. 1. Assume T and F constitute a finite trace situation. For $X \in \mathbb{C}$, we let $\overline{F^+}X := F^+X$. For $f : X \nrightarrow Y$, we define $\overline{F^+}f : F^+X \nrightarrow F^+Y$ as the unique homomorphism from $\overline{F}(\mathrm{id}_{F^+X} + f) \odot J(\iota_X^F)^{-1}$ to $J(\iota_Y^F)^{-1}$.
2. Assume T and F constitute an infinitary trace situation. For $X \in \mathbb{C}$, we let $\overline{F^\oplus}X := F^\oplus X$. For $f : X \nrightarrow Y$, we define $\overline{F^\oplus}f : F^\oplus X \nrightarrow F^\oplus Y$ as the greatest homomorphism from $\overline{F}(\mathrm{id}_{F^\oplus X} + f) \odot J\zeta_X^F$ to $J\zeta_Y^F$.

$$
\begin{array}{ccc}
F(F^+X + Y) \xdashrightarrow{\;\;\dashv\;} F(F^+Y + Y) & \quad & F(F^\oplus X + Y) \xrightarrow{\;\;\;} F(F^\oplus Y + Y) \\
\overline{F}(\mathrm{id}+f)\!\!\uparrow \;\;\; \overline{F(F^+ f + \mathrm{id}_{F^+ X})} & & \overline{F}(\mathrm{id}+f)\!\!\uparrow \;\;\; \overline{F(F^\oplus f + \mathrm{id}_{F^\oplus x})} \\
F(F^+X + X) \;\; = \;\; J(\iota_Y^F)^{-1}\!\uparrow\;\cong & & F(F^\oplus X + X) \;\; =_\nu \;\; J\zeta_Y^F\!\uparrow\;\cong \\
J(\iota_X^F)^{-1}\!\uparrow\;\cong & & J\zeta_X^F\!\uparrow\;\cong \\
F^+X \xdashrightarrow{\;\;\overline{F^+}f\;\;} F^+Y & & F^\oplus X \xrightarrow{\;\;\overline{F^\oplus}f\;\;} F^\oplus Y
\end{array}
$$

In the rest of this section, we check under which conditions $\overline{F^+}$ and $\overline{F^\oplus}$ are functors and form liftings of F^+ and $F^\oplus$. Functoriality of $\overline{F^+}$ holds if and only if for each $f : X \nrightarrow Y$ and $g : Y \nrightarrow W$, $F^+g \odot F^+f$ is the unique homomorphism from $\overline{F}(\mathrm{id} + g) \odot \overline{F}(\mathrm{id} + f) \odot J(\iota_X^F)^{-1}$ to $J(\iota_W^F)^{-1}$. Similarly, functoriality of $\overline{F^\oplus}$ holds if and only if $F^\oplus g \odot F^\oplus f$ is the greatest homomorphism from $\overline{F}(\mathrm{id} + g) \odot \overline{F}(\mathrm{id} + f) \odot J\zeta_X^F$ to $J\zeta_W^F$.

The former always holds by the finality. In contrast, the latter doesn't necessarily hold: a counterexample is $T = \mathcal{D}$ and $F = \{o\} \times (\_)^2$ (see Example C.1 for details). Hence we need an extra assumption to make $\overline{F^\oplus}$ a functor. We hereby assume a stronger condition than is needed for the sake of discussions in Sect. 4.

Definition 3.11 $(\Phi_{c,\sigma})$. Let $c : X \nrightarrow \overline{F}X$ and $\sigma : \overline{F}Y \nrightarrow Y$. We define a function $\Phi_{c,\sigma} : \mathcal{Kl}(T)(X, Y) \to \mathcal{Kl}(T)(X, Y)$ by $\Phi_{c,\sigma}(f) := \sigma \odot \overline{F}f \odot c$.

Definition 3.12. Assume that T and F constitute an infinitary trace situation. Let $\zeta^F : Z \to FZ$ be a final F-coalgebra. We say that T and F satisfy the *gfp-preserving* condition with respect to an $\overline{F}$-algebra $\sigma : FY \twoheadrightarrow Y$ if for each $X \in \mathbb{C}$ and $c : X \twoheadrightarrow FX$, if $l : X \twoheadrightarrow Z$ is the greatest homomorphism from c to $J\zeta^F$ and the function $\Phi_{J\zeta^F,\sigma}$ has the greatest fixed point $m : Z \twoheadrightarrow Y$, then $m \odot l : X \twoheadrightarrow Y$ is the greatest fixed point of $\Phi_{c,\sigma}$.

$$\begin{array}{ccc}
FX & \xrightarrow{\overline{F}l} FZ \xrightarrow{\overline{F}m} & FY \\
c\uparrow & =_\nu J\zeta^F\uparrow\cong \quad =_\nu & \uparrow\sigma \\
X & \xrightarrow{\quad l \quad} Z \xrightarrow{\quad m \quad} & Y
\end{array}$$

We next check if $\overline{F^+}$ and $\overline{F^\oplus}$ are liftings of F^+ and $F^\oplus$. By definition, we have $\overline{F^+}JX = JF^+X$ and $\overline{F^\oplus}JX = JF^\oplus X$ for each $X \in \mathbb{C}$. Let $f : X \to Y$. By definition, $\overline{F^+}Jf = JF^+f$ holds if and only if JF^+f is a unique homomorphism from $\overline{F}(\mathrm{id} + Jf) \odot J(\iota_X^F)^{-1}$ to $J(\zeta_Y^F)^{-1}$. Similarly, $\overline{F^+}Jf = JF^+f$ holds if and only if JF^+f is the greatest homomorphism from $\overline{F}(\mathrm{id} + Jf) \odot J\zeta_X^F$ to $J\zeta_Y^F$.

The former is easily proved by the finality of $J(\iota_Y^F)^{-1}$, while the latter requires an assumption again.

Definition 3.13. Assume T and F constitute an infinitary trace situation. Let $\zeta^F : Z \to FZ$ be a final F-coalgebra. We say that T and F satisfy the *deterministic-greatest* condition if for $c : X \to FX$ in $\mathbb{C}$, if $u : X \to Z$ is the unique homomorphism from c to ζ^F then Ju is the greatest homomorphism from Jc to $J\zeta^F$.

Concluding the discussions so far, we obtain the following proposition.

Proposition 3.14. *1. If T and $F(\_ + A)$ constitute a finite trace situation for each $A \in \mathbb{C}$, the operation $\overline{F^+}$ is a functor and is a lifting of F^+.*

2. If T and $F(\_ + A)$ constitute an infinitary trace situation and satisfy the gfp-preserving condition with respect to an arbitrary algebra and the deterministic-greatest condition for each $A \in \mathbb{C}$, $\overline{F^\oplus}$ is a functor and is a lifting of $F^\oplus$. $\square$

Hence under appropriate conditions, a lifting $\overline{F} : \mathcal{K\ell}(T) \to \mathcal{K\ell}(T)$ of F gives rise to liftings of F^+ and $F^\oplus$. By repeating this, we can define $\overline{F_j^{(i)}}$ for each i and j.

See Sect. D of the extended version [20] for the distributive laws corresponding to the liftings defined above.

Example 3.15. Let $F = \mathsf{A} \times (\_)$ and $T = \mathcal{P}$. As we have seen in Example 3.3, $F^{+\oplus}X \cong (\mathsf{A}^+)^+X + (\mathsf{A}^+)^\omega$. Let $\overline{F}$ be a lifting that is given as in Example 2.7. We can construct a lifting $\overline{F^{+\oplus}}$ using Proposition 3.14, and for $f : X \twoheadrightarrow Y$ in $\mathcal{K\ell}(\mathcal{P})$, $\overline{F^{+\oplus}}f : (\mathsf{A}^+)^+X + (\mathsf{A}^+)^\omega \twoheadrightarrow (\mathsf{A}^+)^+Y + (\mathsf{A}^+)^\omega$ is given by $\overline{F^{+\oplus}}f(w) = \{w'y \mid y \in f(x)\}$ if $w = w'x$ where $w' \in (\mathsf{A}^+)^+$ and $x \in X$, and $\{w\}$ if $w \in (\mathsf{A}^+)^\omega$.

4   Decorated Trace Semantics of Büchi (T, F)-systems

4.1   Definition

Assumption 4.1. Throughout this section, let T be a monad and F be an endofunctor on $\mathbb{C}$, and assume that each homset of $\mathcal{K}\ell(T)$ carries a partial order $\sqsubseteq$. We further assume the following conditions for each $A \in \mathbb{C}$.

1. $F^+, F^{+\oplus} : \mathbb{C} \to \mathbb{C}$ are well-defined and liftings $\overline{F}, \overline{F^+}, \overline{F^{+\oplus}} : \mathcal{K}\ell(T) \to \mathcal{K}\ell(T)$ are given.
2. T and $F(\_ + A)$ satisfy the conditions in Theorem 2.9 with respect to $\overline{F}(\_ + A)$ and $\sqsubseteq$, and hence constitute a finite trace situation.
3. T and $F^+(\_ + A)$ constitute an infinitary trace situation with respect to $\overline{F^+}(\_ + A)$ and $\sqsubseteq$.
4. T and $F^+(\_ + A)$ satisfy the gfp-preserving condition wrt. an arbitrary σ.
5. T and $F^+(\_ + A)$ satisfy the deterministic-greatest condition.
6. The liftings $\overline{F^+}$ and $\overline{F^{+\oplus}}$ are obtained from $\overline{F}$ and $\overline{F^+}$ using the procedure in Definition 3.10 respectively.
7. $\overline{F^+}(\_ + A)$ and $\overline{F^{+\oplus}}(\_ + A)$ are locally monotone.
8. T and F constitute a Büchi trace situation with respect to $\sqsubseteq$ and $\overline{F}$.

Using the categorical data type defined in Sect. 3, we now introduce a *decorated Büchi trace semantics* $\mathsf{dtr}_1(c) : X_1 \nrightarrow F^+(F^{+\oplus}0)$ and $\mathsf{dtr}_2(c) : X_2 \nrightarrow F^{+\oplus}0$.

Definition 4.2 ($\mathsf{dtr}_i(c)$). For a Büchi (T, F)-system $(c, (X_1, X_2))$, the *decorated Büchi trace semantics* is a solution $(\mathsf{dtr}_1(c) : X_1 \nrightarrow F^+(F^{+\oplus}0), \mathsf{dtr}_2(c) : X_2 \nrightarrow F^{+\oplus}0)$ of the following HES (see also Eq. (2)).

$$\begin{cases} v_1 =_\nu J(\beta_{1\,0})^{-1} \odot \overline{F}(v_1 + v_2) \odot c_1 \in (\mathcal{K}\ell(T)(X_1, F^+(F^{+\oplus}0)), \sqsubseteq) \\ v_2 =_\nu J(\beta_{2\,0})^{-1} \odot \overline{F}(v_1 + v_2) \odot c_2 \in (\mathcal{K}\ell(T)(X_2, F^{+\oplus}0), \sqsubseteq) \end{cases}$$

Existence of a solution will be proved in the next section.

4.2   Trace Semantics vs. Decorated Trace Semantics

This section is devoted to sketching the proof of the following theorem, which relates decorated trace semantics $\mathsf{dtr}_i(c)$ and Büchi trace semantics $\mathsf{tr}_i^{\mathsf{B}}(c)$ in [21] via the natural transformation in Definition 3.8.

Theorem 4.3. *For each* $i \in \{1, 2\}$, $\mathsf{tr}_i^{\mathsf{B}}(c) = p_{i\,0}^{(2)} \circ \mathsf{dtr}_i(c)$. $\square$

To prove this, we introduce Kleisli arrows $c_2^\dagger$, $\tilde{\ell}_1^{(1)}$, $\tilde{\ell}_1^{(2)}$ and $\tilde{\ell}_2^{(2)}$. They are categorical counterparts to $f_2^\dagger$, $l_1^{(1)}$, $l_1^{(2)}$ and $l_2^{(2)}$ (see Definition 2.2) for the HES defining $\mathsf{tr}_i^{\mathsf{B}}(c)$ (see Definition 2.13), and bridge the gap between $\mathsf{dtr}_i(c)$ and $\mathsf{tr}_i^{\mathsf{B}}(c)$.

Definition 4.4 $(c_2^{\ddagger}, \tilde{\ell}_1^{(1)}, \tilde{\ell}_1^{(2)}, \tilde{\ell}_2^{(2)})$. We define Kleisli arrows $\tilde{\ell}_1^{(1)} : X_1 \nrightarrow F^+ X_2$, $c_2^{\ddagger} : X_2 \nrightarrow F^+ X_2$, $\tilde{\ell}_2^{(2)} : X_2 \nrightarrow F^{+\oplus} 0$ and $\tilde{\ell}_1^{(2)} : X_1 \nrightarrow F^{+\oplus} 0$ as follows:

- We define $\tilde{\ell}_1^{(1)} : X_1 \nrightarrow F^+ X_2$ as the unique homomorphism from an $F(\_ + X_2)$-coalgebra c_1 to $J(\iota_{X_2}^F)^{-1}$ (see the left diagram in Eq. (4) below).
- We define $c_2^{\ddagger} : X_2 \nrightarrow F^+ X_2$ by:

$$c_2^{\ddagger} := \left(X_2 \xrightarrow{c_2} F(X_1 + X_2) \xrightarrow{\overline{F(\tilde{\ell}_1^{(1)}} + \text{id})} F(F^+ X_2 + X_2) \xrightarrow{J\iota_{X_2}^F} F^+ X_2 \right).$$

- We define $\tilde{\ell}_2^{(2)} : X_2 \nrightarrow F^{+\oplus} 0$ as the greatest homomorphism from $c_2^{\ddagger}$ to $J\zeta_0^{F^+}$ (see the right diagram below).

$$
\begin{array}{ccc}
F(X_1 + X_2) \xrightarrow{\overline{F(\tilde{\ell}_1^{(1)}} + \text{id})} F(F^+ X_2 + X_2) & F^+(X_2) \xrightarrow{\overline{F^+(\tilde{\ell}_2^{(2)})}} F^+(F^{+\oplus} 0)) & \\
\uparrow c_1 \qquad \underset{\tilde{\ell}_1^{(1)}}{=} \qquad \cong \uparrow J(\iota_{X_2}^F)^{-1} & \uparrow c_2^{\ddagger} \qquad \underset{\tilde{\ell}_2^{(2)}}{=}\nu \qquad \cong \uparrow J\zeta_0^{F^+} & (4) \\
X_1 \dashrightarrow F^+ X_2 & X_2 \xrightarrow{\quad\quad} F^{+\oplus} 0 &
\end{array}
$$

- We define $\tilde{\ell}_1^{(2)} : X_1 \nrightarrow F^+(F^{+\oplus} 0)$ as follows:

$$\tilde{\ell}_1^{(2)} := \left(X_1 \xrightarrow{\tilde{\ell}_1^{(1)}} F^+ X_2 \xrightarrow{\overline{F^+ \tilde{\ell}_2^{(2)}}} F^+(F^{+\oplus} 0) \right).$$

We explain an intuition why Kleisli arrows defined above bridge the gap between $\text{tr}_i^{\text{B}}(c)$ and $\text{dtr}_i(c)$. One of the main differences between them is that $\text{tr}_1^{\text{B}}(c)$ is calculated from $l_1^{(1)}(u_2)$ which is the least fixed point of a certain function, while $\text{dtr}_1(c)$ is defined as the greatest homomorphism. The arrow $\tilde{\ell}_1^{(1)}$ fills the gap because it is defined as the unique fixed point, which is obviously both the least and the greatest fixed point.

We shall prove Theorem 4.3 following the intuition above. The lemma below, which is easily proved by the finality of a, shows that not only $\tilde{\ell}_1^{(1)}$ but also $\tilde{\ell}_1^{(2)}$ is characterized as the unique homomorphism.

Lemma 4.5. *The Kleisli arrow* $\tilde{\ell}_1^{(2)} :$ *$X_1 \nrightarrow F^+(F^{+\oplus} 0)$ is the unique homomorphism from* $\overline{F}(\text{id} + \tilde{\ell}_2^{(2)}) \odot c_1$ *to* $J(\iota_{F+\oplus 0}^F)^{-1}$. □

$$
\begin{array}{ccc}
F(X_1 + F^{+\oplus} 0) \xrightarrow{\quad} F(F^+(F^{+\oplus} 0) + F^{+\oplus} 0) \\
\overline{F}(\text{id} + \tilde{\ell}_2^{(2)}) \uparrow \qquad\qquad \uparrow \overline{F}(\tilde{\ell}_1^{(2)} + \text{id}) \\
F(X_1 + X_2) \qquad = J(\iota_{F+\oplus 0}^F)^{-1} \uparrow \cong \\
c_1 \uparrow \qquad\qquad \tilde{\ell}_1^{(2)} \\
X_1 \dashrightarrow F^+(F^{+\oplus} 0)
\end{array}
$$

Together with the definition of $\tilde{\ell}_2^{(2)}$, we have the following proposition.

Proposition 4.6. *For each* $i \in \{1, 2\}$, $\tilde{\ell}_i^{(2)} = \text{dtr}_i(c)$. □

This proposition implies the existence of a solution of the HES in Definition 4.2.

It remains to show the relationship between the $\tilde{\ell}_j^{(i)}$ and $\text{tr}_i^{\text{P}}(c)$. By using that $\tilde{\ell}_1^{(1)}$ is the unique fixed point (and hence the least fixed point), we can prove the following equality for an arbitrary $u_2 : X_2 \nrightarrow F^{\oplus} 0$.

$$l_1^{(1)}(u_2) = \left(X_1 \xrightarrow{\tilde{\ell}_1^{(1)}} F^+ X_2 \xrightarrow{\overline{F^+ u_2}} F^+ F^{\oplus} 0 \xrightarrow{Jp_{1F\oplus0}^{(1)}} F^{\oplus} F^{\oplus} 0 \xrightarrow{\mu_0^{F\oplus}} F^{\oplus} 0 \right)$$

The following equalities are similarly proved using the equality above.

$$l_1^{(2)}(*) = \left(X_1 \xrightarrow{\tilde{\ell}_1^{(2)}} F^+ F^\oplus 0 \xrightarrow{\overline{F+F^\oplus i}}_{F+\oplus 0} F^+ F^\oplus (F^\oplus 0) \xrightarrow{Jp_1^{(2)}}_{F^\oplus 0} F^\oplus F^\oplus 0 \xrightarrow{\mu_0^{F^\oplus}} F^\oplus 0 \right)$$

$$l_2^{(2)}(*) = \left(X_2 \xrightarrow{\tilde{\ell}_2^{(2)}} F^\oplus 0 \xrightarrow{\overline{F^\oplus i}}_{F+\oplus 0} F^\oplus (F^\oplus 0) \xrightarrow{Jp_2^{(2)}}_{F^\oplus 0} F^\oplus F^\oplus 0 \xrightarrow{\mu_0^{F^\oplus}} F^\oplus 0 \right)$$

By the definition of $\mathsf{tr}_i^{\mathsf{B}}(c)$, these equalities imply the following proposition.

Proposition 4.7. *For each $i \in \{1, 2\}$, $\mathsf{tr}_i^{\mathsf{B}}(c) = p_{i0}^{(2)} \circ \tilde{\ell}_i^{(2)}$.*   □

Propositions 4.6 and 4.7 immediately imply Theorem 4.3.

5   Decorated Trace Semantics for Nondeterministic Büchi Tree Automata

We apply the framework developed in Sects. 3 and 4 to *nondeterministic Büchi tree automata* (NBTA), systems that nondeterministically accept *trees* with respect to the Büchi condition (see e.g. [17]). We show what datatypes $F^+(F^{+\oplus}0)$ and $F^{+\oplus}0$, and $\mathsf{dtr}_i(c)$ characterize for an NBTA. We first review some basic notions.

5.1   Preliminaries on Büchi Tree Automaton

Definition 5.1 (Ranked Alphabet). A *ranked alphabet* is a set Σ equipped with an *arity function* $|\_| : \Sigma \to \mathbb{N}$. We write Σ_n for $\{a \in \Sigma \mid |a| = n\}$. For a set X, we regard $\Sigma + X$ as a ranked alphabet by letting $|x| = 0$. We also regard $\Sigma \times X$ as a ranked alphabet by letting $|(a, x)| = |a|$.

Definition 5.2 (Σ-labeled Tree, [7]). A *tree domain* is a set $D \subseteq \mathbb{N}^*$ s.t.: i) $\langle\rangle \in D$, ii) for $w, w' \in \mathbb{N}^*$, $ww' \in D$ implies $w \in D$ (i.e. it is prefix-closed), and iii) for $w \in D$ and $i, j \in \mathbb{N}$, $wi \in D$ and $j \leq i$ imply $wj \in D$ (i.e. it is downward-closed). A *Σ-labeled (infinitary) tree* is a pair $t = (D, l)$ of a tree domain D and a *labeling function* $l : D \to \bigcup_{n \in \mathbb{N}} \Sigma_n$ s.t. for $w \in D$, $|l(w)| = n$ implies $\{i \in \mathbb{N} \mid wi \in D\} = [0, n-1]$. A Σ-labeled tree $t = (D, l)$ is *finite* if D is a finite set. We write $\mathsf{Tree}_\infty(\Sigma)$ (resp. $\mathsf{Tree}_{\mathsf{fin}}(\Sigma)$) for the set of Σ-labeled infinitary (resp. finite) trees. For $w \in D$, the *w-th subtree t_w* of t is defined by $t_w = (D_w, l_w)$ where $D_w := \{w' \in \mathbb{N}^* \mid ww' \in D\}$ and $l_w(w') := l(ww')$. A *branch* of t is a possibly infinite sequence $i_1 i_2 \ldots \in \mathbb{N}^\infty$ s.t. $i_1 i_2 \ldots i_k \in D$ for each $k \in \mathbb{N}$, and if it is a finite sequence $i_1 i_2 \ldots i_k$ then $|l(i_0 i_1 \ldots i_k)| = 0$. We sometimes identify a branch $i_0 i_1 \cdots \in \mathbb{N}^\infty$ with a sequence $l(\langle\rangle) l(i_1) l(i_1 i_2) \cdots \in \Sigma^\infty$.

Remark 5.3. For the sake of notational simplicity, we identify a Σ-labeled tree with a Σ-term in a natural manner. For example, a $\{a, b\}$-term $(a, (b, b))$ denotes an $\{a, b\}$-labeled finite tree $t = (\{\langle\rangle, 0, 1\}, [\langle\rangle \mapsto a, 0 \mapsto b, 1 \mapsto b])$. Moreover, for $\{a, b, c\}$-labeled trees $t_0 = (D_0, l_0)$ and $t_1 = (D_1, l_1)$, we write (c, t_0, t_1) for a tree $t = (\{\langle\rangle\} \cup \{0w \mid w \in D_0\} \cup \{1w \mid w \in D_1\}, [\langle\rangle \mapsto c, 0w \mapsto l_0(w), 1w \mapsto l_1(w)])$.

Definition 5.4 (NBTA). A *nondeterministic Büchi tree automaton* (NBTA) is a tuple $\mathcal{A} = (X, \Sigma, \delta, \mathsf{Acc})$ of a *state space* X, a ranked alphabet Σ, a *transition function* $\delta : X \to \mathcal{P}(\coprod_{n \in \mathbb{N}} \Sigma_n \times X^n)$ and a set $\mathsf{Acc} \subseteq X$ of *accepting states*.

Definition 5.5 [$L^{\mathrm{B}}_{\mathcal{A}}$]. Let $\mathcal{A} = (X, \Sigma, \delta, \mathsf{Acc})$ be an NBTA. A *run tree* over $\mathcal{A}$ is a $(\Sigma \times X)$-labeled tree ρ such that for each subtree $((a, x), ((a_0, x_0), t_{00}, \dots, t_{0n_0}), \dots, ((a_n, x_n), t_{n0}, \dots, t_{nn_n}))$, $(a, x_0, \dots, x_n) \in \delta(x)$ holds. A run tree is *accepting* if for each branch $(a_0, x_0)(a_1, x_1) \dots \in (\Sigma \times X)^\omega$, $x_i \in \mathsf{Acc}$ for infinitely many i. We write $\mathrm{Run}_{\mathcal{A}}(x)$ (resp. $\mathrm{AccRun}_{\mathcal{A}}(x)$) for the set of run trees (resp. accepting run trees) whose root node is labeled by $x \in X$. For $A \subseteq X$, $\mathrm{Run}_{\mathcal{A}}(A)$ denotes $\cup_{x \in A} \mathrm{Run}(x)$. We define $\mathrm{AccRun}_{\mathcal{A}}(A)$ similarly. If no confusion is likely, we omit the subscript $\mathcal{A}$. We define $\mathrm{DelSt} : \mathrm{Run}(X) \to \mathsf{Tree}_\infty(\Sigma)$ by $\mathrm{DelSt}(D, l) := (D, l')$ where $l'(w) := \pi_1(l(w))$. The *language* $L^{\mathrm{B}}_{\mathcal{A}} : X \to \mathcal{P}\mathsf{Tree}_\infty(\Sigma)$ of $\mathcal{A}$ is defined by $L^{\mathrm{B}}_{\mathcal{A}}(x) = \mathrm{DelSt}(\mathrm{AccRun}_{\mathcal{A}}(x))$.

5.2   Decorated Trace Semantics of NPTA

A ranked alphabet Σ induces a functor $F_\Sigma = \coprod_{n \in \mathbb{N}} \Sigma_n \times (\_)^n : \mathbf{Sets} \to \mathbf{Sets}$. In [21], an NBTA $\mathcal{A}$ was modeled as a Büchi $(\mathcal{P}, F_\Sigma)$-system, and it was shown that $L^{\mathrm{B}}_{\mathcal{A}}$ is characterized by a coalgebraic Büchi trace semantics $\mathrm{tr}^{\mathrm{B}}_i(c)$.

Proposition 5.6 ([21]). *For $X, Y \in \mathbf{Sets}$, we define an order $\sqsubseteq$ on $\mathcal{K}\ell(\mathcal{P})(X, Y)$ by $f \sqsubseteq g \overset{def}{\Leftrightarrow} \forall x \in X. f(x) \subseteq g(x)$. We define $\overline{F_\Sigma} : \mathcal{K}\ell(\mathcal{P}) \to \mathcal{K}\ell(\mathcal{P})$ by $\overline{F_\Sigma} X := X$ for $X \in \mathcal{K}\ell(\mathcal{P})$ and $\overline{F_\Sigma} f(a, x_1, \dots, x_n) := \{(a, y_1, \dots, y_n) \mid y_i \in f(x_i)\}$ for $f : X \nrightarrow Y$. It is easy to see that $\overline{F_\Sigma}$ is a lifting of F_Σ. Then we have:*

1. *$\mathcal{P}$ and F_Σ constitute a Büchi trace situation (Definition 2.13) with respect to $\sqsubseteq$ and $\overline{F_\Sigma}$.*
2. *The carrier set of the final F_Σ-coalgebra is isomorphic to $\mathsf{Tree}_\infty(\Sigma)$.*
3. *For an NBTA $\mathcal{A} = (X, \Sigma, \delta, \mathsf{Acc})$, we define a Büchi $(\mathcal{P}, F_\Sigma)$-system $(c : X \nrightarrow \overline{F_\Sigma} X, (X_1, X_2))$ by $c := \delta$, $X_1 := X \setminus \mathsf{Acc}$ and $X_2 := \mathsf{Acc}$. Then we have: $[\mathrm{tr}^{\mathrm{B}}_1(c), \mathrm{tr}^{\mathrm{B}}_2(c)] = L^{\mathrm{B}}_{\mathcal{A}} : X \to \mathcal{P}\mathsf{Tree}_\infty(\Sigma)$.*   □

In the rest of this section, for an NBTA $\mathcal{A} = (X, \Sigma, \delta, \mathsf{Acc})$ modeled as a $(\mathcal{P}, F_\Sigma)$-system $(c : X \to \mathcal{P}F_\Sigma X, (X_1, X_2))$, we describe $\mathrm{dtr}_i(c)$ and show the relationship with $\mathrm{tr}^{\mathrm{B}}_i(c)$ using Theorem 4.3.

We first describe datatypes $F^+_\Sigma(F^{+\oplus}_\Sigma 0)$ and $F^{+\oplus}_\Sigma 0$ referring to the construction of a final coalgebra in Theorem 2.5. We can easily see that $F^+_\Sigma A \cong \mathsf{Tree}^+_{\mathsf{fin}}(\Sigma, A) := \mathsf{Tree}_{\mathsf{fin}}(\Sigma + A) \setminus \{(x) \mid x \in A\}$. Hence for each $i \in \omega$, by a similar characterization to Example 3.3, we have:

$$(F^+_\Sigma(\_ + 0))^i 1 \cong \underbrace{\mathsf{Tree}^+_{\mathsf{fin}}(\Sigma, \mathsf{Tree}^+_{\mathsf{fin}}(\Sigma, \dots \mathsf{Tree}^+_{\mathsf{fin}}(\Sigma, \{*\}) \dots))}_{i} \cong$$

$$\left\{ \begin{array}{c} \xi \in \mathsf{Tree}_{\mathsf{fin}}(\Sigma \times \{\bigcirc, \circledcirc\} \\ + \{*\}) \end{array} \middle| \begin{array}{l} \text{the root node is labeled by } \circledcirc, \text{ and for each branch} \\ \text{whose last component is } *, \circledcirc \text{ appears exactly } i\text{-times} \end{array} \right\}.$$

Therefore $F_\Sigma^{+\oplus}0$, a limit of the above sequence by Theorem 2.5, and $F_\Sigma^+(F_\Sigma^{+\oplus}0)$ are characterized as follows:

Proposition 5.7. *We define* $\mathrm{AccTree}_i(\Sigma) \subseteq \mathrm{Tree}_\infty(\Sigma \times \{\bigcirc, \textcircled{\scriptsize\textbf{0}}\})$ *by:*

$$\mathrm{AccTree}_i(\Sigma) := \left\{ \begin{array}{c} \xi \in \mathrm{Tree}_\infty(\Sigma \times \{\bigcirc, \textcircled{\scriptsize\textbf{0}}\}) \\ +A) \end{array} \middle| \begin{array}{l} \text{the root node is labeled by } \bullet, \text{ and for each} \\ \text{infinite branch } \textcircled{\scriptsize\textbf{0}} \text{ appears infinitely often} \end{array} \right\}.$$

where $i \in \{1, 2\}$ *and* $\bullet$ *is* $\bigcirc$ *if* $i = 1$ *and* $\textcircled{\scriptsize\textbf{0}}$ *if* $i = 2$. *Then* $\mathrm{AccTree}_1(\Sigma) \cong F_\Sigma^+(F_\Sigma^{+\oplus}0)$ *and* $\mathrm{AccTree}_2(\Sigma, A) \cong F_\Sigma^{+\oplus}0$. □

We now show what $\mathsf{dtr}_i(c)$ characterizes for an NBTA with respect to the characterization in Proposition 5.7. Firstly, Assumption 4.1 in the previous section is satisfied.

Proposition 5.8. *Assumption 4.1 is satisfied by* $(T, F) = (\mathcal{P}, F_\Sigma)$. □

By Proposition 5.7, for $i \in \{1, 2\}$, $\beta_{i\,0}$ (see Definition 3.4) has a type

$$\beta_{i\,0} : \mathrm{AccTree}_i(\Sigma) \to \coprod_{n\in\omega} \Sigma_n \times (\mathrm{AccTree}_1(\Sigma) + \mathrm{AccTree}_2(\Sigma)),$$

and is given by $\beta_{i\,A}(\xi) = (a, \xi_0, \ldots, \xi_{n-1})$ if the root of ξ is labeled by $(a, \bullet) \in \Sigma_n \times \{\bigcirc, \textcircled{\scriptsize\textbf{0}}\}$. Using this, we can show the following characterization of $\mathsf{dtr}_i(c)$.

Proposition 5.9. *Let* $\mathcal{A} = (X, \Sigma, \delta, \mathrm{Acc})$ *be an NBTA. We define* $\Omega :$ $\mathrm{Run}(X) \to \mathrm{Tree}_\infty(\Sigma \times \{\bigcirc, \textcircled{\scriptsize\textbf{0}}\})$ *by* $\Omega(D, l) := (D, l')$ *where for* $w \in D$ *s.t.* $l(w) = (a, x)$, $l'(w) := (a, \bigcirc)$ *if* $x \notin \mathrm{Acc}$ *and* $(a, \textcircled{\scriptsize\textbf{0}})$ *if* $x \in \mathrm{Acc}$. *We define a Büchi* $(\mathcal{P}, F_\Sigma)$-*system* $(c : X \nrightarrow \overline{F_\Sigma}X, (X_1, X_2))$ *as in Proposition 5.6.3. Then for* $i \in [1, 2n]$ *and* $x \in X_i$,

$$\mathsf{dtr}_i(c)(x) = \left\{ \Omega(\rho) \in \mathrm{AccTree}_i(\Sigma) \mid \rho \in \mathrm{AccRun}_\mathcal{A}(x) \right\}. \qquad \square$$

We conclude this section by instantiating $p_{i\,A}^{(2)}$ (Definition 3.8) for NBTAs.

Proposition 5.10. *We overload* DelSt *and define* DelSt : $\mathrm{AccTree}_1(\Sigma) + \mathrm{AccTree}_2(\Sigma) \to \mathrm{Tree}_\infty(\Sigma)$ *by* $\mathrm{DelSt}(D, l) := (D, l')$ *where* $l'(w) := \pi_1(l(w))$. *Then with respect to the isomorphism in Proposition 5.7,* $\mathrm{DelSt}(\xi) = p_{i\,A}^{(2)}(\xi)$ *for each* $i \in \{1, 2\}$ *and* $\xi \in \mathrm{AccTree}_i(\Sigma)$. □

Hence Theorem 4.3 results in the following (obvious) equation for NBTAs:

$$\left\{ \mathrm{DelSt}(\Omega(\rho)) \mid \rho \in \mathrm{AccRun}_\mathcal{A}(x) \right\} = L_\mathcal{A}^{\mathrm{B}}(x).$$

6   Systems with Other Branching Types

In this section we briefly discuss other monads than $T = \mathcal{P}$. As we have discussed in Sect. 3.3, the framework does not apply to $T = \mathcal{D}$.

Let $T = \mathcal{L}$ and $F = F_{\Sigma}$. A Büchi $(\mathcal{L}, F_{\Sigma})$-system $(c : X \nrightarrow \overline{F_{\Sigma}}X, (X_1, \dots, X_{2n}))$ is understood as a Σ-labeled deterministic Büchi tree automaton with an exception. In a similar manner to $T = \mathcal{P}$ we can prove that they satisfy Assumption 4.1. The resulting decorated trace semantics has a type $\mathsf{dtr}_i(c) : X_i \to \{\bot\} + \mathrm{AccTree}_i(\Sigma)$. Note that once $x \in X$ is fixed, either of the following occurs: a decorated tree is determined according to c; or $\bot$ is reached at some point. The function $\mathsf{dtr}_i(c)$ assigns $\bot$ to $x \in X_i$ if and only if $\bot$ is encountered from x or the resulting decorated tree does not satisfy the Büchi condition: otherwise, the generated tree is assigned to x. See Sect. E.1 of the extended version [20] for detailed discussions, which includes the case of parity automata.

We next let $T = \mathcal{G}$. A Büchi $(\mathcal{G}, F_{\Sigma})$-system is understood as a *probabilistic Büchi tree automaton*. In fact, it is open if $T = \mathcal{G}$ and $F = F_{\Sigma}$ satisfy Assumption 4.1. The challenging part is the gfp-preserving condition (Assumption 4.1.4). However, by carefully checking the proofs of the lemmas and the propositions where the gfp-preserving condition is used (i.e. Proposition 3.14, Lemma 4.5 and Proposition 4.7), we can show that Assumption 4.1.4 can be relaxed to the following weaker but more complicated conditions:

4'-1. T and $F^+(\_ + A)$ satisfy the gfp-preserving condition with respect to
$$\overline{F^+}(F^{+\oplus}B + A) \xrightarrow{\overline{F_i^+}(\mathrm{id}+f)} \overline{F^+}(F^{+\oplus}B + B) \xrightarrow{J(\zeta_B^{F^+})^{-1}} F^{+\oplus}B \text{ for each}$$
$f : A \nrightarrow B$;

4'-2. T and $F^+(\_ + A)$ satisfy the gfp-preserving condition with respect to an algebra $F^+(F^{\oplus\oplus}A + A) \xrightarrow{J\tau} F^{\oplus}(F^{\oplus\oplus}A + A) \xrightarrow{J(\zeta_A^{F^{\oplus}})^{-1}} F^{\oplus\oplus}A$ where τ is the unique homomorphism from $(\iota_{F^{\oplus\oplus}A+A}^F)^{-1}$ to $\zeta_{F^{\oplus\oplus}A+A}^F$; and

4'-3. T and $F(\_ + A)$ satisfy the gfp-preserving condition with respect to an algebra $F(F^{\oplus}A + F^{\oplus}A + A) \xrightarrow{JF([\mathrm{id},\mathrm{id}]+\mathrm{id})} F(F^{\oplus}A + A) \xrightarrow{J(\zeta_A^F)^{-1}} F^{\oplus}A$.

In fact, only the first condition is sufficient to prove Proposition 3.14 and Lemma 4.5.

We can show that $T = \mathcal{G}$ and $F = F_{\Sigma}$ on **Meas** satisfy the above weakened gfp-preserving condition, and hence we can consider a decorated trace semantics $\mathsf{dtr}_i(c)$ for a Büchi $(\mathcal{G}, F_{\Sigma})$-system $(c : X \nrightarrow \overline{F_{\Sigma}}X, (X_1, X_2))$ and use Theorem 4.3.

Assume X is a countable set equipped with a discrete σ-algebra for simplicity. Then the resulting decorated trace semantics $\mathsf{dtr}_i(c)$ has a type $X_i \to \mathcal{G}(\mathrm{AccTree}_i(\Sigma), \mathfrak{F}_{\mathrm{AccTree}_i(\Sigma)})$ where $\mathfrak{F}_{\mathrm{AccTree}_i(\Sigma)}$ is the standard σ-algebra generated by cylinders. The probability measure assigned to $x \in X_i$ by $\mathsf{dtr}_i(c)$ is defined in a similar manner to the probability measure over the set of run trees generated by a probabilistic Büchi tree automaton (see e.g. [18]).

The situation is similar for parity $(\mathcal{G}, F_{\Sigma})$-systems. See [20, Sect. E.2] for details.

7　Conclusions and Future Work

We have introduced a categorical data type for capturing behavior of systems with Büchi acceptance conditions. The data type was defined as an alternating fixed point of a functor, which is understood as the set of traces decorated with priorities. We then defined a notion of coalgebraic decorated trace semantics, and compared it with the coalgebraic trace semantics in [21]. We have applied our framework for nondeterministic Büchi tree automata, and showed that decorated trace semantics is concretized to a function that assigns a set of trees decorated with priorities so that the Büchi condition is satisfied in every branch. We have focused on the Büchi acceptance condition for simplicity, but all the results can be extended to the parity acceptance condition (see Sect. A of [20] for the details).

Future Work. There are some directions for future work. In this paper we focused on systems with a simple branching type like nondeterministic or probabilistic. Extending this so that we can deal with systems with more complicated branching type like *two-player games* (systems with two kinds of nondeterministic branching) or *Markov decision processes* (systems with both nondeterministic and probabilistic branching) is a possible direction of future work.

Another direction would be to use the framework developed here to categorically generalize a verification method. For example, using the framework of coalgebraic trace semantics in [21], a simulation notion for Büchi automata is generalized in [19]. Searching for an existing verification method that we can successfully generalize in our framework would be interesting.

Finally, it was left open in Sect. 6 if Assumption 4.1.4 is satisfied by $T = \mathcal{G}$ and $F = F_\Sigma$. Investigating this is clearly a future work.

Acknowledgments. We thank Kenta Cho, Shin'ya Katsumata and the anonymous referees for useful comments. The authors are supported by JST ERATO HASUO Metamathematics for Systems Design Project (No. JPMJER1603), and JSPS KAKENHI Grant Numbers 15KT0012 & 15K11984. Natsuki Urabe is supported by JSPS KAKENHI Grant Number 16J08157.

References

1. Adámek, J., Koubek, V.: Least fixed point of a functor. J. Comput. Syst. Sci. **19**(2), 163–178 (1979). https://doi.org/10.1016/0022-0000(79)90026-6
2. Adámek, J., Milius, S., Moss, L.S.: Fixed points of functors. J. Log. Algebraic Methods Program. **95**, 41–81 (2018)
3. Arnold, A., Niwiński, D.: Rudiments of μ-Calculus. Studies in Logic and the Foundations of Mathematics. Elsevier, Amsterdam (2001)
4. Borceux, F.: Handbook of Categorical Algebra, Encyclopedia of Mathematics and its Applications, vol. 1. Cambridge University Press, Cambridge (1994)
5. Ciancia, V., Venema, Y.: Stream automata are coalgebras. In: Pattinson, D., Schröder, L. (eds.) CMCS 2012. LNCS, vol. 7399, pp. 90–108. Springer, Heidelberg (2012). https://doi.org/10.1007/978-3-642-32784-1_6
6. Cleaveland, R., Klein, M., Steffen, B.: Faster model checking for the modal mu-calculus. In: von Bochmann, G., Probst, D.K. (eds.) CAV 1992. LNCS, vol. 663, pp. 410–422. Springer, Heidelberg (1993). https://doi.org/10.1007/3-540-56496-9_32

7. Courcelle, B.: Fundamental properties of infinite trees. Theor. Comput. Sci. **25**, 95–169 (1983). https://doi.org/10.1016/0304-3975(83)90059-2

8. Etessami, K., Wilke, T., Schuller, R.A.: Fair simulation relations, parity games, and state space reduction for Büchi automata. SICOMP **34**(5), 1159–1175 (2005)

9. Ghani, N., Hancock, P., Pattinson, D.: Representations of stream processors using nested fixed points. Log. Methods Comput. Sci. **5**(3) (2009)

10. Hasuo, I., Jacobs, B.: Context-free languages via coalgebraic trace semantics. In: Fiadeiro, J.L., Harman, N., Roggenbach, M., Rutten, J. (eds.) CALCO 2005. LNCS, vol. 3629, pp. 213–231. Springer, Heidelberg (2005). https://doi.org/10.1007/11548133_14

11. Hasuo, I., Jacobs, B., Sokolova, A.: Generic trace semantics via coinduction. Log. Methods Comput. Sci. **3**(4) (2007)

12. Jacobs, B.: Trace semantics for coalgebras. Electr. Notes Theor. Comput. Sci. **106**, 167–184 (2004). https://doi.org/10.1016/j.entcs.2004.02.031

13. Jacobs, B.: Introduction to Coalgebra: Towards Mathematics of States and Observation. Cambridge Tracts in Theoretical Computer Science. Cambridge University Press, New York (2016)

14. Milius, S.: Completely iterative algebras and completely iterative monads. Inf. Comput. **196**(1), 1–41 (2005)

15. Mulry, P.S.: Lifting theorems for Kleisli categories. In: Brookes, S., Main, M., Melton, A., Mislove, M., Schmidt, D. (eds.) MFPS 1993. LNCS, vol. 802, pp. 304–319. Springer, Heidelberg (1994). https://doi.org/10.1007/3-540-58027-1_15

16. Power, J., Turi, D.: A coalgebraic foundation for linear time semantics. Electr. Notes Theor. Comput. Sci. **29**, 259–274 (1999)

17. Thomas, W.: Languages, automata, and logic. In: Rozenberg, G., Salomaa, A. (eds.) Handbook of Formal Languages, pp. 389–455. Springer, Heidelberg (1997). https://doi.org/10.1007/978-3-642-59126-6_7

18. Urabe, N., Hasuo, I.: Coalgebraic infinite traces and Kleisli simulations. CoRR abs/1505.06819 (2015). http://arxiv.org/abs/1505.06819

19. Urabe, N., Hasuo, I.: Fair simulation for nondeterministic and probabilistic Buechi automata: a coalgebraic perspective. LMCS **13**(3) (2017)

20. Urabe, N., Hasuo, I.: Categorical Büchi and parity conditions via alternating fixed points of functors. arXiv preprint (2018)

21. Urabe, N., Shimizu, S., Hasuo, I.: Coalgebraic trace semantics for Büchi and parity automata. In: Proceedings of the CONCUR 2016. LIPIcs, vol. 59, pp. 24:1–24:15. Schloss Dagstuhl - Leibniz-Zentrum fuer Informatik (2016)

Author Index

Printed in the United States
By Bookmasters